Lecture Notes in Earth Sciences 46

Gianni Galli

Temporal and Spatial Patterns in Carbonate Platforms

Springer-Verlag
Berlin Heidelberg GmbH

Author

Dr. Gianni Galli
Via Samacchini 5
I-40141 Bologna

"For all Lecture Notes in Earth Sciences published till now please see final pages of the book"

ISBN 978-3-540-56231-3

Library of Congress Cataloging-in-Publication Data

Galli, Gianni, 1956-
Temporal and spatial patterns in carbonate platforms / Gianni Galli. p. cm. - (lecture notes in earth sciences; 46) Includes bibliographical references.
ISBN 978-3-540-56231-3 ISBN 978-3-540-47486-9 (eBook)
DOI 10.1007/978-3-540-47486-9
1. Rocks, Carbonate. 2. Sedimentation and deposition. I. Title. II. Series.
QE471. 15.C3G35 1993 552' .58-dc20 93-8317

Originally published by Springer-Verlag Berlin Heidelberg New York in 1993

Typesetting: Camera ready by author
32/3140-543210 - Printed on acid-free paper

ACKNOWLEDGMENTS

I would like to express my sincere thanks to all of those who helped me during this 12-yr long work project, started in 1981.

Prof.G.B.Vai (University of Bologna) allowed me to make a fresh start at the beginning by introducing me to the field geology of the Carnic Alps and later acting as a supervisor of my PhD thesis. He also oriented my mental attitude towards a neocatastrophist approach of geological processes.

Prof.R.N.Ginsburg allowed me to study Pleistocene cores of Florida platform in 1987 and put me in the condition of enjoying the fresh breeze of Fisher Island Station during a Fullbright tenure.

Prof.C.G.St.C.Kendall gave access to the Sedpak program at the South Carolina University, probably one of the most powerful simulation programs existing today.

Dr.Mia Van Steenwinkel quite kindly sent me her thesis in 1988 which allowed the development of the sequence stratigraphy approach to the case histories summarized in this work.

I am also indebted to Prof. Jonathan Tennenbaum (Fusion Energy Foundation) who pushed me to develop section V on the relativistic distribution of event horizons.

TABLE OF CONTENTS

Introduction

"The geological history,as expressed by the stratigraphic column,is basically composed of cycles of sedimentation, stratification and magmatism which correlate with relative changes in sea level determined in turn by different types of crustal movements. The classical sequence of stages "transgression - inundation - differentiation - regression - emergence" is believed to reflect the deformation phases of a geotectonic cycle" (Wezel,1988: p.37).

The concept of geotectonic cycle is fundamental in geology because it links tectonics with sedimentary processes.

According to Wezel (1988) the geotectonic cycle is an expression of cyclic variations in the behavior of the crust; more precisely,it is a geodynamic response to the Earth's variations in the rate of rotation (Mörner,1986; Whyte,1977; Carey,1976).Based on a global analysis of geotectonic data, synchronous episodes of intense global swelling, governed by cyclically ordered diastrophic processes, were identified (Wezel,1985;1988).

The process leading to these swells was termed krikogenesis (Wezel, 1988).It basically consists of not steady, localized, migratory vertical movements linked to mantle diapirism and concentrated in single zones.The overlying crust adjusts itself to mantle motions induced by krikogenesis, with the formation of transient troughs and swells ('touche-de-piano' tectonics).This mechanism was individuated in several areas (Wezel,1988).

The history of the Earth is described by six episodes that repeat in the same way in the course of geological time.Their duration progressively decreases:the first cycle has a duration of about 200 million years, the following,younger cycles lasted 150,115,65,45 and 20 m.y.

These cycles can be traced along a time-spiral,its length representing the time (Fig.1). Each cycle of deposition and uplift can be subdivided into the following phases: krikogenesis, inundation, regression and emergence. The last phase is preceeded by a tectogenic phase, linked to an increase in tectonic activity (Fig.2).

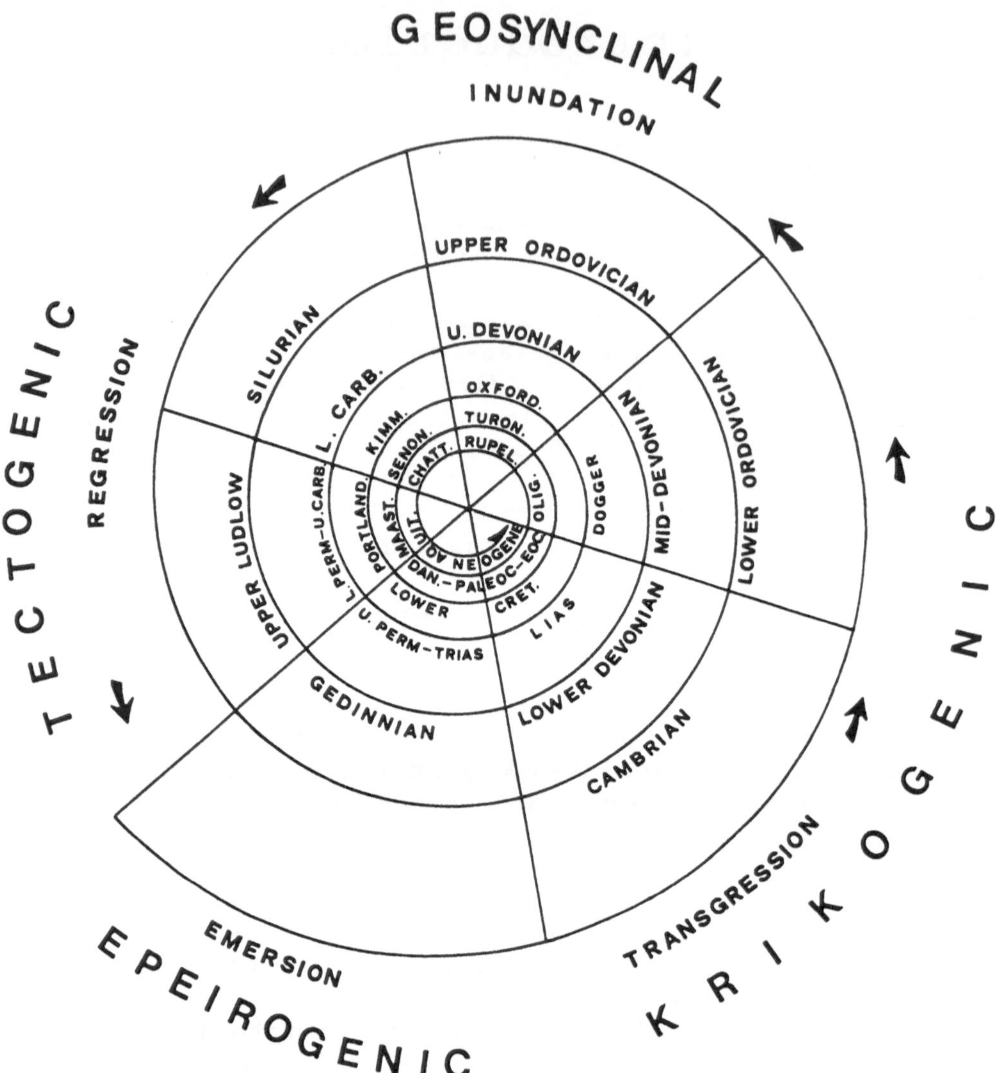

Fig.1 - Time spiral showing the six major sedimentary cycles of Earth history, each separated by a tectogenetic phase (Wezel,1976:p.87).Each cycle is divisible into four main phases.

The geological history is marked by the rhythm of krikogenetic rejuvenation and krikogenetic quiescence periods (Wezel,1988), each punctuated by a recurrent set of changes of different organic and inorganic processes.

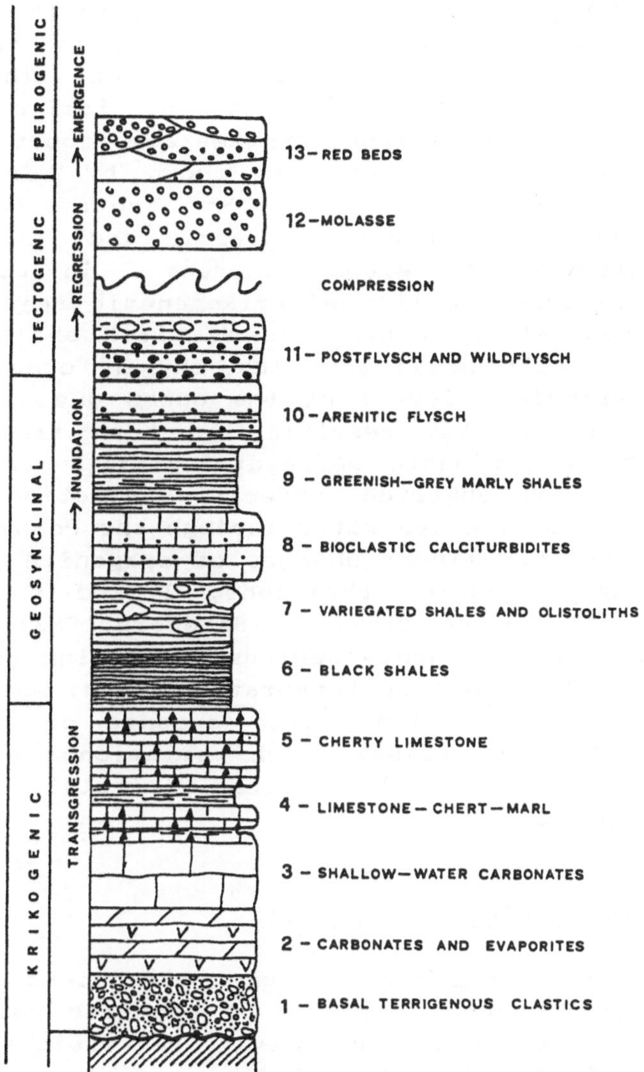

Fig.2 - Geotectonic cycle showing the vertical sequence of lithofacies and correlations with the diastrophic phases (adapted from Wezel,1988).The sequence, consisting of a transgressive hemicycle overlain by a regressive hemicycle, occurs in different situations on various scales, also in carbonate settings ranging from shallow-water to basinal.

Krikogenetic rejuvenation periods

Periods characterized by krikogenetic rejuvenation record the formation of a 'basin-and-swell' morphology (geoanticlinal ridges), high heat flow and volcanicity of shoshonite to tholeite affinity. These periods are associated with episodes of normal magnetic polarity, increases in the rate of sea-floor spreading,global temperature (leading to the deposition of evaporites) and climatic amelioration. There is also an increase in the seismicity at low latitudes (Whyte,1977). Several storm depositional systems are reported from periods of krikogenetic rejuvenation, as shown by Marsaglia and Klein (1983) who listed several storm occurrences dated to the Ordovician, Devonian,Early Jurassic and late Cretaceous, all time intervals of global increases in ocean temperature. Another feature peculiar of krikogenetic rejuvenation seems to be the generalized appearance of black shales ('AOE' events) probably resulting from a low oxygen concentration in the deep-intermediate levels of the ocean due to a reduced water circulation possibly resulting in turn from an increase in temperature. Carbonate sedimentologists working on carbonate mineralogy have observed systematic variations in the isotopic composition of the sea-water through the Phanerozoic. Lowenstam (1963) found non-linear changes of aragonite - high-Mg calcite mineralogy of marine skeletons, linked to changes in the atmospheric carbon dioxide (Sandberg,1983). Fischer (1981) distinguished greenhouse periods containing an elevated amount of carbon dioxide and temperature, and icehouse periods of colder temperature and lesser amounts of carbon dioxide.As shown by Sandberg (1983), such variations seem to be tied to worldwide sea-level fluctuations,ultimately global tectonics and volcanic activity.

Krikogenetic quiescence periods

Periods characterized by krikogenetic quiescence are linked to reverse or frequent inversion of magnetic polarity, a slowing down of tectonic activity, regressions and climatic deteriorations. Mountain uplifts produce a deflection at low latitudes which can be responsible for global cooling, hence,glaciations. These periods are marked by biotic crises (Whyte,1977), increase in shallow seismicity at high latitudes and few storm occurrences.

According to Wezel (1988) and Whyte (1977) the decrease in the rate of Earth's rotation which leads to a transition from krikogenetic rejuvenation to quiescence coincides with a change in the geoidal configuration from an oblate to a prolate shape.The geotectonic cycle is seen as the product of such cyclical variations. The transgressive hemicycle comprising krikogenetic and geosynclinal phases records a period of acceleration in the rate of rotation ('spin maximum' of Whyte,1977). Conversely, the uplift of the geosynclinal pile and molasse deposition records a decrease in the speed of the Earth's rotation ('spin minimum' of Whyte,1977).

Four globally synchronous episodes of krikogenetic rejuvenation were identified: 1) Late Lower Jurassic - Middle Jurassic; 2) Aptian - Senonian; 3) Paleogene; and 4) Pliocene - Quaternary (Wezel,1988). The Cretaceous was a period of long, intense krikogenetic activity.

The "sea-level curve" published by Haq, et al.(1987) which shows relative sea-level fluctuations throughout the Phanerozoic, consists of a sequence of irregular bumps and wiggles. Long-term sea level fluctuations "are most likely related to changes in the volume of ecean basin driven by changes in the rate of sea floor spreading and ridge volume. The shorter term eustatic changes ... are most probably due in large part to changes in polar ice volume and possibly part of some other, as yet unknown mechanism" (Vail and Haq,1988).

In several works Mörner introduced the concept of geoidal eustasy (i.e. Mörner,1986).The effects of short-term geoidal eustasy have been underestimated, as geoidal pulses tend to be correlated with tectonic phases of glacial fluctuations. Conversely, geoidal eustasy is an independent mechanism, controlled by processes seated in the upper mantle.

The study of carbonate sediments offers two advantages: 1) carbonates store a moltitude of physical, chemical, inorganic and organic variations that have occurred throughout the Earth history; 2) carbonate sediments, more than siciclastic sediments, give a fine, precise record of vertical crustal movements (relative sea-level changes by Kendall and Schlager,1981). Therefore, the study of carbonate rocks is especially suitable for defining alternations of krikogenetic periods, and possibly a hierarchy of krikogenetic processes. These may be linked to geoidal eustasy.

This preliminary work has been centered on the following sequential questions:

1) Are there specific carbonate geometries that reflect krikogenetic rejuvenation and quiescence periods?

2) Is there a sequence in the carbonate world encompassing shallow-water and slope sectors equivalent or comparable to the geotectonic cycle?

3) If existent, are such geometries and sequences indicative of short-term geoidal pulses?

4) Are short-term geoidal pulses distributed at random through geological time, or conversely ordered in time by some physical law?

The following time intervals are considered to be pertinent to this study: 1) Ordovician; 2) Upper Devonian- Lower Carboniferous; 3) Anisian-Ladinian;4) Pliensbachian- Aalenian; 5) Valanginian; 6) Cenomanian-Turonian; 7) Upper Maastrichtian; 8) Thanetian; 9) Ypresian; 10) Chattian; 11) Serravallian; 12) Messinian; and 13) Pleistocene.

These time intervals are thought to correspond to more intense short-term krikogenetic rejuvenation periods.They were individuated by an integrated analysis of paleontological and geological data whose description is beyond the scope of this work.

In this work I described some features which I thought to be linked generally to krikogenetic rejuvenation periods; the main key elements of this research are the following:

1) intrashelf ramps;
2) divergent (onlap) and convergent (offlap) sedimentary patterns; and
3) megabreccias.

Part I

Intrashelf ramps are basically sedimentary wedges (miniature ramps) developed within carbonate shallow-water complexes.
Onlap and offlap ramps are respectively divergent and convergent prisms. Rotational subsidence is a main control in the formation of onlap ramps.
Different modalities of formation of onlap ramps are discussed in the selected case histories described below. They comprise Pleistocene, Jurassic (Pliensbachian), and Upper Devonian (Givetian-Frasnian) geological intervals.

Introduction

The profile of some basins is that of a ramp consisting of an irregular, undulating topographic outline with a deeper basinal trough and a shallower, overhanging basin located landward of a sill or inflexion of the topographic gradient.

This ramp outline is developed at various scales. At a regional scale it consists of a miogeosynclinal and a eugeosynclinal trough separated by a ridge of the basement. The ridge, or flexure, as well as the landward area of the miogeosyncline may be the sites of formation of smaller scale ramps.These miniature ramps are most commonly developed in divergent prisms of passive margins, in the areas comprised between the hinge

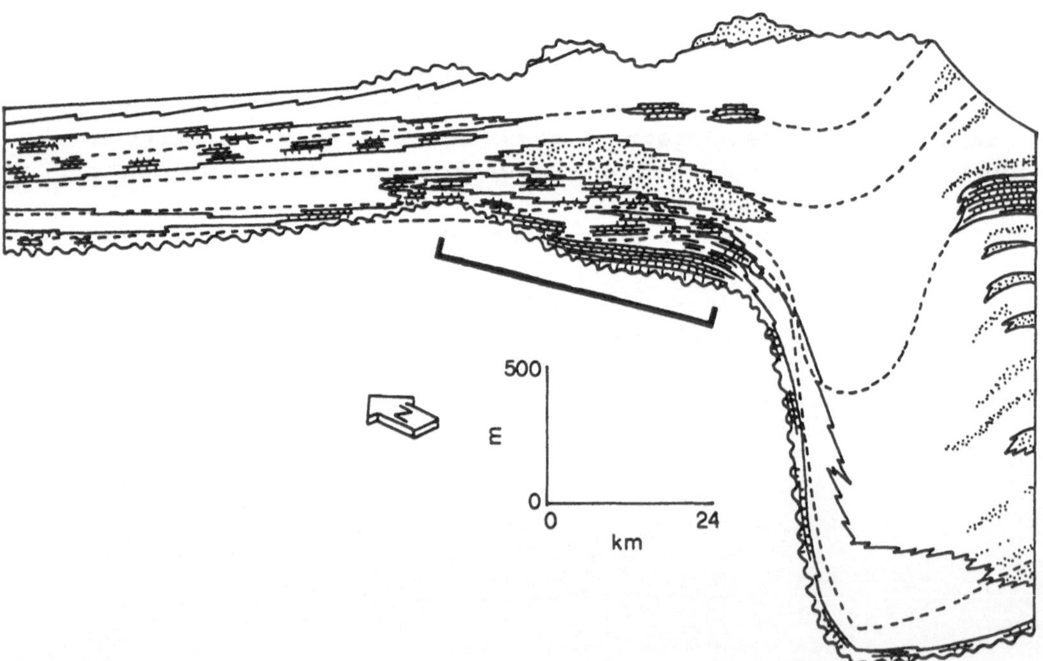

Fig.3 - Cross section of Middle Ordovician sequence from East Tennessee (from Ruppel and Walker,1982). The miniature (intrashelf) ramp occurs in the flexure zone of the sedimentary prism.

and the trough areas,where the time lines begin to diverge from one another.An example occurs in the Middle Ordovician cross section of East Tennessee (Fig.3).

Several modern carbonate shallow-water settings record a complex bathymetry consisting of peritidal banks and islands surrounded by open marine sediments. A number of analogous fossil 'facies mosaics' record a complex interbedding of different, unrelated lithologies. Laporte (1967) and Anderson (1971) interpreted one of these facies mosaics within the Manlius Formation as a complex array of high- and low-tidal flat and lagoonal deposits resulting from a transgression in an epeiric sea (Fig.4).

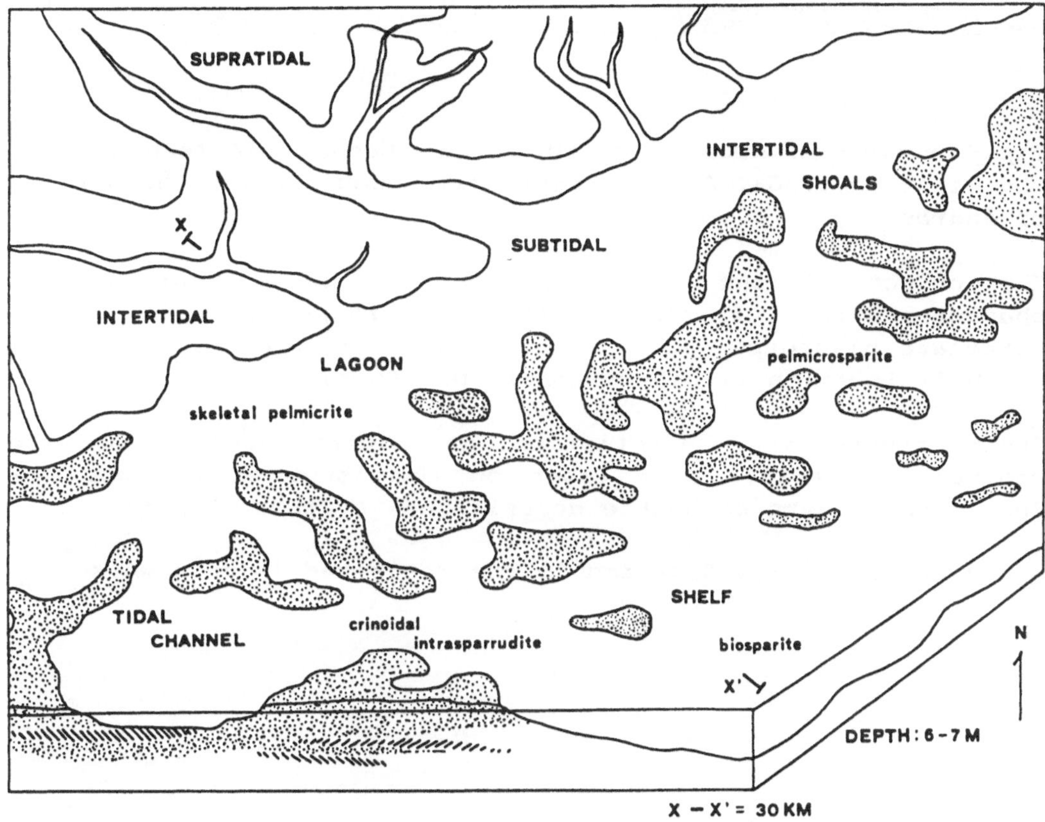

Fig.4 - Reconstructed depositional environment for the Coeymans and Manlius facies, Lower Devonian, New York (from Anderson,1971).Recent reappraisals of the Manlius and Coeyman Formations by Goodwin, et al.(1986) did not disagree with this former sedimentological interpretation.

In some of these situations the present-day or ancient lagoonal floor ranges in depth from a few meters (1.5 to 2 m) to as much as 25 - 35 m.Between shallow and deeper floor there is an inclined surface, 10 to 50 km wide (20 km on the average) that is comparable to a ramp (Ahr,1973), in spite of the smaller size and width. A modern example of this topographic feature is found within the Bahama platform, in the area comprised between the Bimini Bank and the Andros Bank (Eberli and Ginsburg,1981; Fig.5).

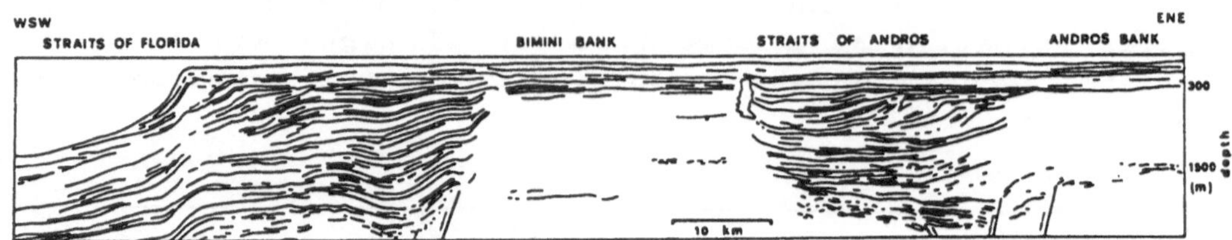

Fig.5 - Seismic cross section of the Bahama platform (Eberli and Ginsburg,1980). An intrashelf ramp occurs below the Straits of Andros.

The object of the following chapters is to outline the characteristics of these miniature ramps located within carbonate platforms. They are termed *intrashelf ramps* because of their location within carbonate platforms.

Their recognition, description and interpretation of the controlling factors were based on the study of a number of fossil and modern carbonate depositional systems (Fig.6).

Study areas # 2,3,4,5,14 and 15 are described in this work.

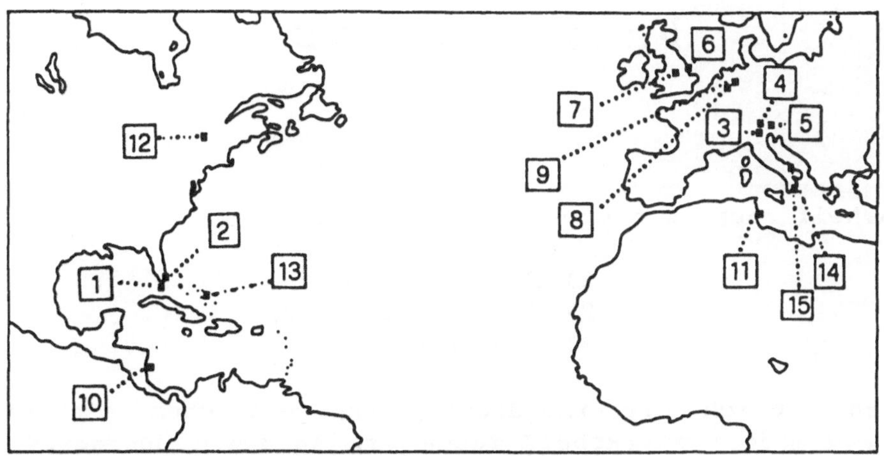

Fig.6 - Locations of study areas.

1): Florida Bay, Holocene, U.S.A.;
2): Fort Thompson Formation, Pleistocene, U.S.A.;
3): 'Calcari Grigi' Formation,Jurassic, Italy;
4): Dürrenstein Formation, Carnian (Triassic), Italy;
5): Devonian platform,Givetian-Frasnian (Devonian), Italy;
6): Cadeby Formation,Permian, England;
7): Asbian - Brigantian platform,Derbyshire, England;
8): Dinantian platform,Upper Carboniferous, Belgium;
9): Devonian platform, Givetian-Frasnian, Belgium;
10): Belize platform,Holocene, British Honduras;
11): Pleistocene carbonates,eastern Tunisia;
12): Manlius Formation, Lower Devonian, U.S.A.;
13): Exuma Cays, Holocene, Bahamas;
14): Cretaceous carbonates, southern Gargano, Italy;
15): Capo Rizzuto carbonates, Pleistocene, Italy.

Facies belts

An intrashelf ramp is composed of three sectors:

1) *Shallow ramp*;
2) *Intermediate ramp*;
3) *Deep ramp*.

Some trends which are immediately apparent when examining ancient and modern intrashelf ramps are landward decreases in the detritus/matrix ratio and faunal diversity. Petrographic trends in most cases are not linear, as a result of a complex topography, variations in width of facies belts and a broad spectrum of variations in the relationships between the three sectors.More linear trends and a gradational transition between shallow and deep ramp reflect regular and negligible topographic variations along the sloping surface; in these simpler situations the intermediate ramp is commonly poorly developed.

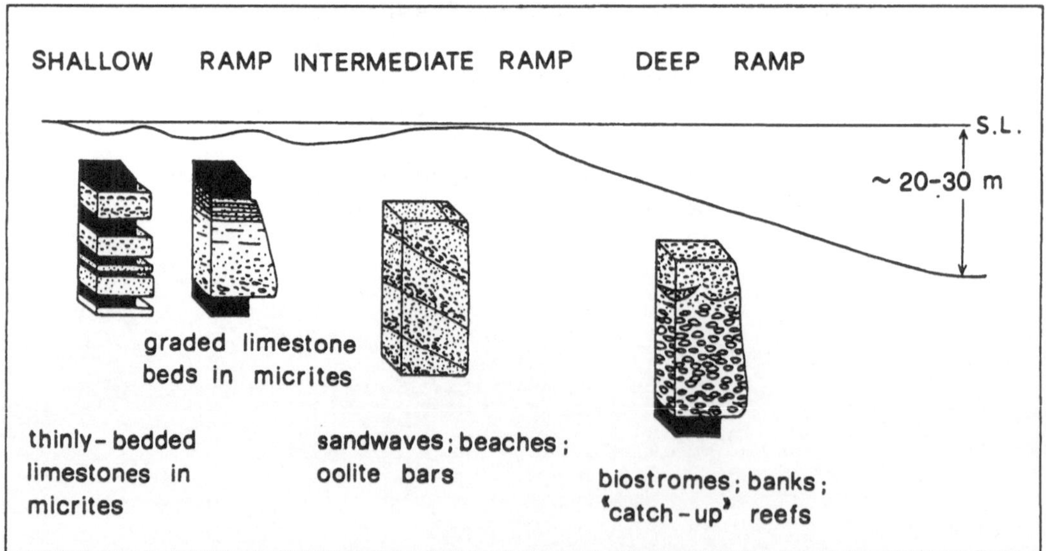

Fig.7 - Profile and main facies types of an intrashelf, onlap ramp.

This chapter is a summary of the main characters of these three sectors which may be recognized on stratigraphic profiles. A general, representative profile of the main lithofacies types in relation to the bathymetry is shown in Fig.7.

Shallow ramp

The shallow ramp is the shallowest site of the ramp. The lagoonal floor is approximately located 2 m below the sea-level. The bathymetry is made uneven by a complex framework composed of peritidal banks and islands surrounded by open marine or lagoonal sediments. This environmental situation is common in nearly all rimmed shelves where the rapid Holocene eustatic sea-level rise (1m/1500 yrs.) determined a drowning within the euphotic zone and the formation of scattered depocenters that have kept pace with the rising sea-level.This process occurred also in pre-Holocene times (Read,1985: p.16; Grotzinger,1986;Pratt and James,1986).

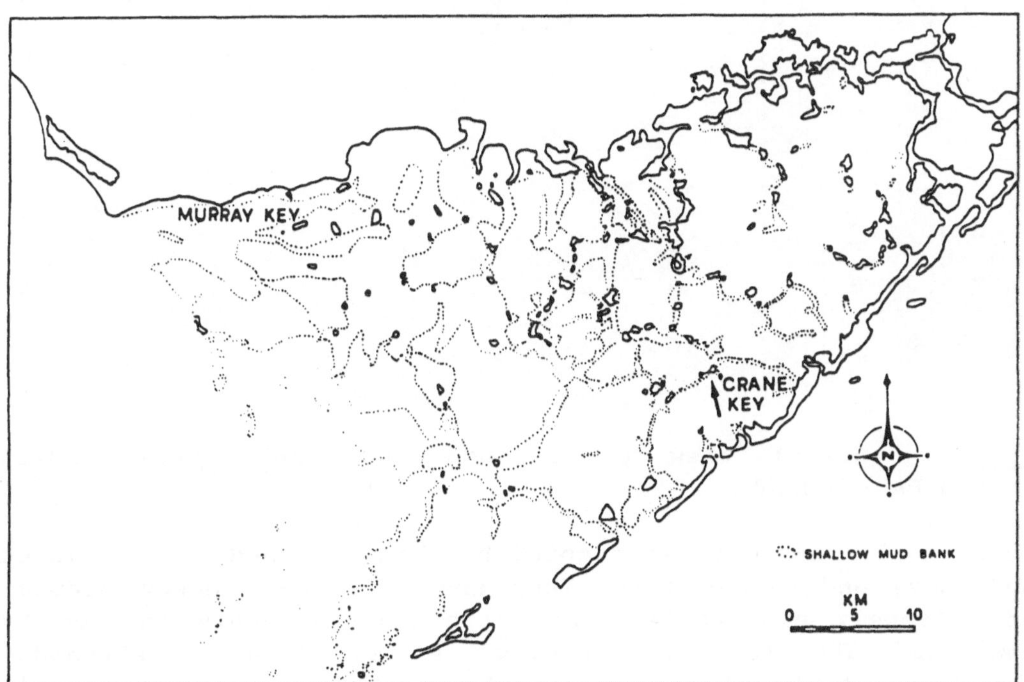

Fig.8 - Distribution of carbonate banks and islands within Florida Bay (Enos and Perkins,1979;Ginsburg,1956).

Progradation mechanisms of depocenters and the dynamics of infilling of the lagoonal depressions are poorly understood. Laporte suggested some possible causes leading to the formation of these mosaics: 1) migration of tidal channels; 2) migration of islands; and 3) short-term sea-level changes.

Shoals range from calcarenite to calcirudite in grain size and have an oncolite, bioclast or intraclast composition. Their width is difficult to reconstruct from ancient examples. Modern shoals, such as those occurring in the Florida platform, are hundred of m to 2 km wide, semicircular or elongated normal to the strike of the three sectors of the ramp (Fig.8; Enos and Perkins,1979).Similar sizes were estimated by Anderson (1971) in the reconstruction of the Manlius Formation (Fig.4). In some cases shoals are narrower in extent, such as those occurring in the Bahama platform (Fig.9).

Fig.9 - Aerial view of islands and banks near Andros Island,Bahama Bank.

Intershoal areas are represented by thinly bedded, dark colored mudstones and wackestones. They are in several cases nodular and strongly bioturbated. They are also probably incised by channels. The faunal diversity is low (algae, ostracods, thin-shelled bivalves, calcispheres), typical of poorly oxygenated environments.The very shallow intertidal mud and lithoclasts are draped by encrusting organisms, such as worms. Freshwater ponds may be present in the landwardmost areas and ponds, as is shown by the occurrence of plane-spiraled gastropods and snails and remains of root traces and plant debris. This situation occurs in Florida Bay (Fig.10), close to the junction to the land.

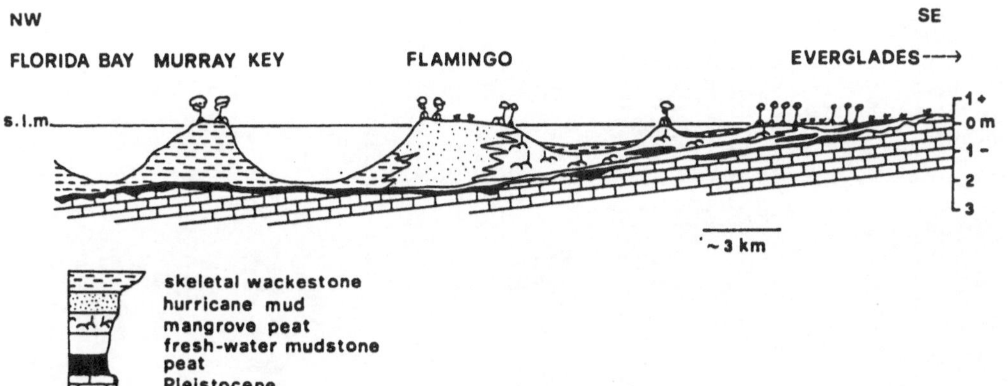

Fig.10 - Schematic profile of the recent sediments close to the Everglades (Scholl,1963).

These areas may be the sites of formation of pisoids and vadose diagenetic features which may also form on top of shoals. Radial oolite horizons may occur in the more sheltered, hypersaline intershoal areas.

Shoals and intershoal areas are organized into two main facies associations:

1) *thin-bedded alternations*; and
2) *thick-bedded alternations*

Thin-bedded alternations

These alternations consist of a few cm to 20 cm thin, calcarenite - calcisiltite layers interbedded with black, cm-thin argillaceous calcilutites which contain vegetal debris and plant remains. Bases and tops of calcarenite and calcisiltite beds are sharp. Physical sedimentary structures are gutter casts at the bases of coarser grained beds, pinch-outs within calcilutites and very thin mm to cm thin lenses of shell debris, or cm-thick mud storm horizons (Fig.11).

Calcarenite and calcisiltite mud layers frequently display a vertical thickening-upward trend which is accompanied by an increase in thickness and number of storm horizons and grain size (Fig.12). In other cases thicker beds are homogeneous mudstones with rare ostracods.

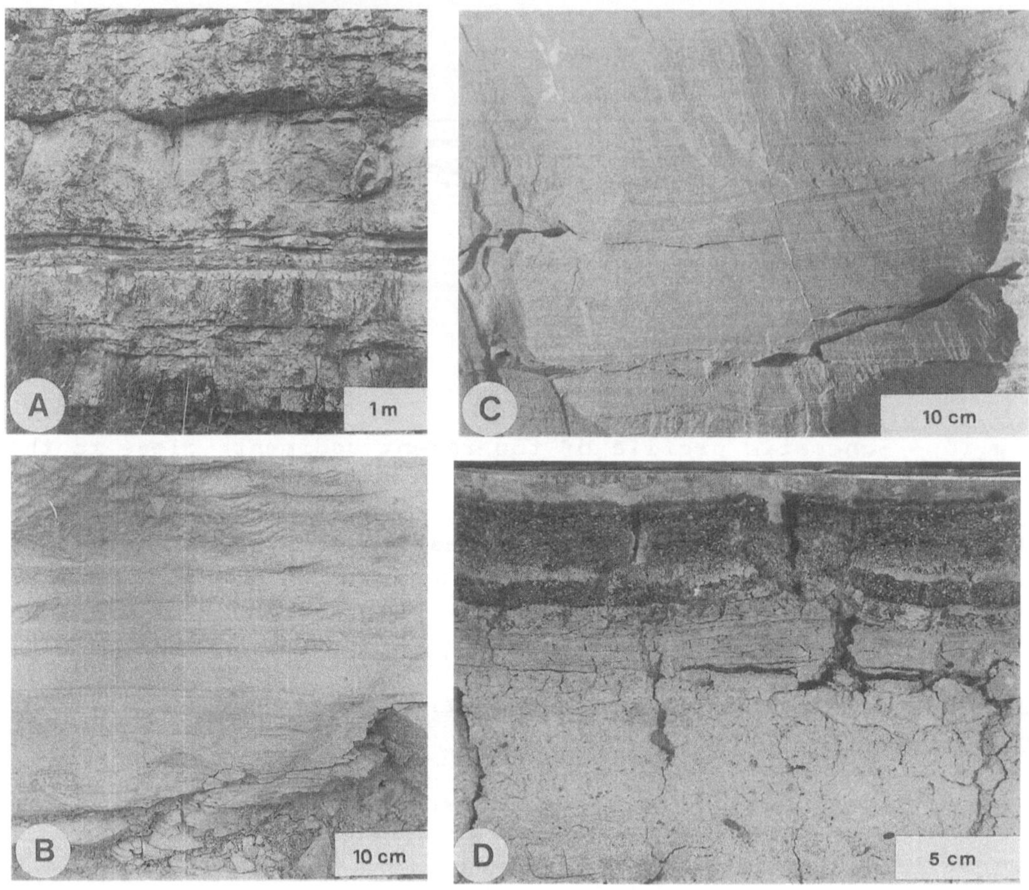

Fig.11 - Thin-bedded alternations. A,B: mudstones intercalated with marls ('Calcari Grigi' Formation, sections #5 and #19). C: very thin, storm-generated lamination (Givetian-Frasnian platform, Belgium). D: Hurricane- and winter-storm generated layers (Crane Key, Florida Bay).

There are basically two end-members of thicker beds:

1) calcarenites with fragments of bioclasts and shell debris;
2) structureless mud.

There is a lateral transition, over short distances, between these two end members, accompanied by a set of structure and compositional variations, as is shown in Fig.13.

A modern analogue of these marlstone - limestone alternations occurs close to the landwardmost termination of Florida Bay, close to the land (Fig.10). There, mud layers are aragonite and contain sparse detritus of mollusks, foraminifers and

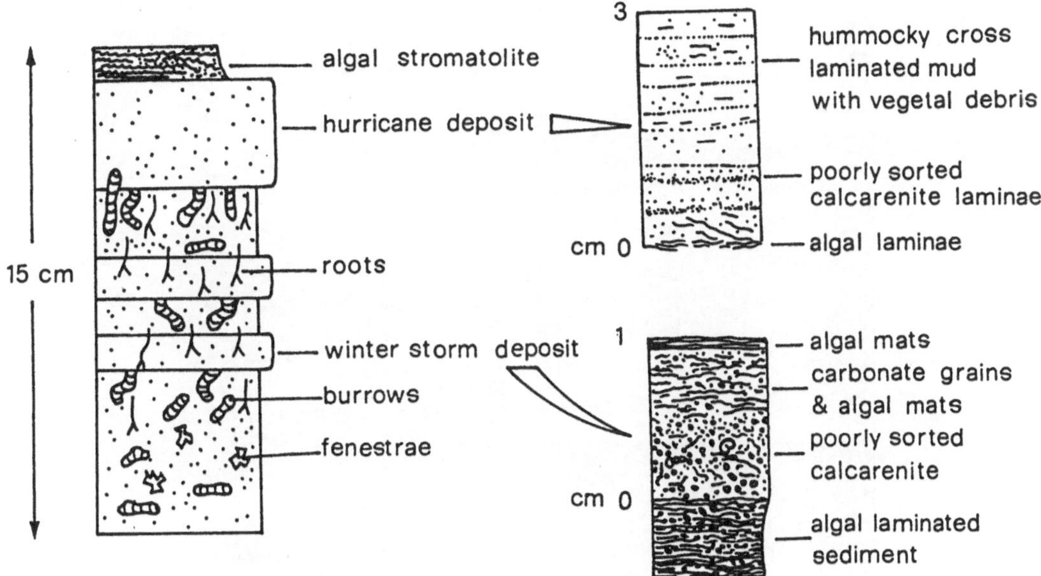

Fig.12 - Details of recent thin-bedded alternations from Florida Bay (Galli,1990).

calcareous algae. The mud contains mangrove plant debris, angiosperms and peat.

Sharp bases, lack of bioturbation and lags at the bases of individual interbeds evidence for 'event' deposition. These thin calcarenite-calcisiltite interbeds are analogous to interpreted distal tempestites described by Brett (1983),Aigner (1985) and Van Steenwinkel (1988). Their location and stratigraphic relationships with other facies types suggest that they formed in a very shallow setting. They are interpreted as the finer tails of onshore storms directed towards progressively shallower zones. Interbeds are also similar to hurricane-generated thinly bedded grainstones interbedded with organic-rich layers described by Wanless, et al.(1988) from Caicos platform.
Mud layers represent hurricane-generated deposits; they are similar to mud layers deposited by hurricanes Donna and Betsy in Florida Bay (Pray,1966;;Galli,1990;Fig.12).

Thick-bedded alternations

These are characterized by fining-upward calcarenites and

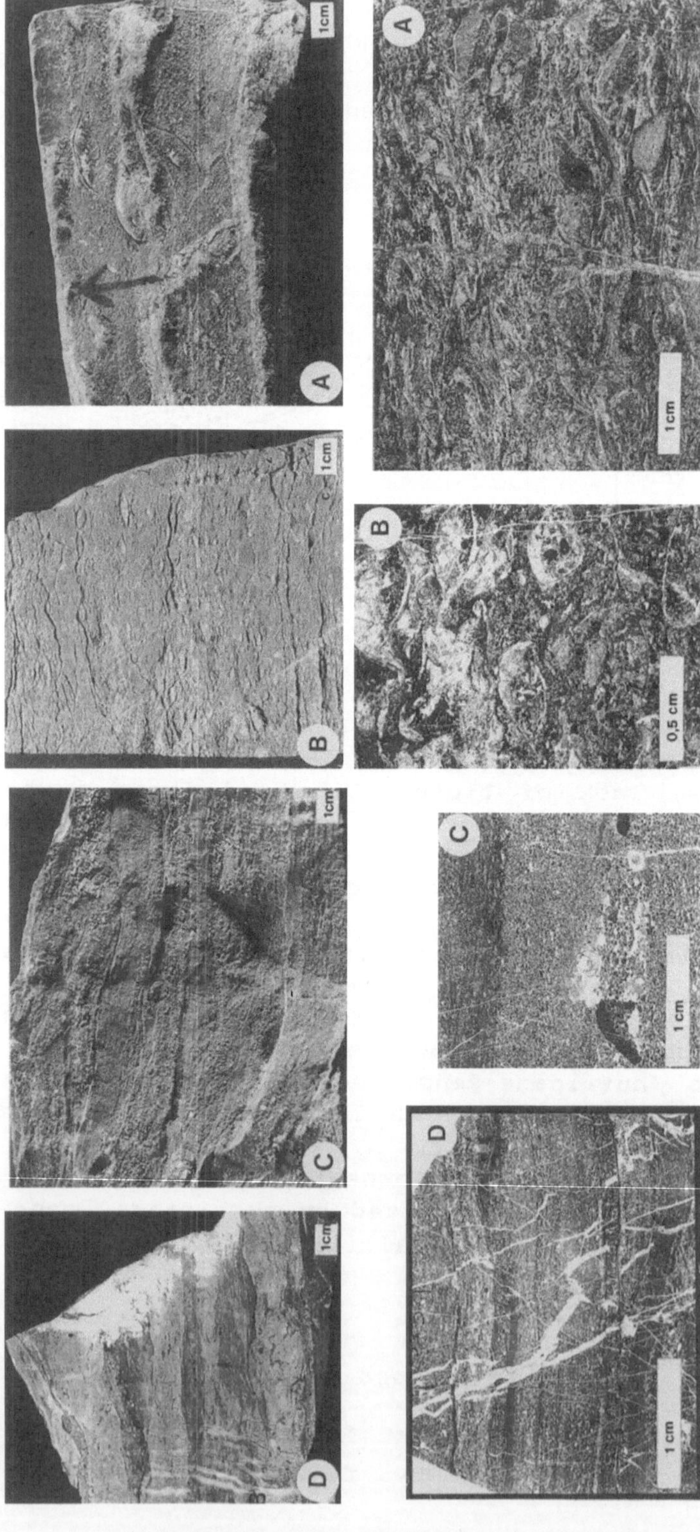
A
1cm
B
1cm
C
1cm
D
1cm
A
1cm
B
0,5 cm
C
1cm
D
1cm

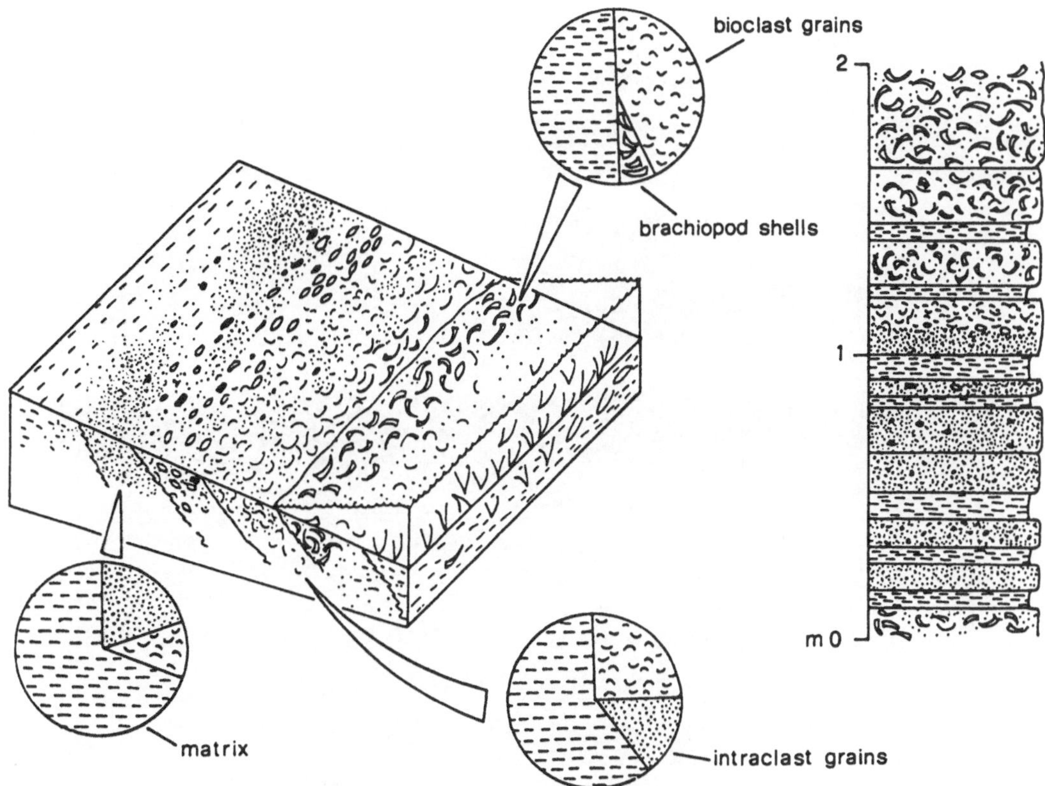

Fig.13 - Fining-leftward grain size trends and compositional changes of thin-bedded alternations (Devonian platform, Carnic Alps):brachiopod coquina (A)-->bioclast and intraclast grains (B)-->peloidal detritus with sparse bioclasts interbedded with mud laminae (C)-->mud laminae with sparse detritus interbedded with mudstones (D). These trends result from an onshore storm current (towards the left). The sketch shows a reconstruction of the paleoenvironment and the corresponding stratigraphic log.

---

calcisiltites, normally between 0.70 and 2m thick, interbedded with thin calcareous black mudstones containing scarse fossils. Thick beds are characterized by the following modal tripartite division (Fig.14):

1) a basal shell lag;
2) an intermediate mudstone-wackestone unit; and
3) a parallel-laminated, well sorted upper grainstone unit.

The upper unit may be missing. The lower contact above black mudstones is planar to erosional, as is shown by erosional

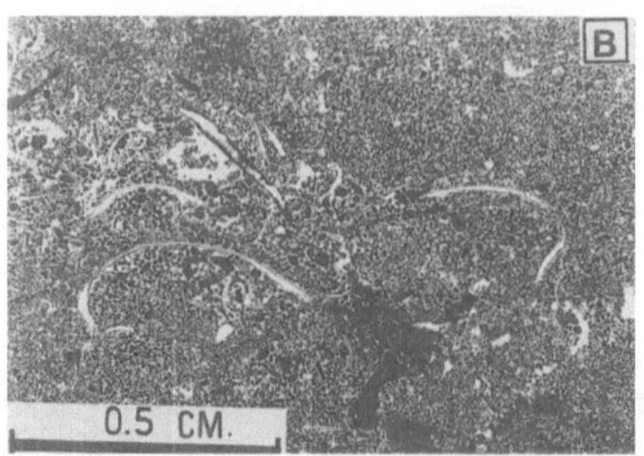

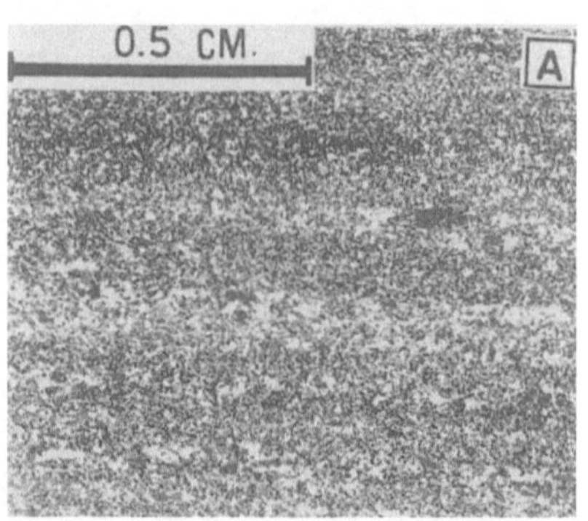

Fig.14 - Thick-bedded alternation ('Calcari Grigi' Formation:section #25). This composite bed consists of a scoured contact above lime mudstones, a basal lag zone composed of *Lithiotis* shells and other bioclasts (B), an intermediate wackestone unit consisting of amalgamated dm-thick layers, and an upper, hummocky-cross bedded unit with well sorted grainstones (A).

traces and gutter casts. The lower shell lag is poorly sorted and contains lithoclasts eroded from the underlying muddy interface, in addition to various types of unsorted bioclasts. Imbrications within more densely packed units contribute to the development of a better sorting. The transition to the intermediate unit is sharp; the muddy unit appears homogeneous; sedimentary structures were possibly present (such as faint

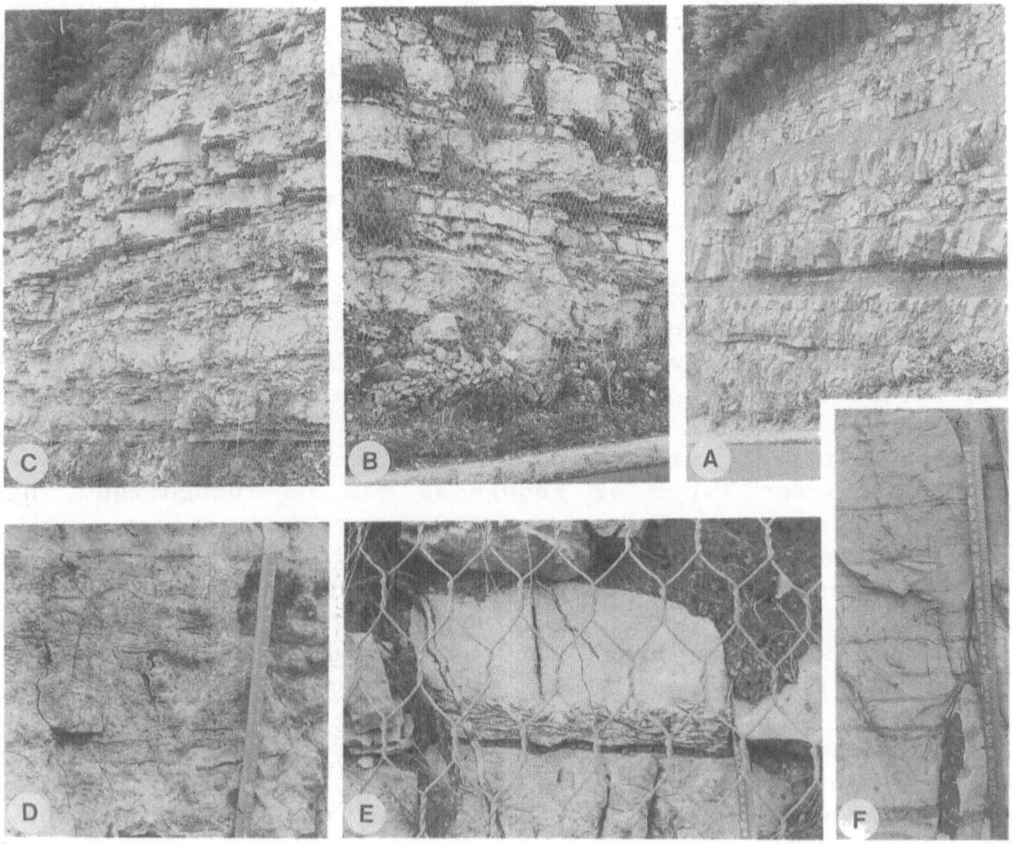

PROXIMAL DISTAL

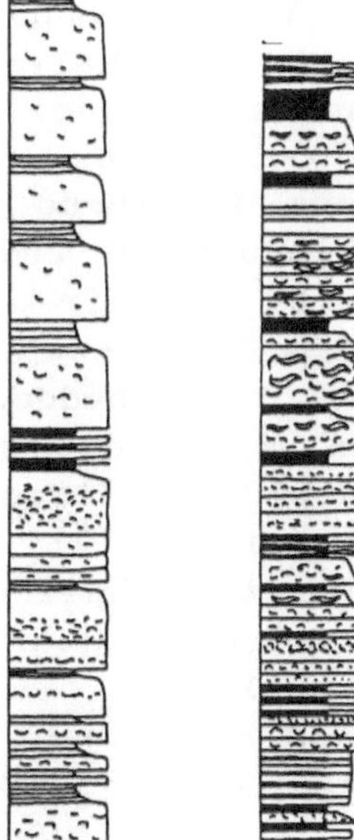

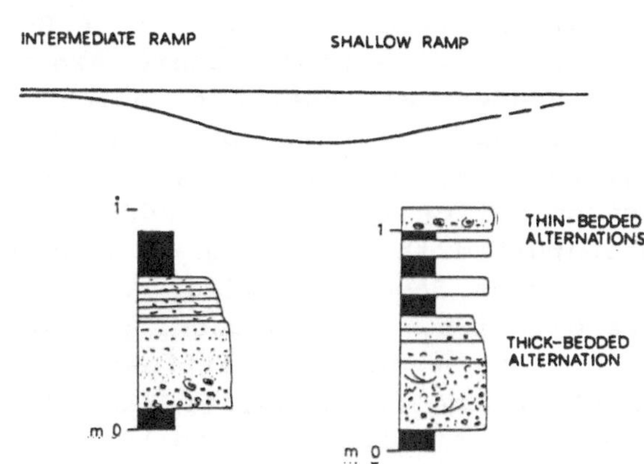

undulations: Fig.14) but were destroyed by non figurative bioturbation which is of varying intensity. The transition to the overlying unit is diffuse. The thinner upper unit is characterized by well sorted, laminated grainstones with hummocky, to wave ripple lamination, or parallel lamination.

This sequence is similar to other nearshore storm deposits described by Kreisa (1982), Brenchley, et al.(1979) and Kumar and Sanders (1976), among others.Storms and hurricanes are mainly responsible for the formation of these sequences. High flow regimes can be inferred from a number of features, such as spar calcite cement, shelter porosity cement, amalgamated beds, well sorted laminae and hummocky cross-lamination.

Based on different proportions of the intermediate and lower units, different types of sequences can be recognized. They represent a spectrum of landward to seaward transitions, with a landward tendency towards a decrease in shell percentage. Fig.15, taken from the 'Calcari Grigi' Formation, illustrates changes in bedding styles going from the intermediate ramp to marsh environments typical of the landwardmost site of the shallow ramp.

This trend is analogous to that occurring in Florida Bay; the Holocene platform, 36 km wide, does not exceed 12 m in depth. The floor which gently dips towards south and southeast

---

<u>Fig.15</u> - <u>A,B,C</u>: changes in bedding styles of thick-bedded and thin-bedded alternations from areas closer to the shallow ramp (right) to the intermediate ramp (left) in the 'Calcari Grigi' Formation (sections #19 : A,B-->#6 : C). Closer to the intermediate ramp (C) there are no mudstone interbeds within thin-bedded alternations which are distinguished from the underlying thick-bedded alternations by their small thicknesses. In more distal positions lime mudstone become predominating and thick-bedded alternations, interbedded with bioturbated thicker sets of lime mudstones, stand out clearly on outcrop faces. Proximal and distal thin-bedded alternations are organized into thickening-upward trends (cf.Fig.12). <u>D,E,F</u>: closer views of compositional changes from the shallow ramp (D,E) to the intermediate ramp (F), where shallow-water forms of *Lithiotis* are replaced by thin-shelled mollusks.
The sketches below show proximal (section #6) and distal (section #19) alternations, and facies types ('Calcari Grigi'Formation).

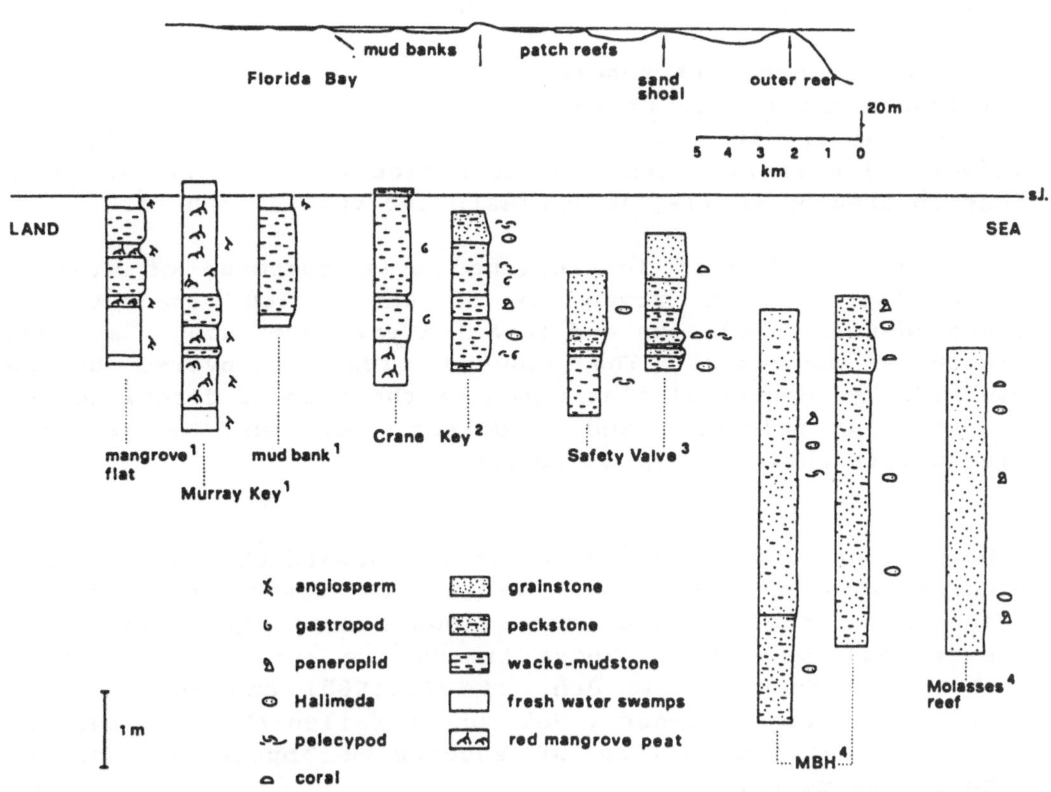

Fig.16 - Bathymetric profile of the Florida platform (Enos,1977) and stratigraphic logs (1: Craighead,1969;2:Enos and Perkins,1979;3:Aigner,1985;4:Enos,1977).

(Atlantic Ocean) is complicated by topographic relieves constituted by carbonate shoals, banks and patch-reefs.

Sedimentation took place during a phase of relative sea-level rise (Scholl,et al.,1969). The flooding of the platform initiated 8500 years ago. A slowing in the sea-level rise took place 5500 years ago (30 cm/1000 years).

The cross section of Fig.16, oriented northwest- southeast, normal to the strike of the platform, reflects a transition from open to semirestricted environments (Enos,1977).

The cycles actually developing in Florida Bay are asymmetric, transgressive sequences capped by intertidal and supratidal deposits (Enos and Perkins,1979). They consist of:

1) a basal coarse mollusk grainstone and packstone (shallow marine bay);

2) a mud bank containing few scattered organisms; and
3) an upper island deposit consisting of algal laminated sediments and cm-thick storm layers.

These banks are intercepted by onshore directed storms (hurricanes and winter storms) and in section show a windward side and a leeward side dipping downcurrent (Fig.17). The windward side is steep and characterized by a nip and a beach ridge covered by shells, intraclasts and vegetal debris.

These keys consist of low-inclined sets composed of stacked graded layers each given by a basal lag (packstones, wackestones with mollusks and foraminifers) overlain by mollusk foraminifer mudstones. The leeward side is covered by an intertidal - supratidal flat which is the site of deposition of fine-grained peloidal and bioclast detritus. A similar structure is shown in Fig.17 (below).

In both modern and ancient situations landward changes in cycle thicknesses, composition and grain-size trends reflect the action of onshore-directed storm flows decreasing in intensity towards shallower areas. These trends are analogous to those described by Bourroulh-le Jan, et al.(1985) and to the case history reported by Aigner (1985) who detailed the structure of spillover lobes consisting of successive inputs of onshore directed sediments.

Proximality-distality trends

The decrease of storm effects towards deeper, offshore water was demonstrated by Aigner (1985) in modern and ancient situations dominated by a progradational, regressive trend. It shows that: "The percentage of sand, storm layer thickness and grain size, as well as the degree of amalgamation show a continuously decreasing trend from shallow to deeper water, while bioturbation generally decreases" (Aigner,1982,1985).

This concept is not applicable in the cases examined further below where the various parameters display opposite trends.

A distal position for thin-bedded alternations is indicated by the following features:

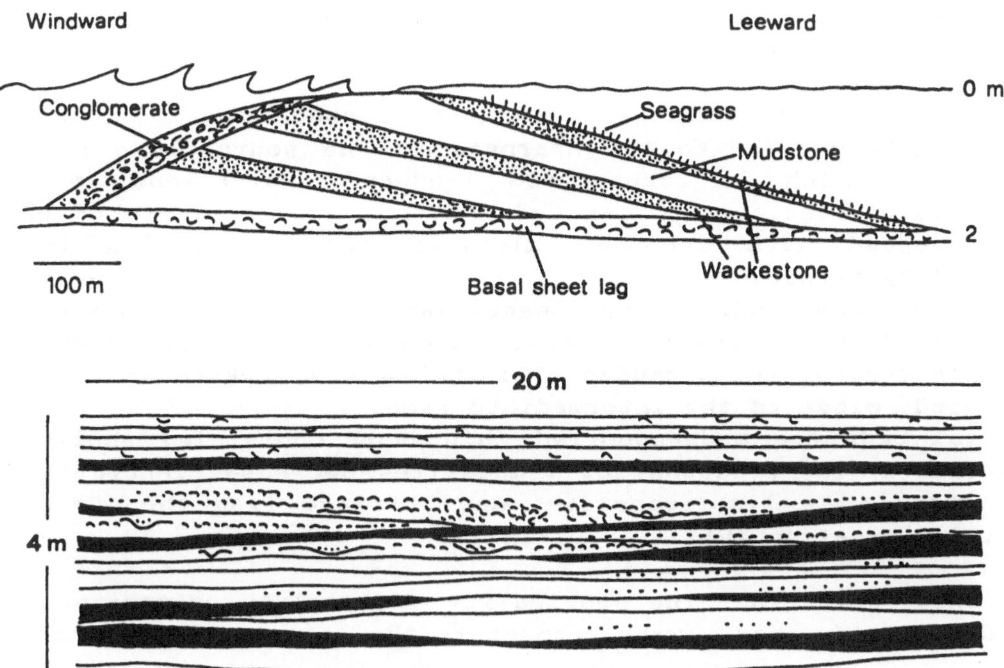

Fig.17 - Above: cross section of a mud mound from Florida Bay (Upper Cross Bank: Bosence,1988). Below: front view of an ancient analogue from the 'Calcari Grigi' Formation (section #6) formed by a stacked lenses of skeletal wackestones interbedded with thin lime mudstones.

1. lack of basal units;
2. flat bases;
3. scarcity of amalgamations;
4. small thicknesses and fine grained size of material.

A proximal position for the thick-bedded alternations is shown by the following:

1. occurrence of complete storm sequences (tripartite division);
2. erosional surfaces;
3. thick basal units with lags and a grainstone texture;
4. amalgamations of sequences; stacking of basal units.

Storm deposits here are a main result of an onshore sediment transport in a shallow, nearshore water, as a direct effect of wind-drift currents. There are two possible causes for this: 1) the deep ramp is not deep enough to favour offshore sediment transport due to gradient currents and bottom return flows; and 2) the overall trend is transgressive.

'Proximal' and 'distal' here are used in a relative meaning: they refer to the nearness to the source area of the material which is transported landward. Variations in the amounts of bioclasts versus mud content, and thickness of beds interbedded with lagoonal micrites reflect proximality - distality trends.

Proximal beds occur in more deeper lagoonal areas which are the source of the bioclast material transported landward; they are bioclast-rich, thick-bedded grainstones and packstones located seaward, close to the intermediate ramp.

Distal beds are mud-supported thin beds containing scattered bioclasts. The frequency of tempestites decreases landward, as is shown in a 'isotempestite' map obtained from the 'Calcari Grigi' Formation (Galli,1990:Fig.2).

In these situations the proximality - distality criterions are 180° out of phase with respect to the proximality-distality

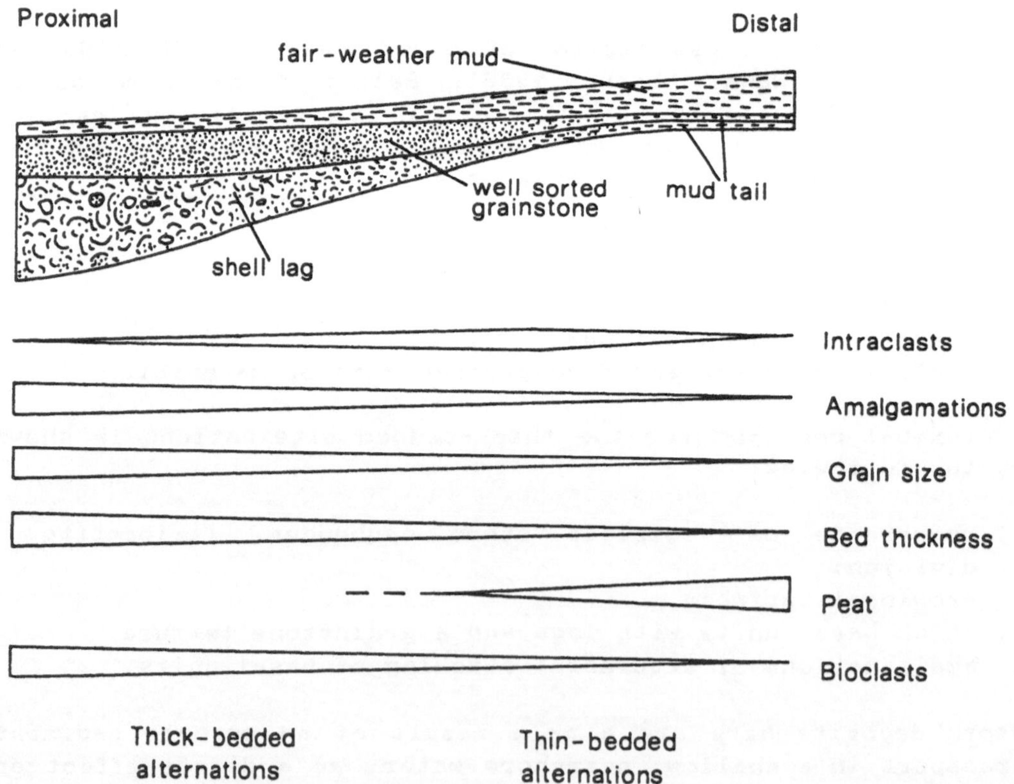

Fig.18 - Proximality-distality trends in the intermediate - shallow ramp.

trends described by Aigner (1985), and others, mainly because the situations encountered are characterized by transgressive, retrogradational trends, whereas the trends described by Aigner (1985) apply to progradational,regressive systems.In order to avoid misleading interpretations,the use of the proximality - distality criterion must be confronted with other independent petrographic and geological data.

Intermediate ramp

The intermediate ramp is located between the lower and the upper limit of wave action. This is a high-energy area characterized by a rough topography and intercepted by waves and currents which are mainly responsible for a complex and lenticular geometry. The large size of bedforms, several m in wavelength, points to a deposition as sand blankets. Common sedimentary structures are erosional features, channels and megaripples migrating under episodic high-energy conditions.

The distribution of sediment in close proximity to the intermediate ramp is controlled by the underlying topographic surfaces.There are different belts: namely 1) skeletal sands; and 2) ooid-rich sands. They form respectively 1) high-energy skeletal beaches; and 2) sandwaves. These two belts may occur

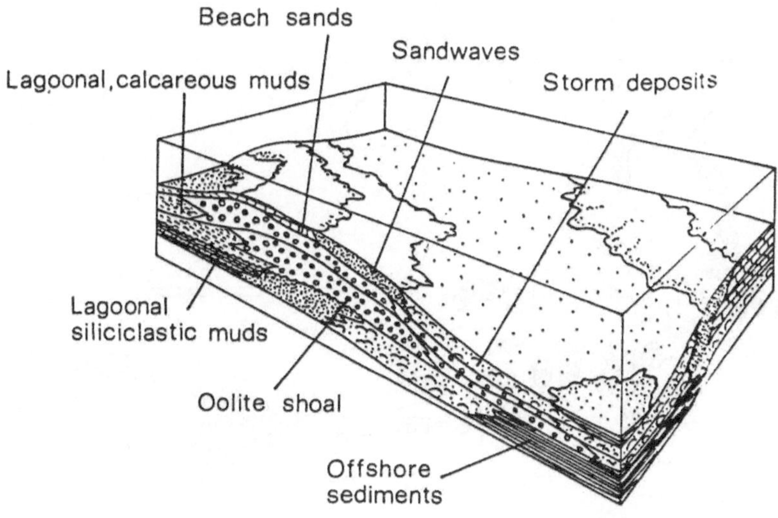

Fig.19 - Example of facies assemblages and facies distribution in the intermediate ramp (Dinantian platform: from Van Steenwinkel,1988:Fig.6.13).

in the same depositional system; for example Van Steenwinkel (1988) described a situation where beaches are located landward with respect to sandwaves (Fig.19).

In the Florida platform the intermediate ramp is a complex area occupied by a discontinuous alignment of nearshore banks and small tidal deltas located close to the tidal inlets and to the transition from Florida Bay and the Gulf of Mexico (Fig.20). The dynamics of banks described by Aigner (1985) takes place by increments of sediment volumes transported by hurricanes. The 'event' accretion dynamics leads to a landward transition from beaches and shell islands,to skeletal banks and mud banks (Aigner,1985: Fig.18).The windward side of lobes and subaqueous dunes is enriched in shells; bioclast beaches may form on their top; the leeward side conversely is muddier in composition. Onshore storm floods also produce spillover lobes (Ball,1967).

A situation similar to that described by Aigner (1985) from Florida Bay occurs in the Devonian platform (Carnic Alps, Italy) where shoreface bars, storm bars and beach bars resulted

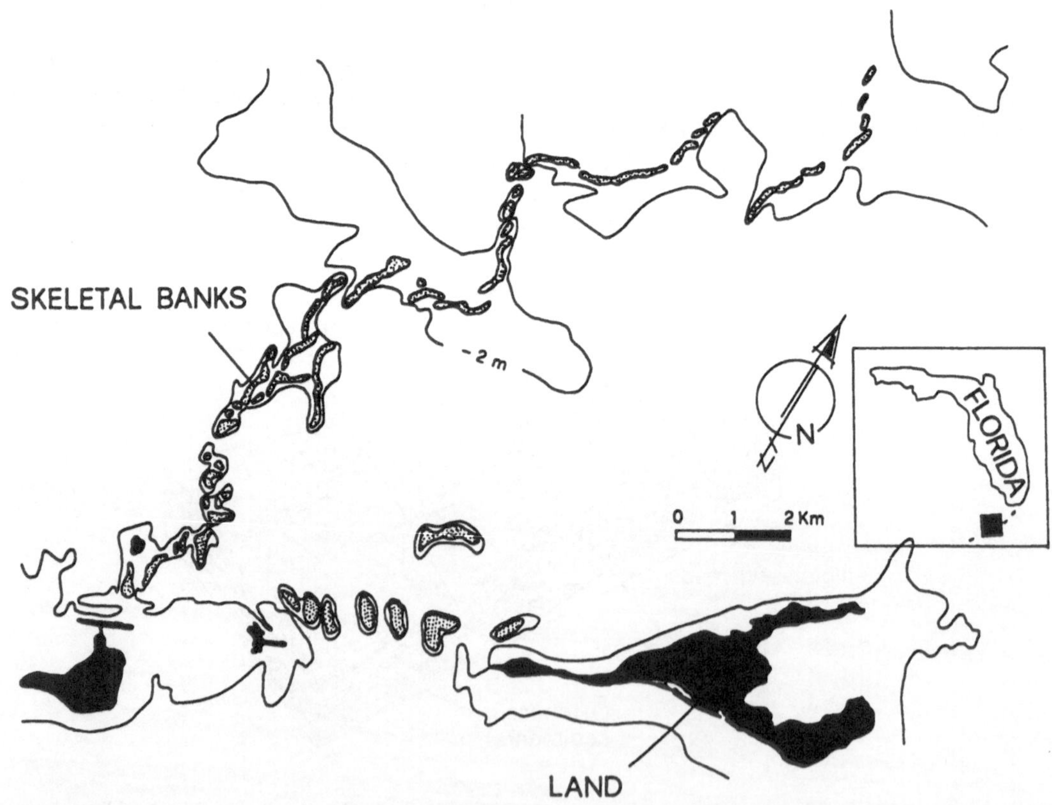

Fig.20 - Skeletal banks of Florida Bay, in the intermediate ramp (from Ginsburg,R.N.,with permission).

from the piling up of shells in a strandline environment. The composition of the beach was built up by stepwise assemblages of sand bodies of different provenance, such as open lagoons and reef flats (Galli,1986; Fig.21).

Open lagoons are associated with sandwaves and bars located in the intermediate ramp. They are constituted by skeletal sands and populated by thin- and thick-shelled mollusks, other than gastropods, crinoids, foraminifers and corals. The open lagoon facies are generally transitional to the shallow ramp and located at both landward and seaward sides of sandwave bodies. More sheltered areas are bioturbated. Open lagoons located seaward are the sites of formation of m-thick bars, whereas those located towards the shallow ramp, on the other side of the barrier, more commonly display fining-upward sequences and bioclast lags transitional to those of the shallow ramp.

In the Bahamas skeletal sands are less than 2 km wide; they are wider (10 km) along the northern margin and extend from the upper slope across the outer margin to water depths of about 10 m. They terminate abruptly against ooid sands and islands.

In the Jurassic 'Calcari Grigi' Formation the main lithofacies of the intermediate ramp are constituted by oolite grainstones and packstones, and skeletal wackestones, with subordinate amounts of nodular wackestones and mudstones. Here the intermediate ramp is dominated by oolite sandwaves, such as that exemplified in Fig.22: the section may be divided into a lower and an upper member. The basal member is a poorly sorted lithofacies with surficial oolites and lumps and an admixture of skeletal grains which constitute the nuclei of the ooids, intraclasts, peloids and coated grains. The upper lithofacies is a well sorted grainstone with tangential ooids. Sedimentary structures are massive bedding at the base, hummocky-cross bedding in the middle and wide, cross-bedded shallow channel fills reminiscent of the swaley cross stratification (Leckie and Walker,1982) at the top.Individual fining-upward sequences within hummocky cross-bedded sets contain internal sequences of structures comparable to the 'b-c-d' Bouma sequence: planar lamination (Walker,et al.,1983) vertically grading to trough-cross lamination with normal graded sets (cf. Dott and Bourgeois,1982). The basal massive bedded oolite grainstones and packstones are interpreted as shoals and sandwaves situated in a storm-dominated area, at a shallow water depth. Ooid sand migration took place under the influence of storms.

Van Steenwinkel (1988) interpreted similar 'oolitic lump facies' from a Carboniferous ramp in Belgium as a transgressive

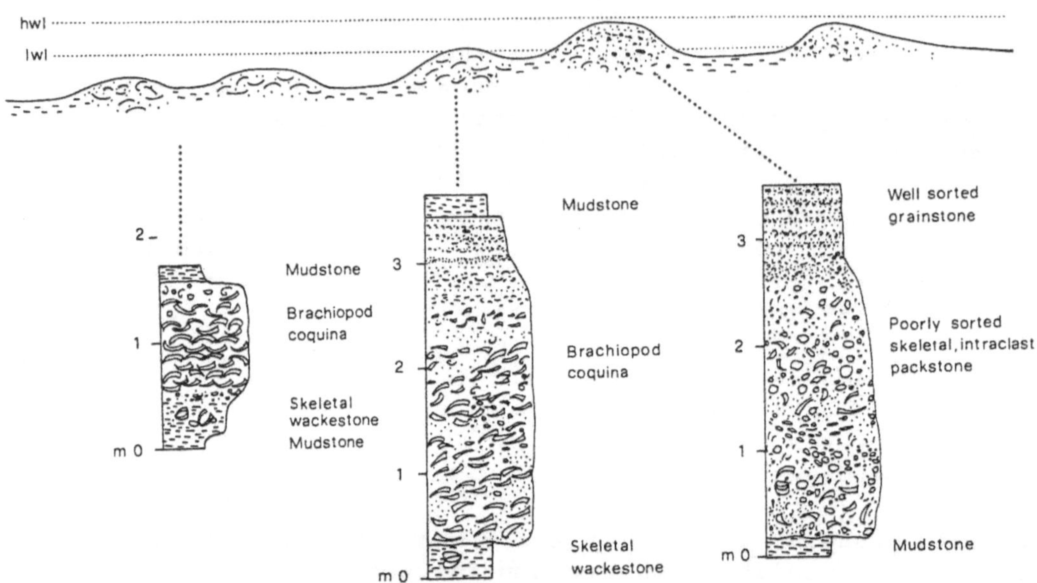

Fig.21 - Bathymetric profile and facies types of the intermediate ramp occurring in the Devonian carbonate platform, Italy (Galli, 1986).

lag deposit formed during an incipient drowning. Brief episodes of bottom agitation alternated with longer, quiet periods characterized by a slow sedimentation rate. She compared such grapestone lumps with the grapestone lumps occurring at Lily Bank (little Bahama Bank) and at Cat Cay Platform (Great Bahama Bank) where oolitically coated sands migrated landward in response to an increase in the sea-level rise and subsequently became inactive and the site of sea-grass growth and grapestone formation. Migration takes place mainly during large storms and hurricanes.

Oolitic sand shoals are common along the edges of several Bahama areas (Cat Island platform, Joulters Cay, Berry Islands, and so on). They occur as active ooid shoals marginal to the open sea, and as stabilized blanket sheets of oolite sand flats which are gradational with other platform sediments (Multer, 1977). Skeletal admixtures are greatest in deeper water areas. Modern oolite sandwaves migrate landward in response to a slow sea-level rise. Subaqueous dunes are elongated parallel to the

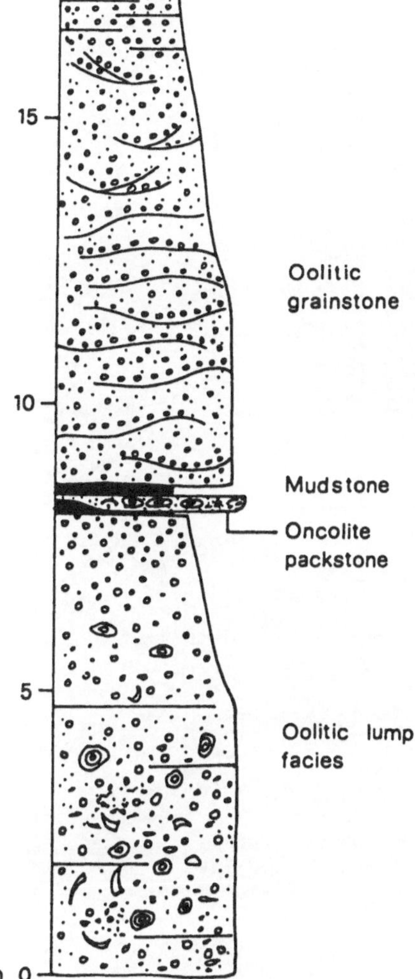

Fig.22 - Oolitic sequence occurring in the 'Calcari Grigi' Formation, Venetian Alps (section #22).

direction of prevailing currents. Active shoals overlie the inactive mixed oolite facies and become stabilized by sea-grass growth.

Deep ramp

The deep ramp is a deep lagoon; the lagoonal floor, located 5 to 15 as much as 20 m below wave base, is typically populated by oligotypic, high-density and low-diversity mollusks. Bivalves form tightly packed accumulations embedded in a wackestone

Fig. 23 - Shell accumulations and bedding styles in the deep ramp. A: *Lithiotis* shells ('Calcari Grigi' Formation, section #12);B: mollusk coquina and corals in the Pleistocene Fort Thompson Formation;C:brachiopod coquina,Devonian platform, Italy; D: *Lithiotis* bank with thin interbeds of skeletal wackestones ('Calcari Grigi' Formation, section #30); E: 9 m thick *Lithiotis* bank ('Calcari Grigi' Formation,section #25): observe cross bedding and scours at the base and traces of multiple scours at the top.

matrix (Fig.23A,B,C). Mudstones alternate with fossiliferous wackestones. These fossils and organisms are also found in other sectors of the ramp, but they do not form big volumes of sediment as in the deep ramp; here they become arranged as catch-up reefs, biostromes and banks reaching 3 as much as 7 - 9 m in thickness (Fig.23 D,E).

Walkden and Gutteridge (1984,1987) and Gutteridge (1987) described different styles of mud mounds from the late Brigantian Eyam Limestone in the Derbyshire carbonate platform (Fig.24). The mounding styles (tabular - dome-shaped - laterally and vertically accreted mounds) were determined by the water depth of deposition. Laterally accreted mounds are usually found in the shallow ramp, whereas vertically accreted mounds occur in the deep ramp.In contrast to the laterally accreted mounds, flank facies of vertically accreted mounds are poorly developed;the sediment surrounding bases of vertically accreted mounds display some cross bedding evidencing for an initial shallow water of deposition. The core, composed of peloidal mud, does not show any zonation which rules out a change in the water depth throughout deposition, except at the very top, where vadose diagenetic features point to periods of subaerial exposure.

Analogous bank styles occur in the Jurassic shallow ramp in the Venetian Alps (Fig.24). Small mounds, 1 m thick to 5 m wide, together with multiply truncation scours infilled with *Lithiotis* shells, occur in the shallow ramp. Banks occurring as steeply inclined, sigmoidal to undulating beds, may correspond to the flank facies of laterally accreted *Lithiotis* banks developed close to the intermediate ramp. The growth of thick *Lithiotis* banks occurred in deeper, subsiding areas of the lagoon, as is suggested by the lack of mechanical structures within banks, a high faunal density, uniformity, large, 'in situ' shell sizes, and occurrence of cross bedding only at the base and top of the banks (Fig.23).

The *Lithiotis* banks were interpreted by Bosellini (1972) to have been deposited within a system of migrating tidal point bars (Fig.25). According to that interpretation, B-type mounds formed at the bottom of point bars, C and D-type mounds were interpreted as point bars or as mechanical deposits formed in the bends of meandering channels or between larger channels.The model put forth by Bosellini (1972) is wrong (cf. Wright,1984; Galli,1990).Nevertheless, his profile of point bars may be taken as the profile of an intrashelf ramp: hence, B-type mounds formed in the deep ramp; C,D, E types developed in the intermediate ramp, whereas channels and scours were produced in the shallow ramp.

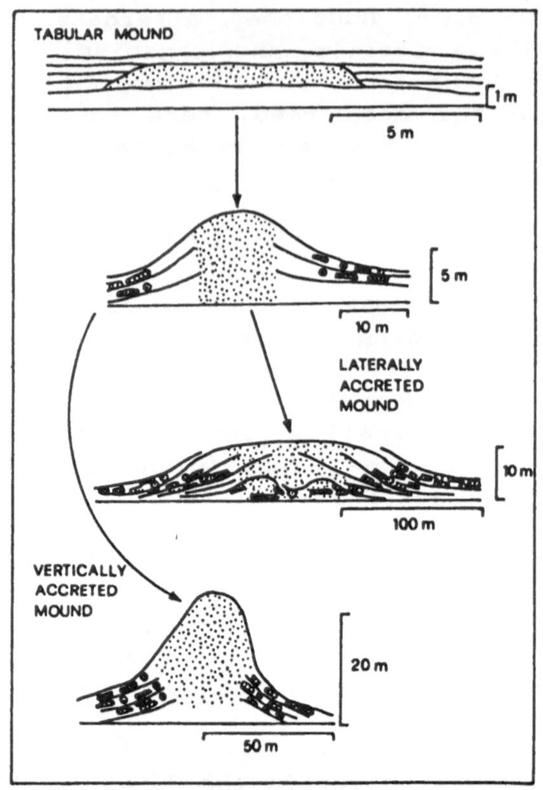

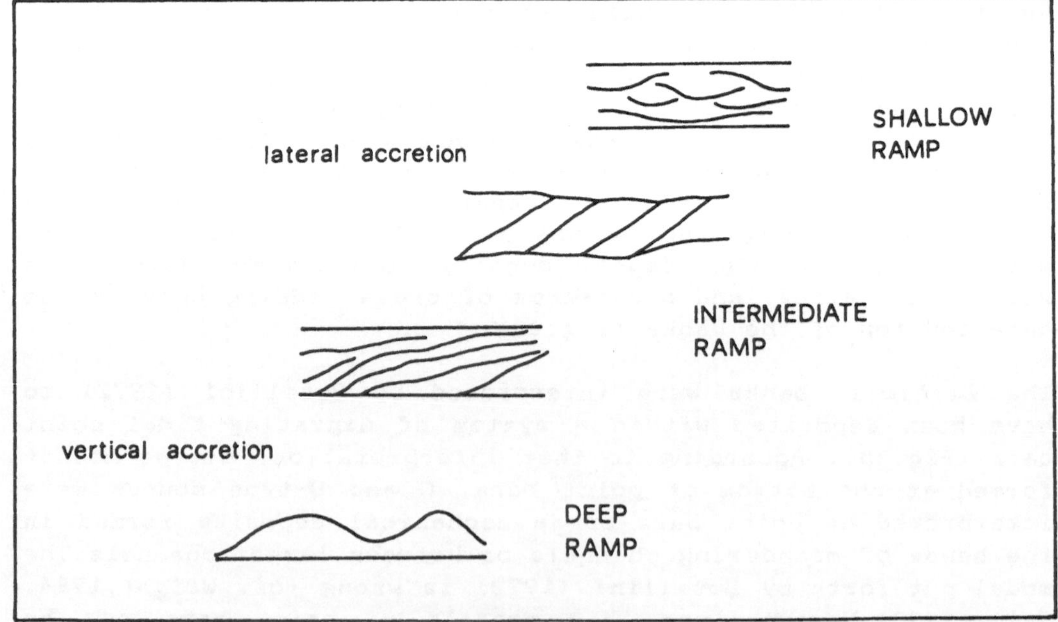

Fig.24 - Top : mud mounds in the Asbian-Brigantian platform, England (Walkden and Gutteridge,1987). Bottom: types of *Lithiotis* banks.

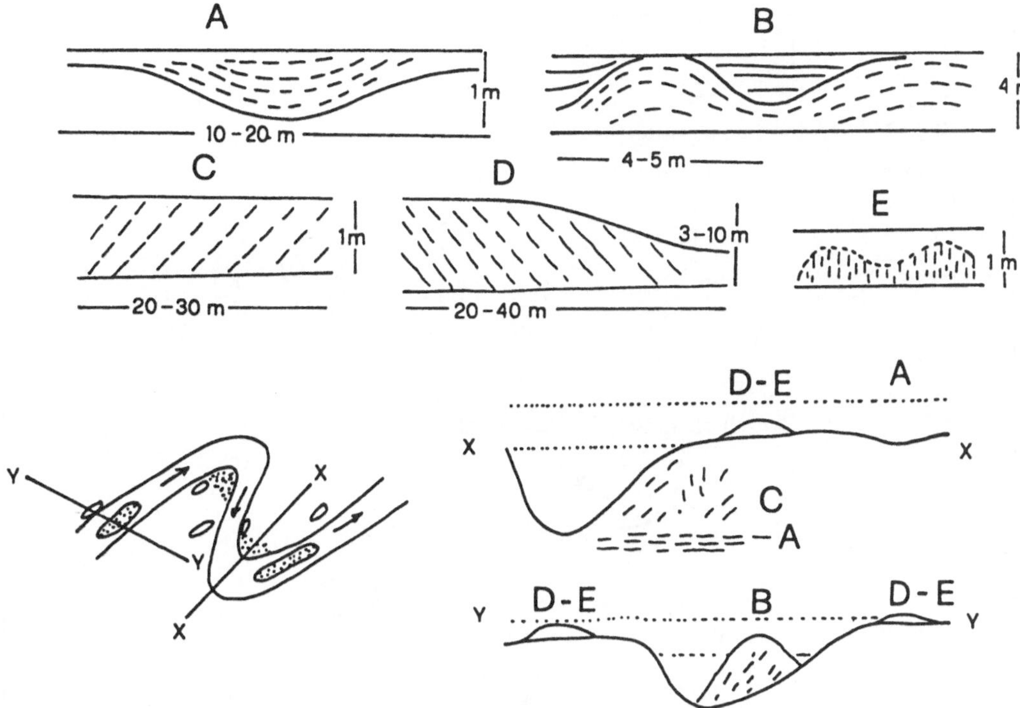

Fig.25 - Paleoenvironmental model of Bosellini (1972). He mistakenly interpreted *Lithiotis* banks to have formed within a system of migrating tidal point bars. However, tidal facies are absent in the study area and there is not much evidence for migration of tidal channels in modern carbonate environments. The model is still applicable if the profile of the 'point bar' is considered to be the profile of an intrashelf ramp.

The Jurassic intrashelf ramp occurring in the 'Calcari Grigi' Formation was subjected to a strong episodic hydraulic action (Galli,1990). Channel morphologies reflect the former depth (Fig.26).Undulating bedforms and flattened,wide channel outlines occur in the deep ramp. Deeply incised,tighter channel outlines cutting shoals are commonly found in the shallow ramp.

The distribution of the types of channel outlines occurring in the study area of the Calcari Grigi' Formation (Fig.26) follows the variations in bathymetry of the ramp. Within shallowing-upward sequences, wide channels occur at the base (Fig.26:A), whereas semicircular scours are found towards the top (Fig.26:B).

This is also apparent from an examination of vertical sequences of channel morphologies (Fig.27).

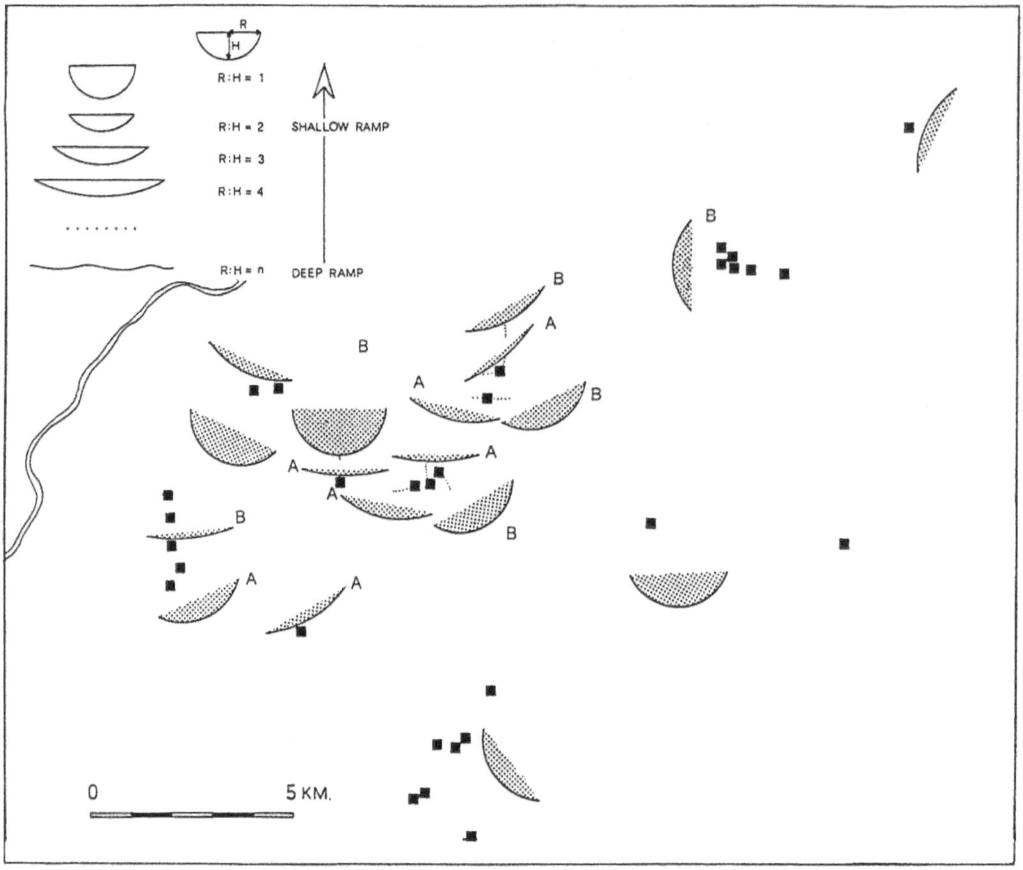

R
H
R:H = 1
R:H = 2
R:H = 3
R:H = 4
R:H = n
SHALLOW RAMP
DEEP RAMP
A
B
0
5 KM.

SHALLOW RAMP
DEEP RAMP.
N

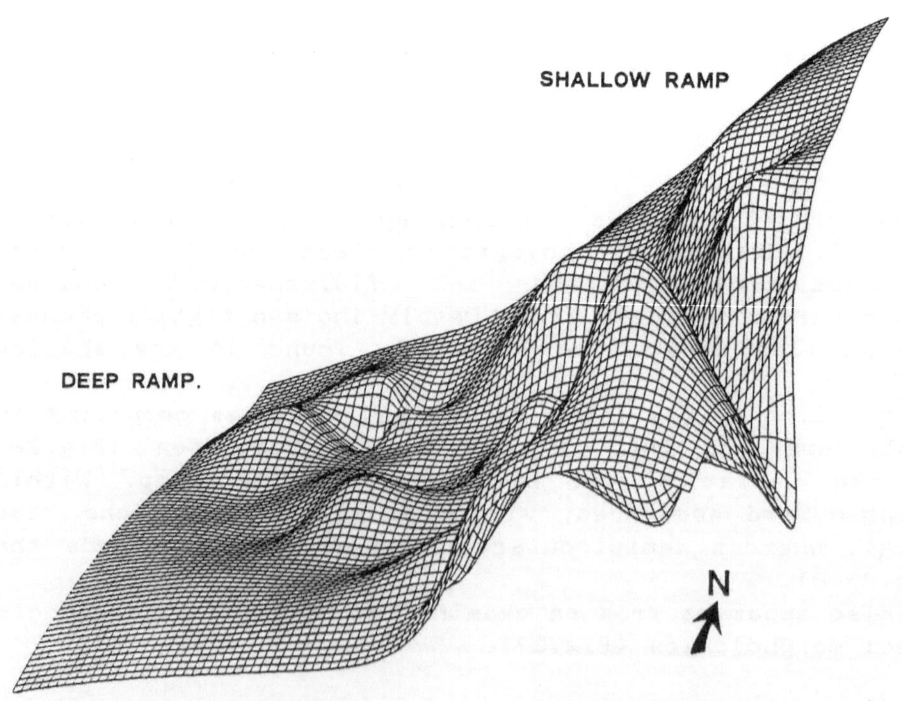

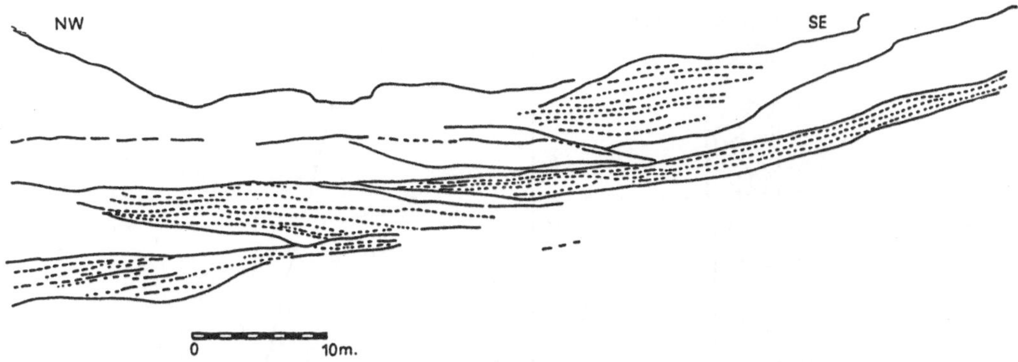

Fig.27 - Upward changes in channel morphologies from undulated to elliptical reflect a shallowing upward trend,confirmed by vertical changes in lithofacies ('Calcari Grigi' Formation: section #27).

Banks occurring in the deep ramp are analogous to a number of 'catch-up' reefs described in the literature (Fig.28), as is indicated by the vertical decrease in size of fossils and an increase in the proportion of lithoclasts. The uppermost parts contain fossils more indicative for a shallower setting, scours, channels and wave-generated structures reflecting a deposition above wave base.Trends displayed by these sequences record transitions from deep, quiet water stages, to a shallower,agitated water stage that corresponds to variations in the steepness of the curve of change in sea-level. Initially, the space available for sedimentation was infilled by a mechanism of vertical sediment growth (vertical aggradation); later, by lateral migration (progradation), once the sedimentary interface caught-up with the sea-level.

In the Florida platform the deep ramp, described by Perkins (1977), consists of a landward area characterized by a semirestricted water circulation, and a seaward area, where circulation is more limited. The composition and grain size of the seawardmost side of the deep ramp reflects the proximity to the reefs. Sediments become finer and of more varied composition towards the intermediate ramp where the inclined

---

Fig.26 - A:Variations of ellipticity of channel outlines as a function of the height:amplitude (H:R) ratios with depth.B: schematic representation of the outlines of channels and scours, in the study area of the 'Calcari Grigi' Formation, related to the reconstructed bathymetry. Circular scours are mainly confined to the shallow ramp; flattened channels occur preferentially in the other sectors.

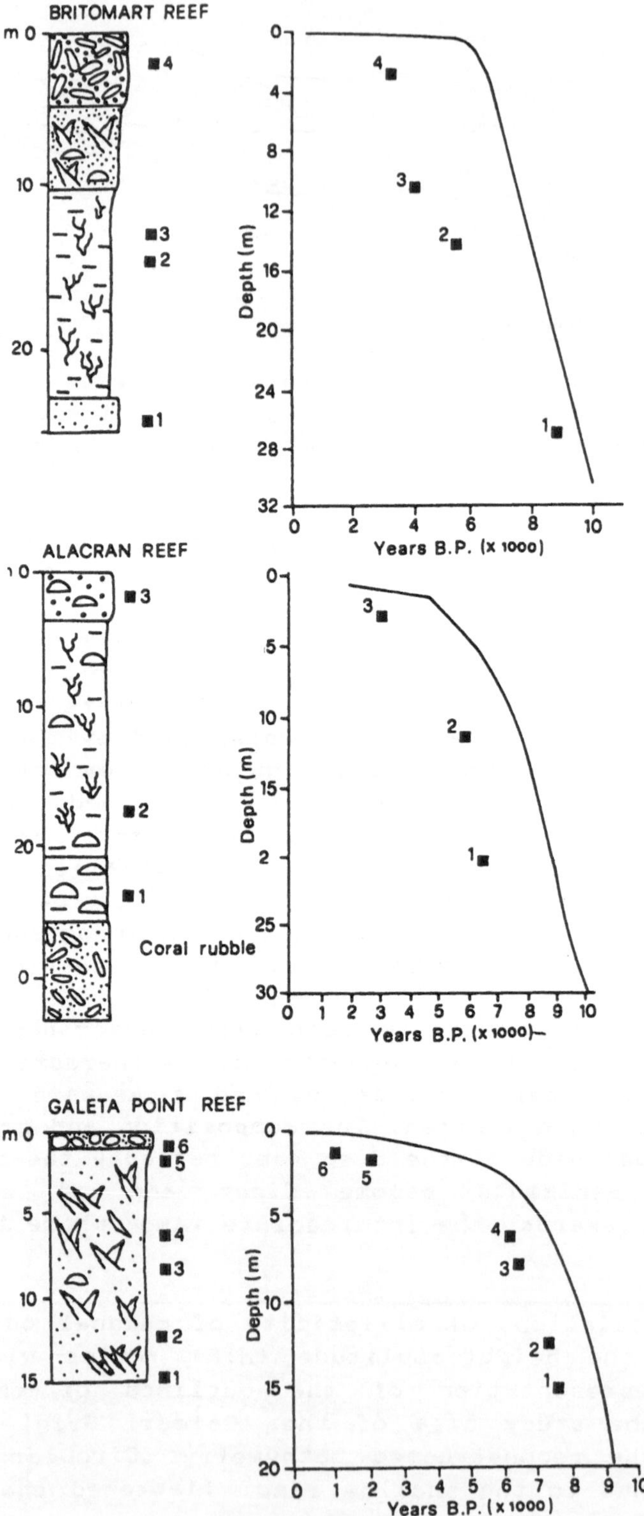
BRITOMART REEF
m 0
10
20
4
3
2
1
Depth (m)
Years B.P. (x 1000)
ALACRAN REEF
Coral rubble
Depth (m)
Years B.P. (x 1000)
GALETA POINT REEF
m 0
5
10
15
6
5
4
3
2
1
Depth (m)
Years B.P. (x 1000)

surface is covered by patch reefs and poorly sorted, bioturbated wackestone banks. These form a compound, elongated, discontinuous wedge parallel to the edge of the platform. They are also asymmetric, with the steepest side located landward. Inner structures close to the intermediate ramp consist of accretionary sets inclined at a low angle (1° - 5°) and striking parallel to the edge of the platform (Enos,1977).They may be a modern analogue of ancient, inclined beds connecting the intermediate and deep ramp (Fig.29).

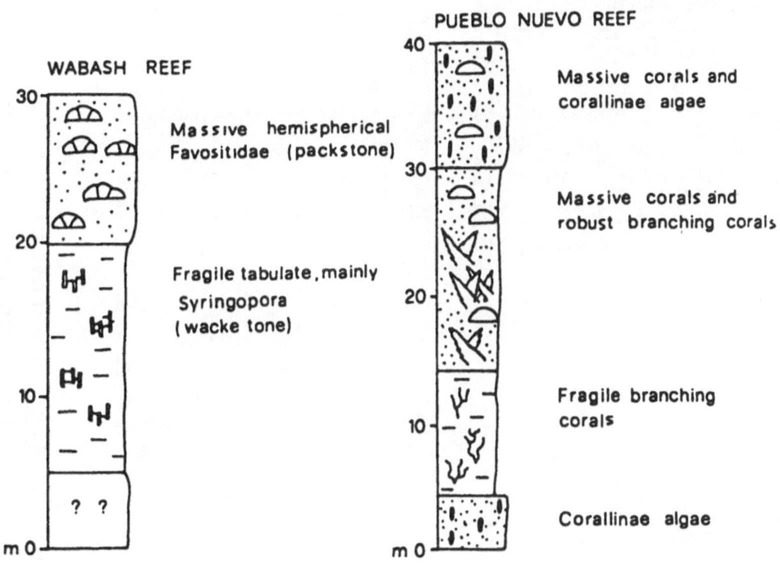

Fig.28 - Examples of 'catch-up' reefs. Britomart Reef,Great Barrier Reef: Johnson, et al.(1984); Galeta Point Reef, Panama: MacIntyre and Glynn (1976);Alacran Reef, Isla Perez,Mexico: MacIntyre, et al.(1977); Wabash Reef,Devonian, Indiana: Lowenstam (1950); Pueblo Nuevo Reef,Oligocene, Mexico: Frost (1977). The stratigraphy and evolution of the Holocene examples shows a strict dependance upon the relative sea-level curve. A similar control is inferred for the older examples.

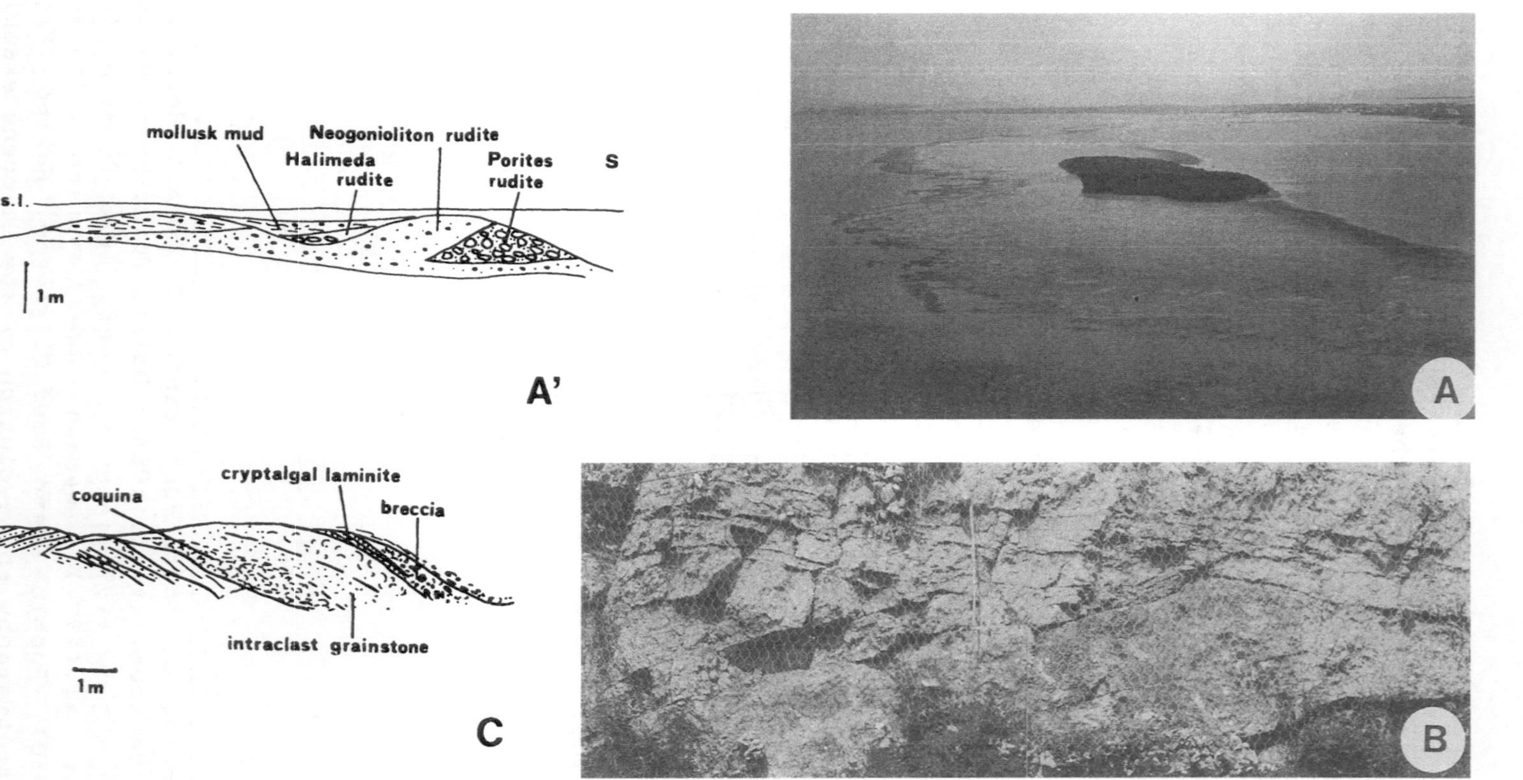

Fig.29 - Examples of inclined beds. A,A':Tavernier Key (Florida Bay) (Bosence,1988).B: 'Calcari Grigi' Formation:section #5. C: Devonian platform, Italy (Galli,1986). These complex bedforms resulting from hydrodynamic instability mark the lateral linkage between deep and intermediate ramp.

Geometries of intrashelf ramps

Intrashelf ramp geometries fall into two fundamental categories (Fig.30) which correspond respectively to divergent and convergent patterns described further below:

1) onlap ramps;
2) offlap ramps

The recognition of these two architectures was attendant on the application of 'event' correlation between stratigraphic sections and logs (Ager,1981; Dixon,et al.,1981; Matthews, 1984;Aigner,1985).The use of physical bodies and stratigraphic horizons as chronostratigraphic tools within shallow-water carbonate platforms offers a better stratigraphic resolution than biostratigraphic criteria because the time interval of formation of physical markers such as storm deposits is shorter than the rate of evolution of shallow-water organisms living in a platform (Sommerville,1979).

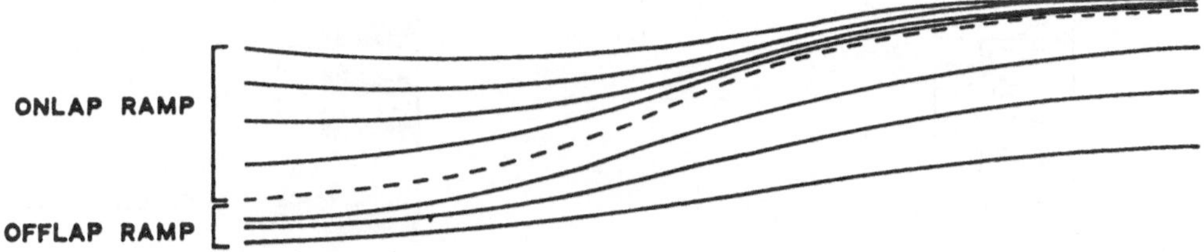

Fig.30 - Geometries of intrashelf ramps.The onlap ramp is generally developed below the offlap ramp and is bounded by flooding surfaces, therefore corresponding to a genetic stratigraphic sequence (Galloway,1989).

Onlap geometry

The onlap geometry is constituted by a 20 ÷ 60 m thick, divergent prism. Isochronous lines converge towards a *hinge* zone. As a result, facies and cycles thin towards the hinge. Facies associations in the hinge zone comprise shallow ramp facies associations (thin- and thick-bedded alternations). The inflection in the topographic profile, which is the site of

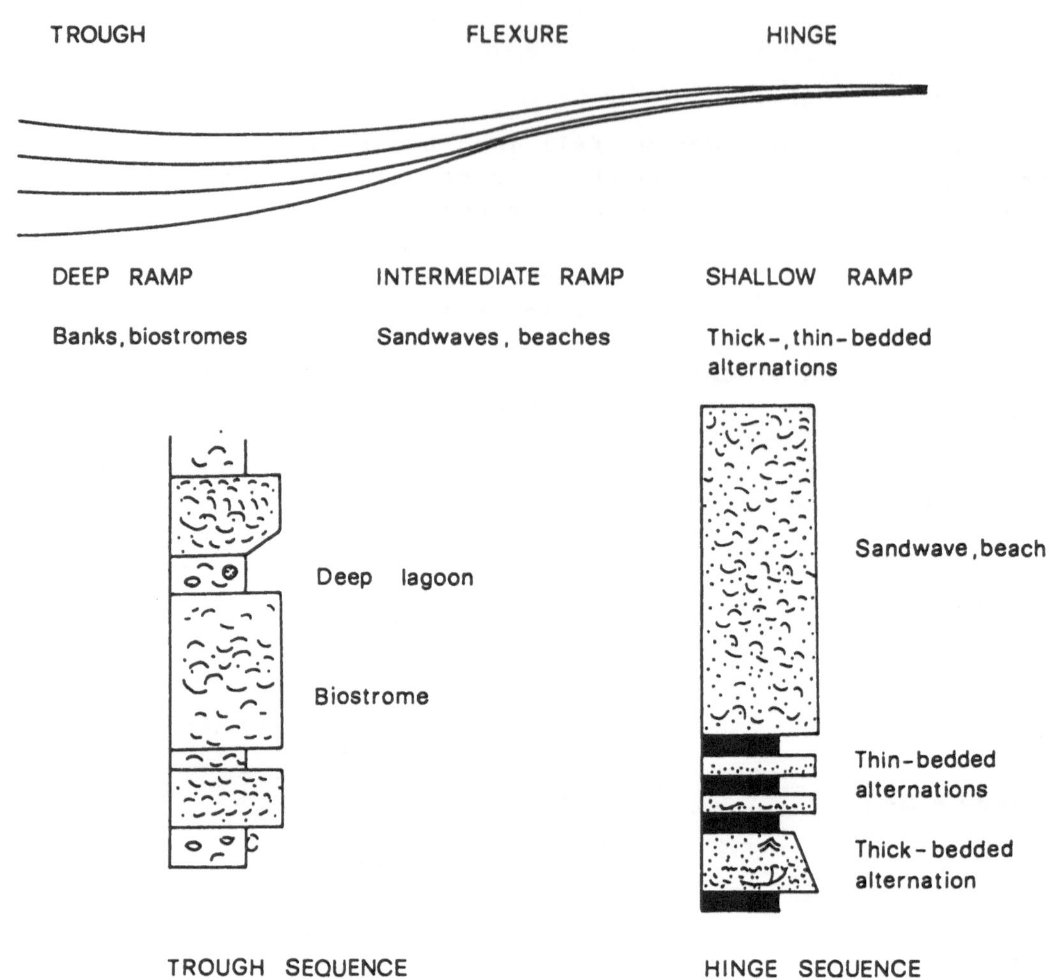

Fig.31 - Sectors, facies associations and modal sequences of an onlap ramp.

development of the intermediate facies association, is the *flexure*. In the *trough* zone, the thickest sector of the onlap ramp, lithofacies are typical of the deep ramp association.

Sedimentation takes place mainly by aggradation. In an onlap geometry facies are developed over progressively wider areas (Krumbein and Sloss, 1963), from bottom to top of stratigraphic sections, and result from a transgressive trend. A consequence of the environmental shift towards the hinge is the formation of deepening-upward sequences and megasequences (trends produced by stacking of individual sequences). The deepening-upward trend may not go to completion as a result of short-term sea-level falls, which determine a shallowing-upward trend at the very top. Upward faunal variations reflect changes from restricted to more open lagoonal environments.

An onlap geometry corresponds to IV and V order eustatic cycles (1 x 10 years). It may represent a depositional sequence (Vail, et al.,1977) because is a 'stratigraphic set composed of a concordant succession of genetically linked system tracts, delimited at the base and top by unconformities (or related paraconformities)'.Sequence boundaries are constituted by a transgressive surface at the base and the maximum flooding surface at the top: in which case the onlap ramp geometry corresponds to a 'genetic stratigraphic sequence' defined as 'a package of sediments recording a significant episode of basin margin outbuilding and basin filling bounded by periods of widespread basin-margin flooding' (Galloway,1989).

A complete *hinge sequence* is a deepening-upward sequence consisting from bottom to top of the following (Fig.31):

1) thick-bedded alternations (shallow-ramp ; hinge zone);
2) thin-bedded alternations (shallow-ramp; hinge zone);
3) sandwaves and bars; beaches (intermediate ramp; flexure);
4) banks and biostromes (deep ramp).

The overposition of thin-bedded alternations (located landward) above thick-bedded alternations (located seaward) suggests that the deepening recorded by the hinge sequence is not a continuous process. The hinge sequence therefore records two major deepening episodes': the first at the base of the sequence, following a sea-level fall; and the second at the base of sandwaves and beaches, or biostromes and banks of the intermediate or deep ramp.In the intervening period,a depositional regression may take place: in fact, carbonate progradation is a rapid process (Kendall and Skipwith,1968). In Florida platform the mainland shoreline is expected to migrate seaward during the actual decrease in the relative sea-level rise (Scholl, et al.,1969). According to Enos & Perkins (1979) mangrove flats will eventually overlie the Florida Bay mud banks; the situation at that time will resemble that occurring in the lower part of the deepening-upward sequence (cf. Fig.16 with Fig.31) where thin-bedded alternations overlie thick-bedded alternations.

A well known deepening-upward sequence similar to the hinge sequence is the "Lofer cyclothem" described by Fischer (1966).Another example of hinge sequence comes from the Devonian carbonate platform occurring in the Carnic Alps, Italy (Fig.32). The completeness of the record of the hinge sequence

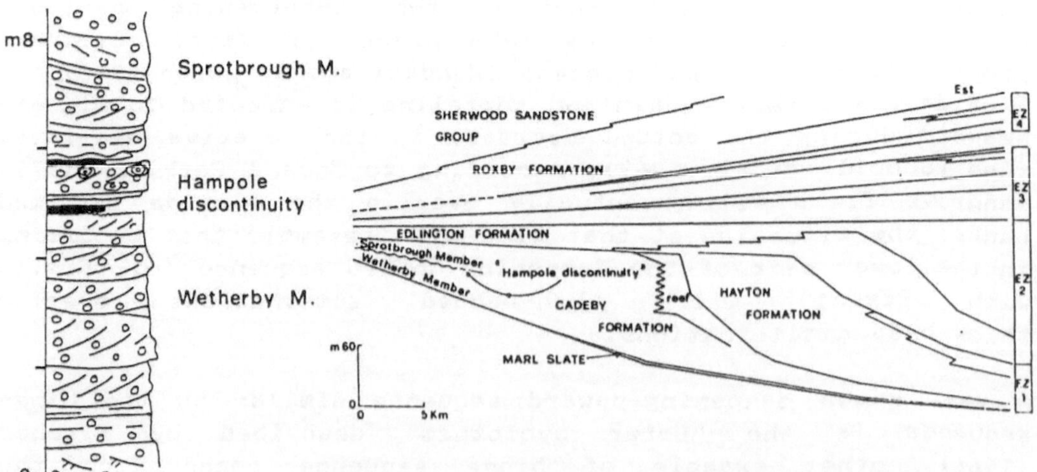
m8
0
Sprotbrough M.
Hampole discontinuity
Wetherby M.
SHERWOOD SANDSTONE GROUP
ROXBY FORMATION
EDLINGTON FORMATION
Sprotbrough Member
Wetherby Member
"Hampole discontinuity"
CADEBY FORMATION
reef
HAYTON FORMATION
MARL SLATE
Est
EZ 4
EZ 3
EZ 2
EZ 1
m 60
0
0
5 Km

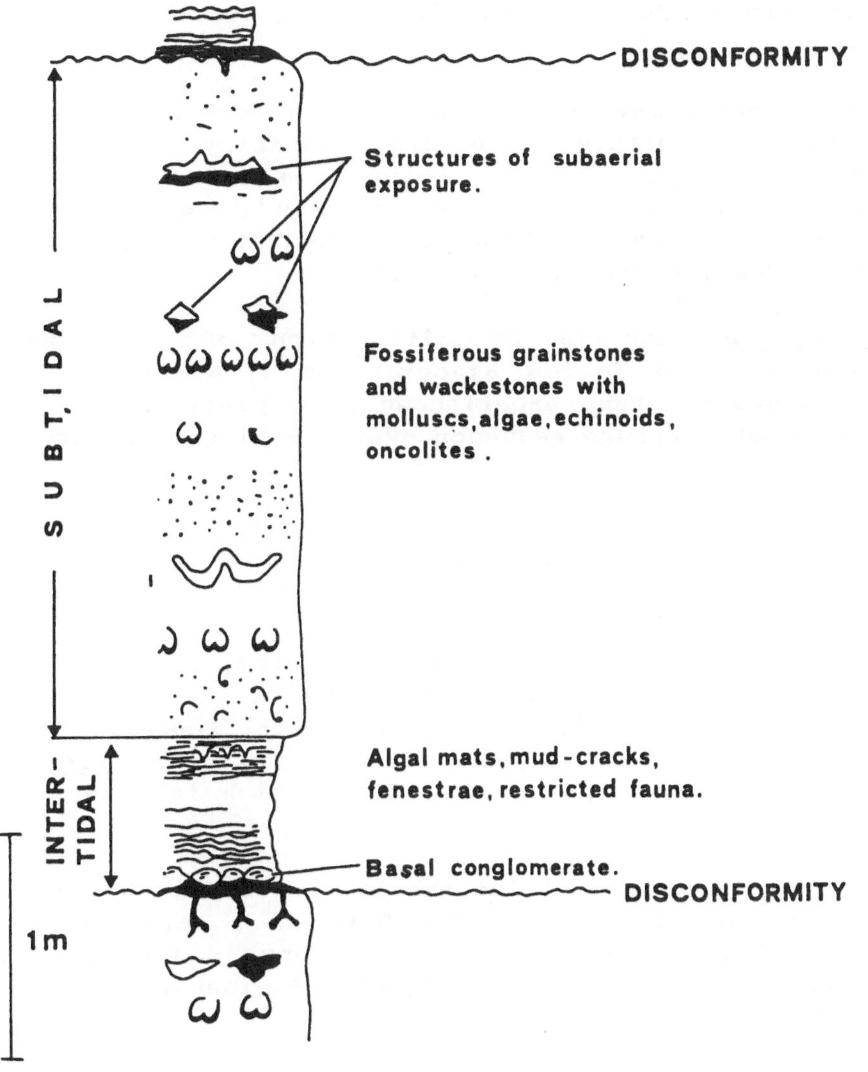

Fig.32 - Deepening-upward sequences analogous to the hinge sequence.A: Example of hinge sequence (Devonian platform, Carnic Alps) consisting of thick-bedded alternations (intraclast grainstones/packstones passing to ostracod-calcisphere - *Amphipora* mudstones which are overlain by a thick bank containing abundant *Amphipora* (deep lagoon). B: Deepening - upward sequence from the English Zechstein (New Micklefield Quarry) consisting of intertidal flat facies ("Hampole Beds") overlain by ooid sandwaves characterized by horizontally bedded grainstones at the base and large scale bedding increasing in size upwards.(The sketch below is from Smith, et al.,1986). C:Lofer cyclothem (Fischer,1966).

depends, among other factors, of the rapidity of the sea-level variations.

A hinge sequence may correspond to the landwardmost termination of clinoforms within carbonate platforms and marks the separation or discontinuity between wedges within carbonate prisms. An example of such a discontinuity resembling a hinge sequence is the "Hampole beds" described by Smith, et al.(1986) from the English Zechstein (Fig.32).

A *trough sequence* is an alternation of deep ramp and intermediate ramp facies associations (Fig.31). A long-term deepening-upward trend experienced by a platform leads to the deposition of a trough sequence over the hinge sequence.

Offlap geometry

The offlap geometry is a sedimentary prism, of the type shown in Fig.30.The shallow ramp zone is the thickest sector of the offlap ramp geometry. The inflection in the topographic profile is the *flexure*. The deep ramp coincides with the *trough zone*. Time lines converge towards the *hinge* which is located in the deep ramp.

Examples of this type of intrashelf ramp geometry include the intrashelf ramp occurring in the Derbyshire carbonate platform (Walkden,1982;Fig.33) and the Pleistocene in the Great Bahama Bank studied by Beach (1982) and Beach & Ginsburg (1980), shown in Fig.34.

The formation of an offlap geometry is determined largely by autocyclic,progradational processes and eustatic variations. In an offlap geometry deep ramp facies are developed over progressively narrow areas ,from bottom to top of stratigraphic sections, with the production of shallowing- and coarsening-upward sequences and megasequences.Upward faunal variations reflect changes from open to semirestricted environments. Tectonic subsidence is not an essential control in the formation of offlap structures; nevertheless, the action of a rotational subsidence around the hinge located in the deep ramp is not excluded.

Short-term sea-level changes (10^4-10^5 years) tuned in the Milankovitch band are especially apparent in the shallow-ramp zone; a tidal flat may form in the shallowest areas of the

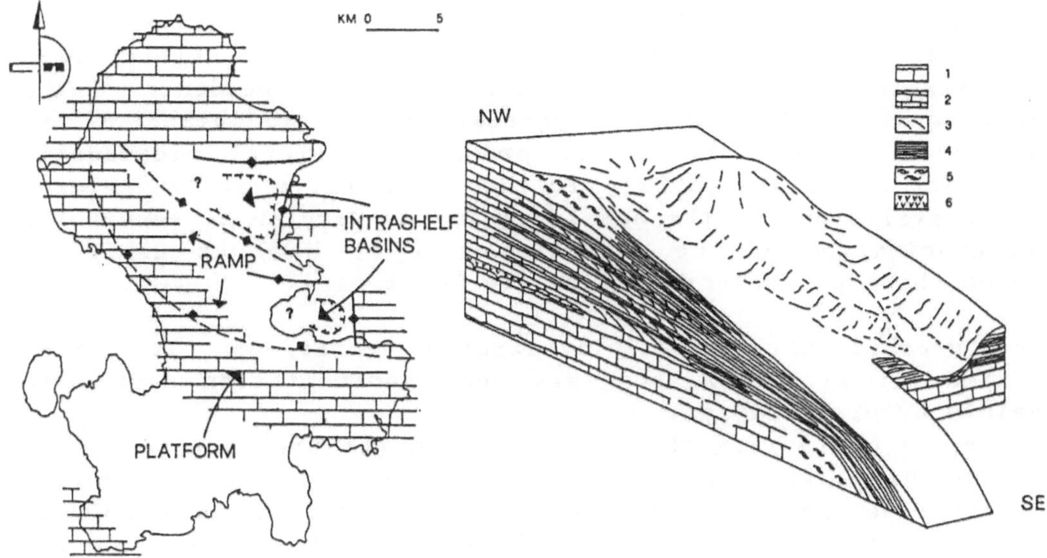

Fig.33 - Offlap ramp (Asbian - Brigantian platform, Derbyshire, England). 1: massive-bedded grainstones and packstones; 2: medium-bedded grey packstones and wackestones; 3: grainstone shoals; 4: thin-bedded dark shaley wackestones and mudstones; 5: knoll reef and marginal reef; 6: lava horizon (from Walkden,1982: right; and Gutteridge,1987: left).

shallow ramp; the deep ramp records the action of storms.A study on storm processes in an ancient, offlap ramp occurring in the Dürrenstein Formation,Dolomites (Galli,1989) showed that proximality trends are analogous to those described by Aigner (1985): they are opposite to those recorded in the onlap ramp geometries (see above).

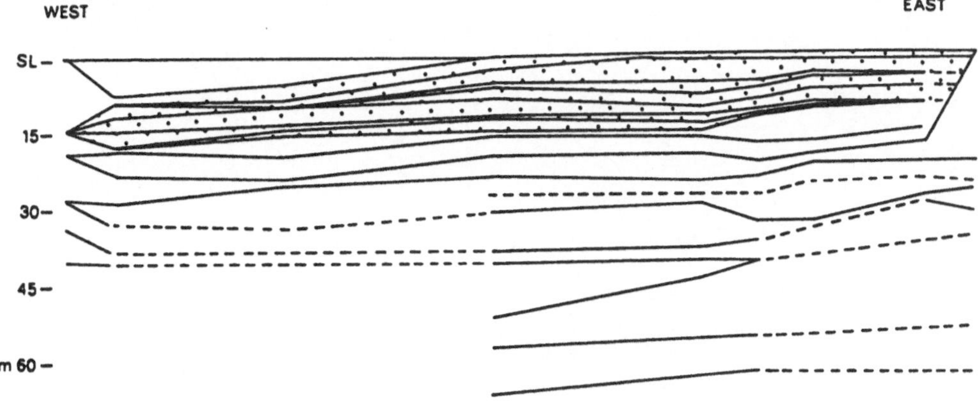

Fig.34 - Offlap ramp (dotted area) in the Bahama Bank (Beach and Ginsburg,1980)

The Dürrenstein Formation (Triassic, eastern Dolomites) was mistakenly interpreted by Bosellini (1984) to be a product of vertical sediment aggradation occurred during a stillstand of the sea-level and absence of differential subsidence. Rather than a vertical sediment aggradation, a stillstand of the sea-level is more likely to produce a lateral progradation (Kendall & Schlager,1981). Furthermore, there is no evidence for a stillstand of the sea-level in the Carnian,as result from an examination of the eustatic curve by Haq, et al.(1987), which shows a first-order slow sea-level rise.

The depositional model, illustrated in Fig.35 (Galli,1989; Bonaga, et al.,1989), suggests that sedimentation took place mainly by progradation.
The modal cycle (Fig.35) displays opposite trends with respect to the hinge sequence of the onlap ramp, where shallow-ramp lithofacies underlie the biostromes and banks of the deep ramp. It consists of a compound shallowing-upward sequence composed of a fining- and thinning-upward sequence, deposited below wave base, in the deep ramp, overlain by a coarsening-upward sequence deposited above wave base, in the shallow-ramp zone.A consequence of the lateral progradation, which is the main process operating in the offlap ramp, was the formation of shallowing-upward megasequences,the shallow-ramp lithofacies and sequences overlying the deep ramp lithofacies.

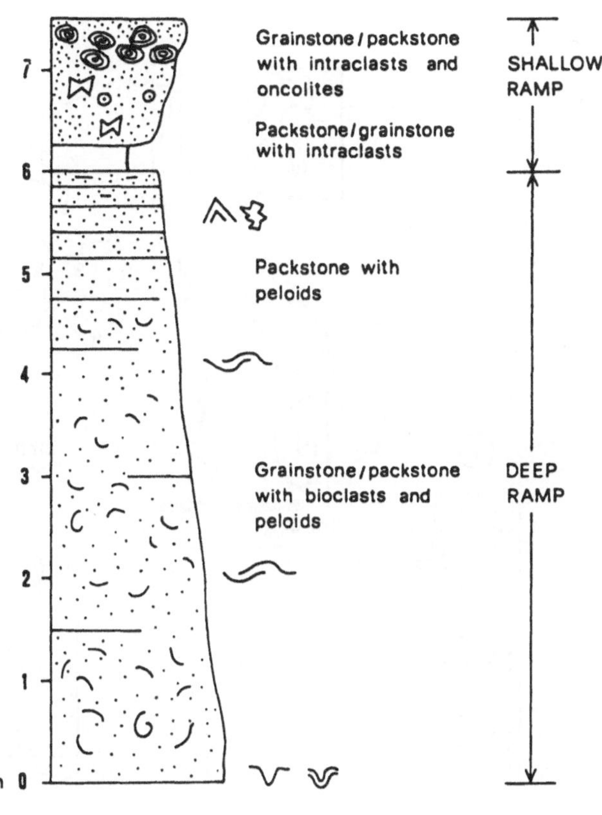
Grainstone / packstone with intraclasts and oncolites
Packstone/grainstone with intraclasts
Packstone with peloids
Grainstone / packstone with bioclasts and peloids
SHALLOW RAMP
DEEP RAMP
m 0
1
2
3
4
5
6
7

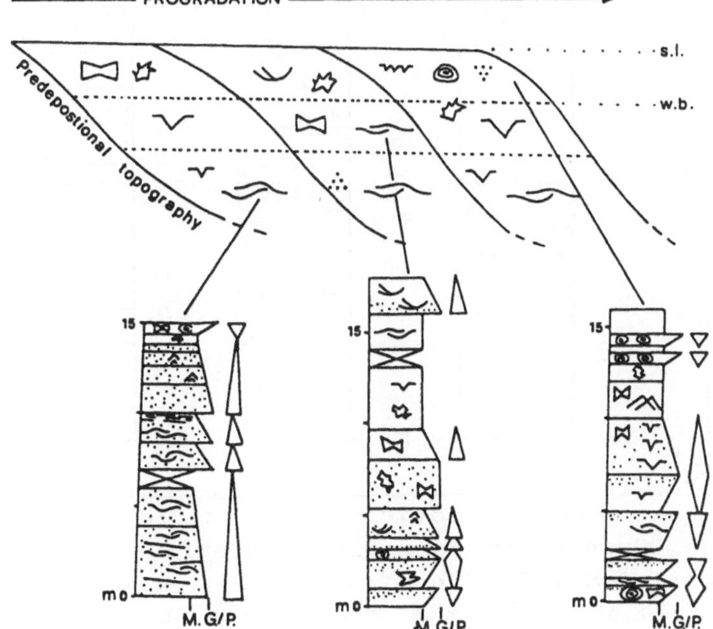
PROGRADATION
s.l.
w.b.
Predepositional topography
15
mo
M.G/P.

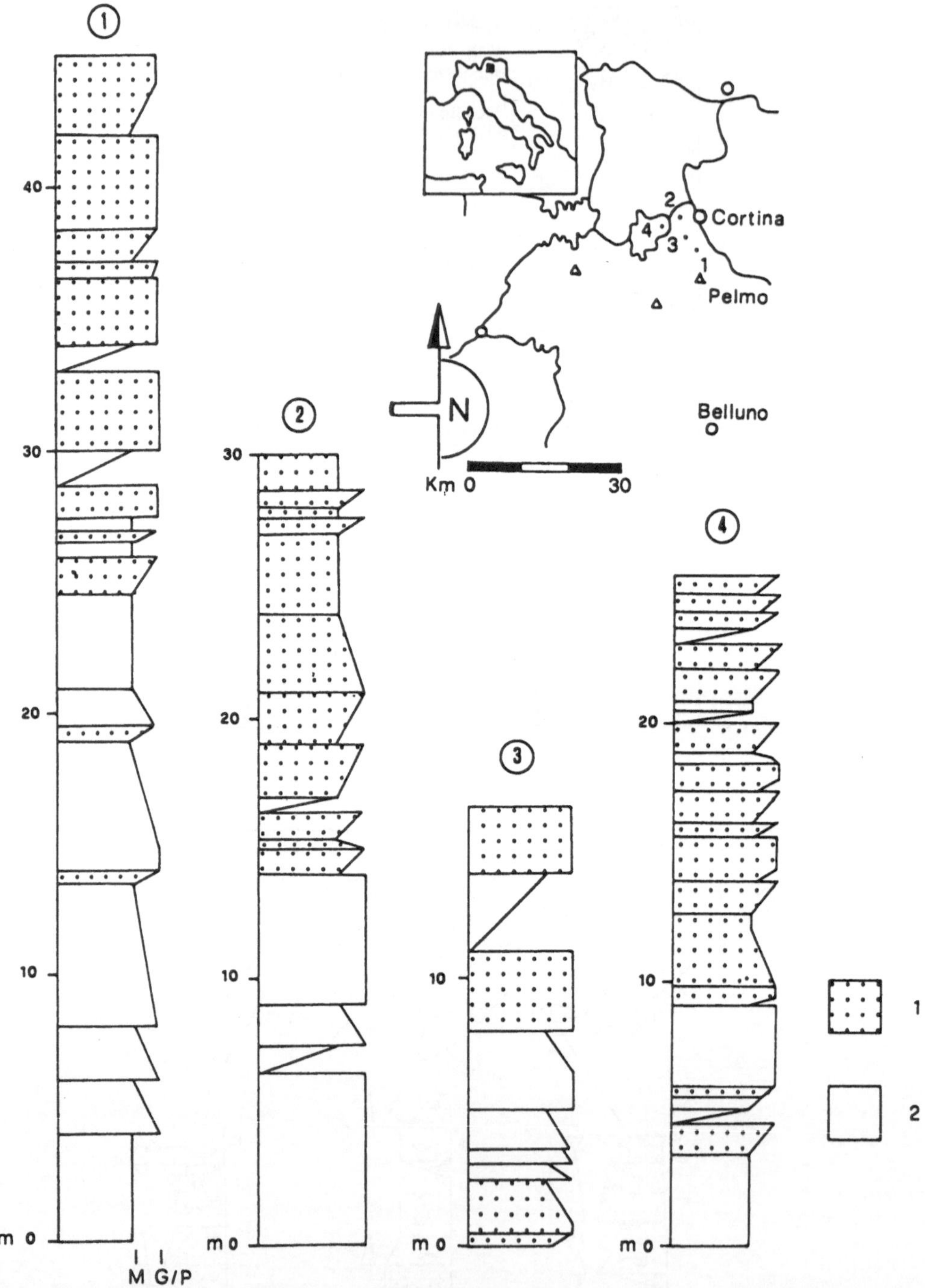

Fig.35 - Stratigraphic sections,modal cycle and depositional model of the Dürrenstein Formation, Dolomites. 1: Shallow ramp facies associations; 2: deep ramp facies (from Galli,1989 and Bonaga, et al.,1989).In the study area the deposition took place by lateral progradation.

Sequence stratigraphy

The sequence stratigraphy depositional model by Vail,et al.(1977;1984) showing depositional sequences and facies tracts is shown in Fig.36.

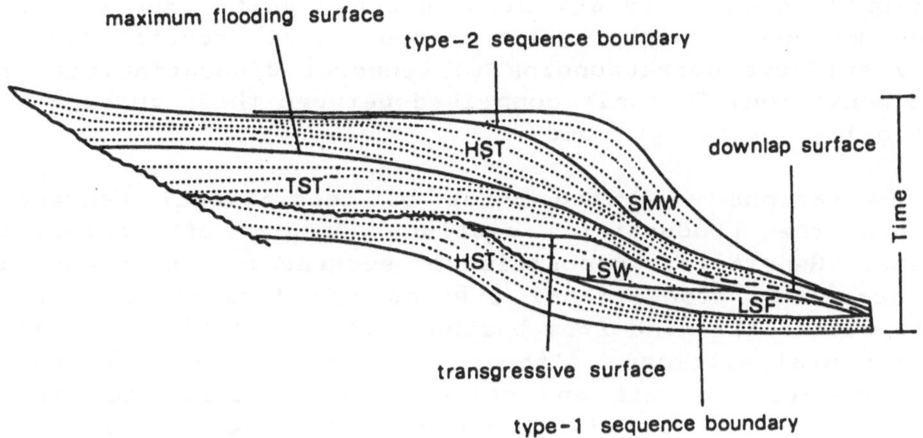

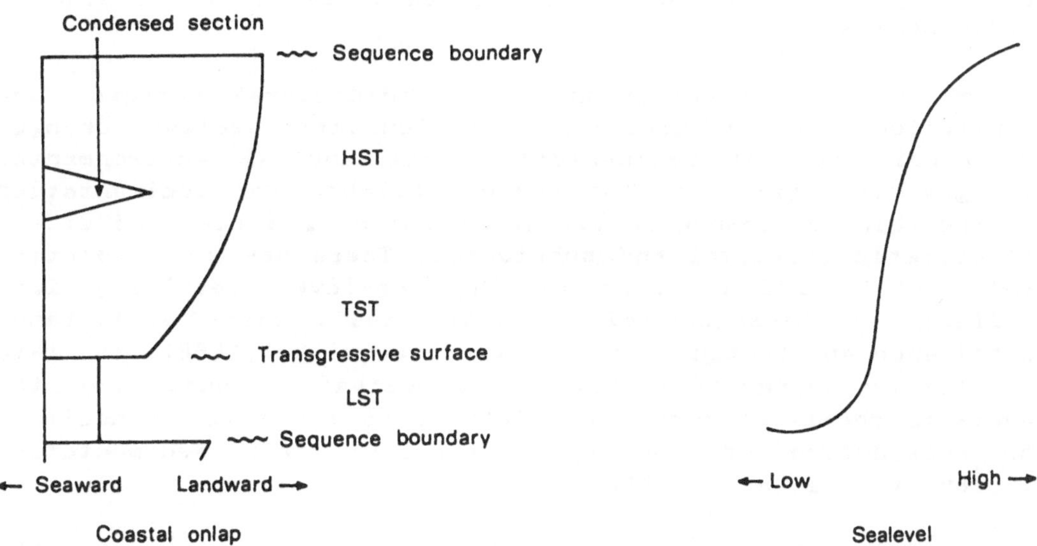

Fig.36 - Depositional sequence (from Swift,et al.,1987) and passive margin model (from Vail,1984;1987).

A depositional sequence is defined as a succession of facies tracts formed in response to a relative sealevel cycle.It is a succession of conformable, genetically-related strata, bounded below and above by unconformities and their correlative conformities (Vail,et al.,1977). A facies tract is "a linkage of contemporaneous depositional systems (Brown & Fischer,1977), each linked to a specific segment of the eustatic curve".A depositional sequence is any stratigraphic unit, from a few m to 1000 m thick, with vertical boundaries constituted by physical surfaces corresponding to temporal discontinuities in the sedimentation. The unit comprised between these surfaces is a genetically homogeneous body.

There are various gerarchic types of depositional sequences which are the product of distinct orders of geological phenomena. Generally, depositional sequences are sigmoidal bodies formed by depositional systems and formations passing from the sea to the land from basinal, slope, platform, paralic and continental settings. Although difficult to define, owing to the detailed analysis and correlations required for their identification, depositional sequences are true natural units which record natural geological processes, such as transgressions, regressions and relative sealevel changes.

The depositional sequences described here formed on a thermally subsiding margin. The hinge is located landward of the area of sedimentation.

Their facies distribution and sedimentary pattern are controlled by the interaction of : 1)eustatic sealevel change; 2) subsidence; 3) sedimentation rate and 4) environmental changes (Schlager,1991).The space available for sedimentation is the relative change in sea-level which is a combined effect of eustatic sea-level and subsidence. There may be a relative fall, stillstand or rise of the sea-level resulting from different combinations of sea-level fall, rise, stillstand, subsidence and tectonic uplift (Van Steenwinkel,1988). The rate of relative change of sealevel is a derivative function of the eustatic sea-level curve (Fig.37): it is the rate of addition or subtraction of the space available for sedimentation (Posamentier,et al.,1988).

The most important changes in sedimentation take place when the rate of sealevel rise or fall are highest.
Subaerial unconformities are produced when the space available for sedimentation is taken away; they occur as a result of a rapid sea-level fall. They form sequence boundaries.
A type-1 sequence boundary is "characterized by a subaerial

exposure and concurrent subaerial erosion, associated with stream rejuvenation (incised valleys), a basinward shift of facies, a landward shift in coastal onlap, and onlap of overlying strata" (Vail,1987). It is accompanied by the development of a lowstand facies tract (Haq,et al.,1987).
A slower rate of sealevel fall, less than or equal the rate of basin subsidence at the platform margin, produces a type-2 sequence boundary. "It is marked by subaerial exposure and a downward shift of coastal onlap landward of the depositional shoreline break, but lacks both subaerial erosion associated with stream rejuvenation and basinward shift in facies" (Vail,1987). In other words, the whole shelf may not be exposed.

At a later stage, when the regional subsidence outrans the slowing rate of sealevel fall, new space available to sedimentation is created by the relative rise in sealevel and a prograding lowstand wedge facies tract accumulates between the shelf edge and the lowstand fan. Later, the lowstand wedge facies tract overlies the lowstand fan.

In a carbonate setting a fall in the sealevel determines non deposition, submarine erosion, cementation and the formation of lithoclast beds. The carbonate 'factory' ceases to work (Droxler & Schlager,1985). Falls of relative sealevel are associated with karst, soil development on the platform (Kendall and Schlager, 1981). According to Sarg (1988) lowstand facies tracts associated with a type-1 unconformity lead to significant slope front erosion and shedding of large volumes of coarse talus into the basin.Lowstand shedding is common in siliciclastic systems; it should be an exception rather than a rule in carbonate systems. In this work it is proposed that deposition of megabreccias (lowstand fan facies) at specific time intervals was triggered by global episodes of geoidal deformation.

A rapid relative sealevel rise determines a transgression, because the sedimentation rate is no longer sufficient to fill-up the space. A transgressive surface develops above a lowstand facies tract. When lowstand deposits are lacking, the transgressive surface coincides with a facies boundary (ravinement surface: Stamp,1922). The formation of a transgressive (retrogradational) facies tract takes place along the steepest part of a rising sealevel curve. A condensed section is deposited seaward of the depositional area because there the sedimentation rate is too low.Younger units are progressively thinner upward and basinward as a result of basin starvation.

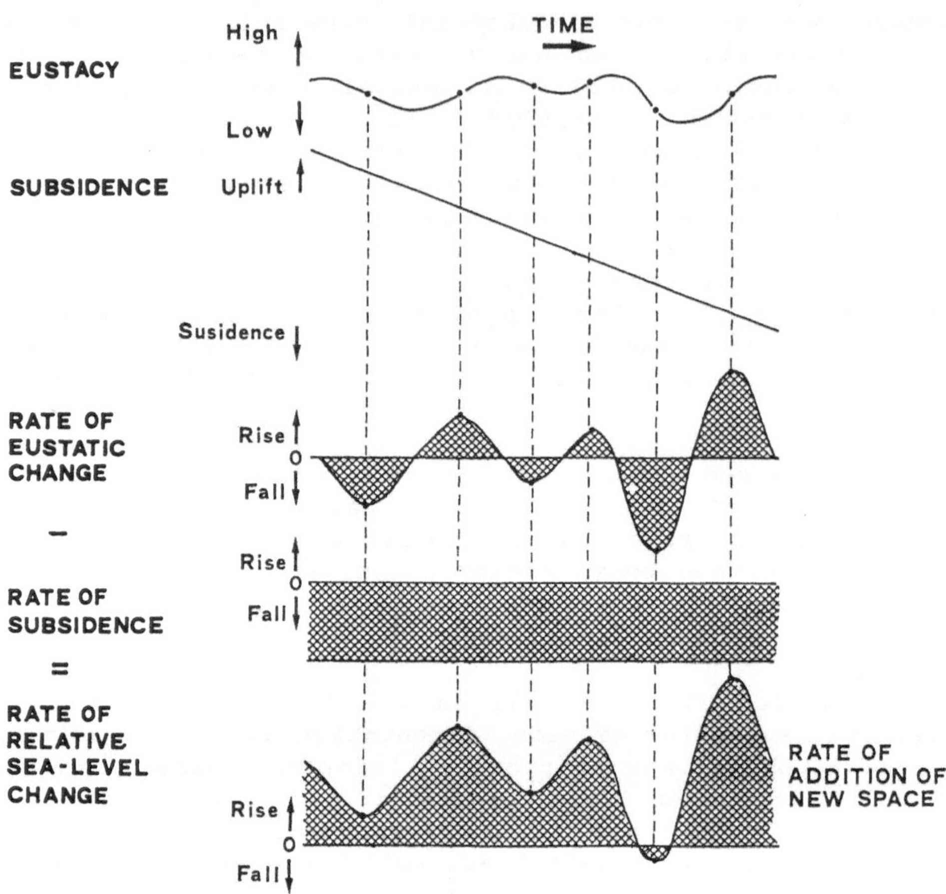

Fig.37 - Relative sealevel as a function of eustacy and subsidence (from Posamentier,et al.,1988).

The maximum flooding surface, or downlap surface, separates the transgressive facies tract from the overlying highstand facies tract. It marks the maximum landward shift of the transgressive facies tract.

As a result of a gradual slowing of the relative sealevel rise,or generally during a decrease in the rate of relative change in sealevel, there is a slow subtraction of the space available to sedimentation and sediments are forced to aggrade and prograde above the transgressive facies tract with the formation of a highstand facies tract. It is possible to distinguish an early aggradational, and a late progradational facies tract.The upper surface of the highstand facies tract may be a type-2 or a type-1 sequence boundary, depending of the rapidity of the sealevel fall.

When a rapid sea-level rise changes into a slow sea-level fall, or a slow fall changes into a slow rise: more generally during a maximum increase in the rate of relative change of sea-level, the newly added space is infilled with an aggradational - progradational complex (shelf margin facies tract) above the highstand facies tract and a type-2 unconformity (Vail,1987). This tract is characterized by stacked sequences with an increasing deepening tendency; it represents the filling-up of topography before drowning. It is overlain by the transgressive facies tract (Haq, et al.,1987).

The relative sea-level is the end-product of two variables: eustasy and subsidence. It is the space available for sedimentation (Posamentier,et al.,1988). The rate of relative sea-level change results from the difference between rate of eustatic change and the rate of subsidence: it is the rate of change of space available to sedimentation. It is a main controlling factor in the depositional pattern as is demonstrated by Van Steenwinkel (1988).As shown above, across the curve of change in the rate of the relative sea-level several tracts may by distinguished that correspond to increases or slowings of the sea-level change leading to regressive or transgressive phases.According to the terminology derived from seismic stratigraphy (Haq,et al.,1987), the facies tracts that form the depositional sequences discussed in the case histories described below are the following:

1 - transgressive facies tract;
2 - shelf margin facies tract;
3 - highstand facies tract;
4 - lowstand facies tract.

Fig.38 summarizes as a reference the facies tracts and associated discontinuity surfaces, in relation to the rate of change of the relative sea-level.

Transgressive facies tract

During this tract a rapid increase in the rate of the relative sea-level determines a drowning : sedimentation does not keep pace with the sea-level rise and 'is taken by surprise' (Kendall and Schlager,1981). Either black micrites or paper shales form during this tract. Beds composed of fast-growing, monotypic epibenthic faunal assemblages may also form, provided the organisms are capable of keeping pace with the rising

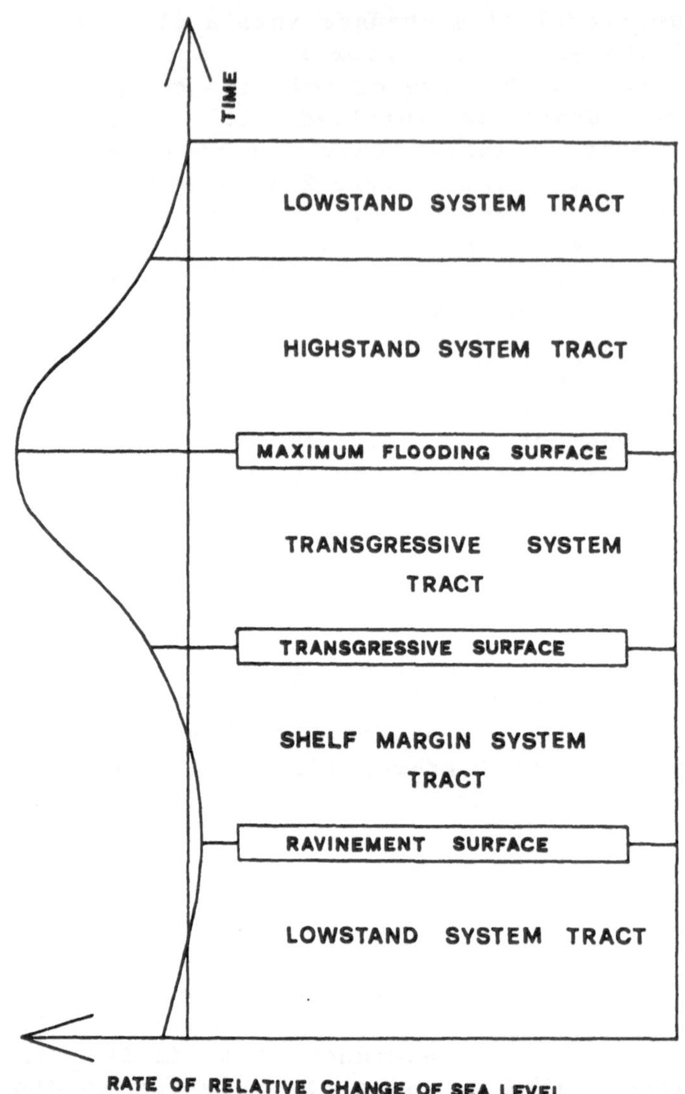

Fig.38 - Relative sea-level, facies tracts and discontinuity surfaces (adapted from Van Steenwinkel,1988 and Posamentier, et al.,1988).

sea-level. Goldhammer,et al.(1990) identified in the Latemar depositional sequence a transgressive facies tract composed of a series of aggradational, thickening - upward amalgamated, subtidal cycles. They also reported syndepositional marine diagenesis associated with this tract.
This tract is also characterized by a transgressive lag containing a condensed fauna and fragments of the underlying lithofacies ('lag and hiatal bypassed concentrations' by Kidwell,1991).

Shelf margin facies tract

This tract is characterized by a slow increase in the rate of relative sea-level rise which determines an incipient drowning. Tangential oolites and skeletal bodies form during this tract: they may also undergo a landward migration resulting in spillover lobes, washovers, or storm bars and high-energy beaches. The generation of oolite shoals in response to the sea-level rise curve is detailed by Hine (1977) who showed that oolites form during a period of gradual rise or slowing of the sea-level. Progressive increase in the rate of sea-level rise may lead to a deepening and to the formation of lumps above oolites and eventually to skeletal beds.In the case history studied by Van Steenwinkel (1988) the onset of the shelf margin facies tract was more or less pronounced and led to different types of sediments,depending of the local sediment supply and bathymetry: grapestone lumps and cryptocrystalline grains similar to those occurring at Lily Bank (Hine,1977), lithoclast beds or crinoidal sands.

Highstand facies tract

During the highstand facies tract the rate of relative sea-level rise decreases. This leads to a reduction in the space available to sedimentation, hence, to a reduction in the sedimentation rate. Sediments undergo a gradual progradation. The net effect is a relative depositional regression which leads to the formation of a shallowing-upward sequence. Within the highstand facies tract an early and a late highstand phase may be distinguished.The gradual decrease in the rate of addition of new space results in an overall thinning of m-scale cycles in the early highstand facies tract.Condensed sequences capped by textures of subaerial diagenesis are produced during the late highstand,in response to a sea-level fall.

Lowstand facies tract

This tract is marked by a relative fall of the sea-level which may stop the lateral progradation and trigger the formation of relict lithoclast beds and vadose pisolite horizons, under vadose meteoric conditions and subaerial diagenesis.Little if any sediment is preserved, because no accomodation space is created. Lateral migration is prevented by the cementation which tends to link together grains, and by the reduced sedimentation rate. This tract is represented by diagenetic caps of shallowing-upward sequences formed during the highstand system and shelf margin system tracts.

A peculiar feature of the stratigraphic columns described in this work is the stacking of unrelated, different lithofacies delimited by discontinuity surfaces. These separate the system tracts and are the product of changes in the rate of relative sea-level rise. Each is associated with a change in the trend along the segment of the relative sea-level curve (Fig.38).

The main surfaces of discontinuity between system tracts are the following:

1 - ravinement surface;
2 - transgressive surface;
3 - surface of maximum flooding.

Ravinement surface

A ravinement surface (Stamp,1922) is a former subaerial surface transformed into a marine surface by erosional shoreface retreat.It is a flooding surface above a former subaerial surface. This discontinuity surface underlies the shelf margin system tract and results from a slow rate of sea-level rise.

Transgressive surface

A transgressive surface is a planar surface overlain by the transgressive system tract. It marks a rapid increase in the rate of sea-level rise and occurs within marine lagoonal facies, separating different lithofacies. The transgressive surface may itself represent the transgressive system tract when the sea-level rise is geologically instantaneous. In such cases this surface is directly overlain by the highstand system tract.

Maximum flooding surface

The maximum flooding surface marks the maximum level reached by the sea during a sea-level rise and overlies the highstand facies tract. It marks the transition between transgressive and highstand facies tracts, often delimiting thickening- and thinning-upward trends. This may correspond to a change in sedimentation trend from vertical sediment growth to lateral facies migration. This surface in some instances is represented by a coquina, or skeletal lag.

These discontinuity surfaces are constituted in several cases by skeletal concentrations. Kidwell (1991) operated a classification of skeletal concentrations into some broad groups (event, composite, hiatal and lag concentrations). She also showed as coquinae are distributed through depositional sequences.According to that classification, hiatal,condensed concentrations may form transgressive surfaces; hiatal, starved concentrations represent maximum flooding surfaces; ravinement surfaces are constituted by lag concentrations.

The existence of different facies tracts separated by discontinuities was highlighted by Goodwin and Anderson (1985) according to which the greatest part of the stratigraphic column is episodic. Their punctuated aggradational cycle ('PAC'),typically 1 to 5 m thick and covering a time interval comprised between 100.000 years (with an average of about 50.000 years),is regarded as the result of geologically instantaneous sea-level rises (1m/1000 years) followed by vertical growth of sediment during a relative stability of the sea-level. Such a situation may represent a transition from a transgressive to a highstand facies tract, according to the terminology derived from the sequence stratigraphy (Haq,et al.,1987).

Mechanisms of formation of onlap ramps

Whereas the eustatic curve is given by a regular sinusoidal function in time, the curve of subsidence may be more irregular, even over limited stratigraphic intervals. Subsidence is the summation of three distinct subsidences (Posamentier,et al.,1988) :1) rotational subsidence; 2) loading subsidence; and 3) tectonic subsidence.

Models of sedimentation assuming a constant rate of subsidence are in reality a simplification which is negligible when considering volumes of sediments formed over large time intervals; short-term spasmodic changes in subsidence rate are probably important for m-scale cycles. The model proposed by Cisne (1987) regards the Lofer cyclothems (Fischer,1966) as tectonic cycles; they would result from stick-slip faulting which is typical of passive continental margins; according to Cisne (1987) these movements would record changes of the litospheric flexure.Dip-slip movements along normal faults would produce abrupt changes in the depth of the sedimentary interface and m-scale depressions successively infilled by sediment aggradation.Cisne (1987) hypothesized subsidence pulses with a recurrence time of about 40.000 years,considered to be characteristic of normal faults.

Differential subsidence had an important control on the stratigraphic pattern of depositional sequences of onlap ramps exemplified by the case histories described below.A tectonic subsidence control is suggested by the following features:

1) lack of regularity of the curves of relative change in the sea-level (i.e. 'Calcari Grigi' Formation).

2) Increases in the number of sequences from the deep ramp to the shallow ramp and their thinning-out towards the hinge which rules out an onlap relation (i.e. Fort Thompson Formation).

3) A predepositional tilting is excluded by the difficulty in reconciling the thickness of sediment in the trough area with the limited bathymetric range of survival or adaptability of shallow-water organisms and fossils found in the deep ramp.

Two models are discussed (Fig.39).

1 - The onlap ramp is produced by a rotational subsidence around a hinge line located in the inner ramp;

2 - The rotational subsidence around the hinge takes place in combination with progressive tilting responsible for the shift of the flexure towards the hinge.

In the first model a flexure can be missing and facies trends vary gradually normal to the strike. In the second model the shift of the flexure determines a stepwise migration of the intermediate ramp facies associations towards the hinge.The first model is not new: it was suggested by Lowenstam (1950) and later by Meissner (1972).

The effects of the rotational subsidence can produce great complications as it is not expected to be characterized by a constant rate through space or time: in both models the creation of new space induced by rotational subsidence is greatest in the trough area and so negligible close to the hinge that sedimentation basically takes place in response to eustatic shifts.

It follows that interpreted eustatic cycles, such as Milankovitch rhythms, and tectonic cycles may occur along the same stratigraphic interval, respectively in the hinge and trough areas.This may lead to some contrasting interpretations because two workers studying the two respective areas,will be forced to apply their (correct) interpretation (tectonic vs. eustatic) to the other area.Such a case is represented by the Triassic Dolomites, considered further below.

Another factor of complication of the environmental setting and lithofacies distribution results from a differential deformation between the hinge and the flexure with the production of a secondary trough in the intervening area.

The accentuation of the flexure relief, consequent to the deformation, may change the flexure into a *sill*, which is the site of formation of:

1) complex facies transitions;
2) condensed sequences;
3) very rapid facies changes; and
4) high-energy structures and deposits such as crinoidal sands and hummocky sequences.

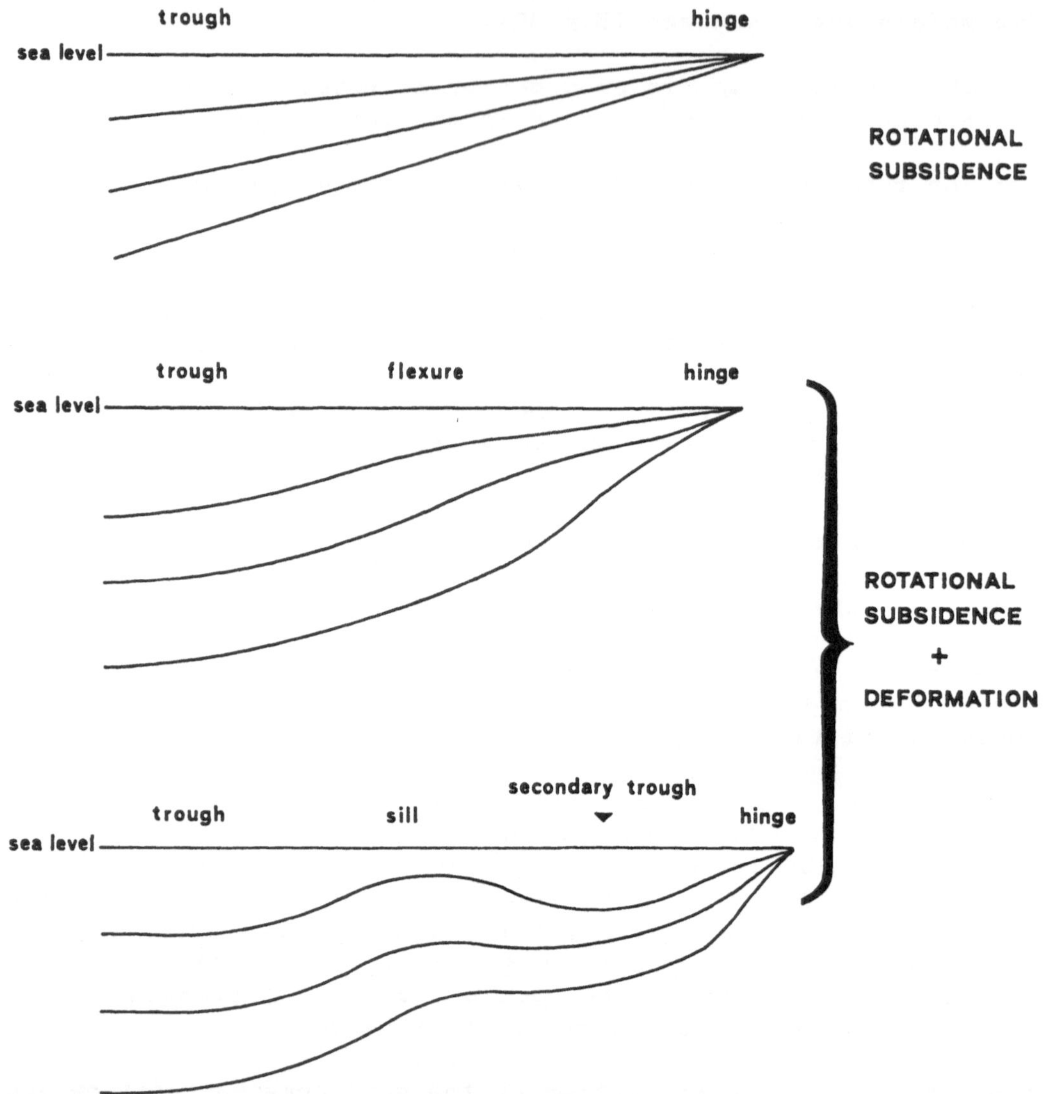

Fig.39 - Types of onlap ramp geometries.

A *sill sequence* is a shallowing-upward sequence formed by the stacking of different,unrelated shallow-water facies.

Sill areas in some cases record the abrupt (channelized) vertical transition from deep lagoonal deposits (deep ramp) to very shallow deposits, such as the example shown in Fig.27 where deep ramp deposits are scoured and infilled with a variety of shallow-water lithofacies.
Another example of sill is recognisable in the cross section of the Pleistocene Florida platform (Fig.40:cf. Fig.39: bottom) by

Perkins (1977).Here, a tectonic control on sedimentation is mainly inferred from a lack of parallelism between time lines.An elevated area, intermediate between the Florida Bay and the outer platform, forms the present-day Florida Keys.A comparable, however more complicated situation occurs in the Middle Ordovician of Tennessee, where the platform lagoonal areas are separated by several elongated, relieved areas (Fig.40: Lower Ordovician islands,shelf edge reef tract,oolite shoals & reefs: Walker, 1974) which may represent sill areas.

In the carbonate world most of the morphology and physiography of elevated features such as reefs, banks, patch-reefs show various degrees of inheritance from the preexisting topography.

The Belize lagoonal reefs for example provide an excellent spectrum of such relationships where shelf atolls and linear reefs mirror Pleistocene foundations (Choi and Ginsburg,1982; Choi and Holmes,1982; Purdy,1974).

A comparison between the two styles of the depositional assemblages examined in the case histories below indicate that the modalities of formation of intrashelf ramps:(simple rotation; rotation and tilting: Fig.39) dictate the type of control exerted by the predepositional topography in the two facies.

Where the inclined surface of the ramp is produced only by a stepwise, simple rotation associated with eustatic shifts, the control played by the antecedent topography is strong (Fig.41 A) and the stratigraphic column consists of a well defined cyclicity, shown in the Pleistocene Fort Thompson Formation (see below).

Conversely, a deformation accompanied by a rotation determines discontinuous variations in the topography that complicate the environmental facies distribution (Fig.41 B).

Complex facies mosaics are expected to form especially in this last situation, other than by autocyclic shifts and variations in the coexisting syndepositional topography (Pratt and James,1986);the predepositional topography is not as important as in the previous case; it may play an important role in the flexure when it is transformed into a sill. The cyclicity is less evident; rather, it is a statistical cyclicity accompanied by random facies transitions which are also a function of the intensity and modality of the deformation.

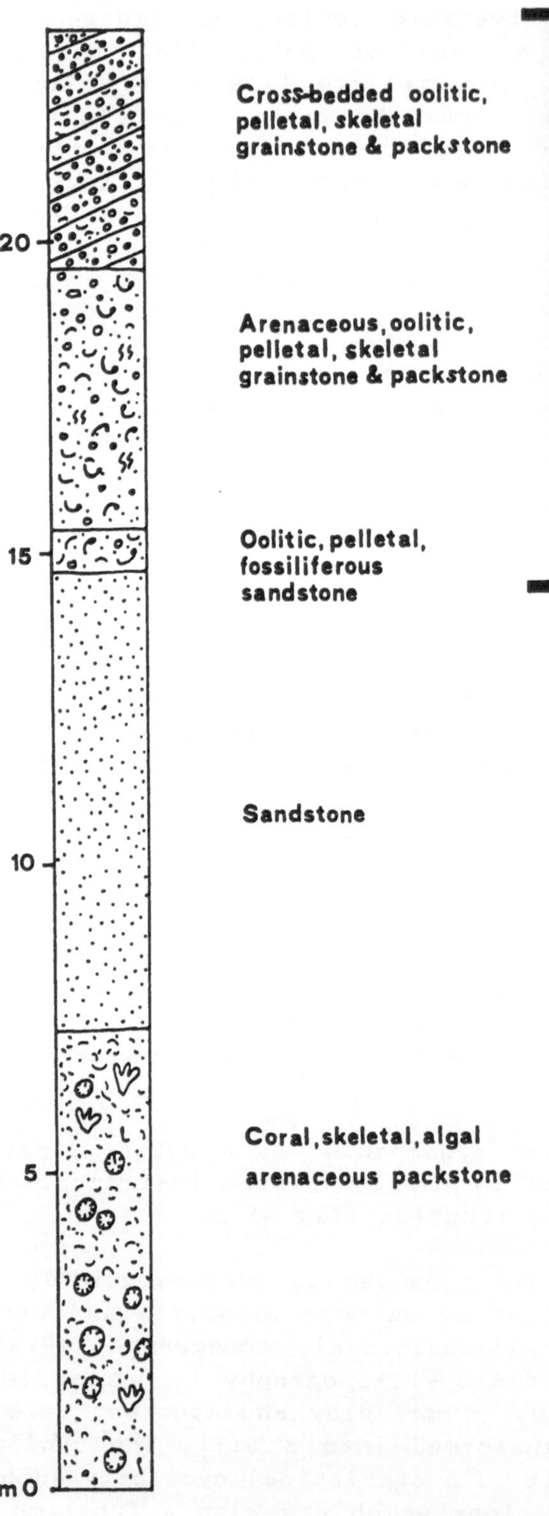

Cross-bedded oolitic, pelletal, skeletal grainstone & packstone
Arenaceous, oolitic, pelletal, skeletal grainstone & packstone
Oolitic, pelletal, fossiliferous sandstone
SILL SEQUENCE
Sandstone
Coral, skeletal, algal arenaceous packstone
20
15
10
5
m 0

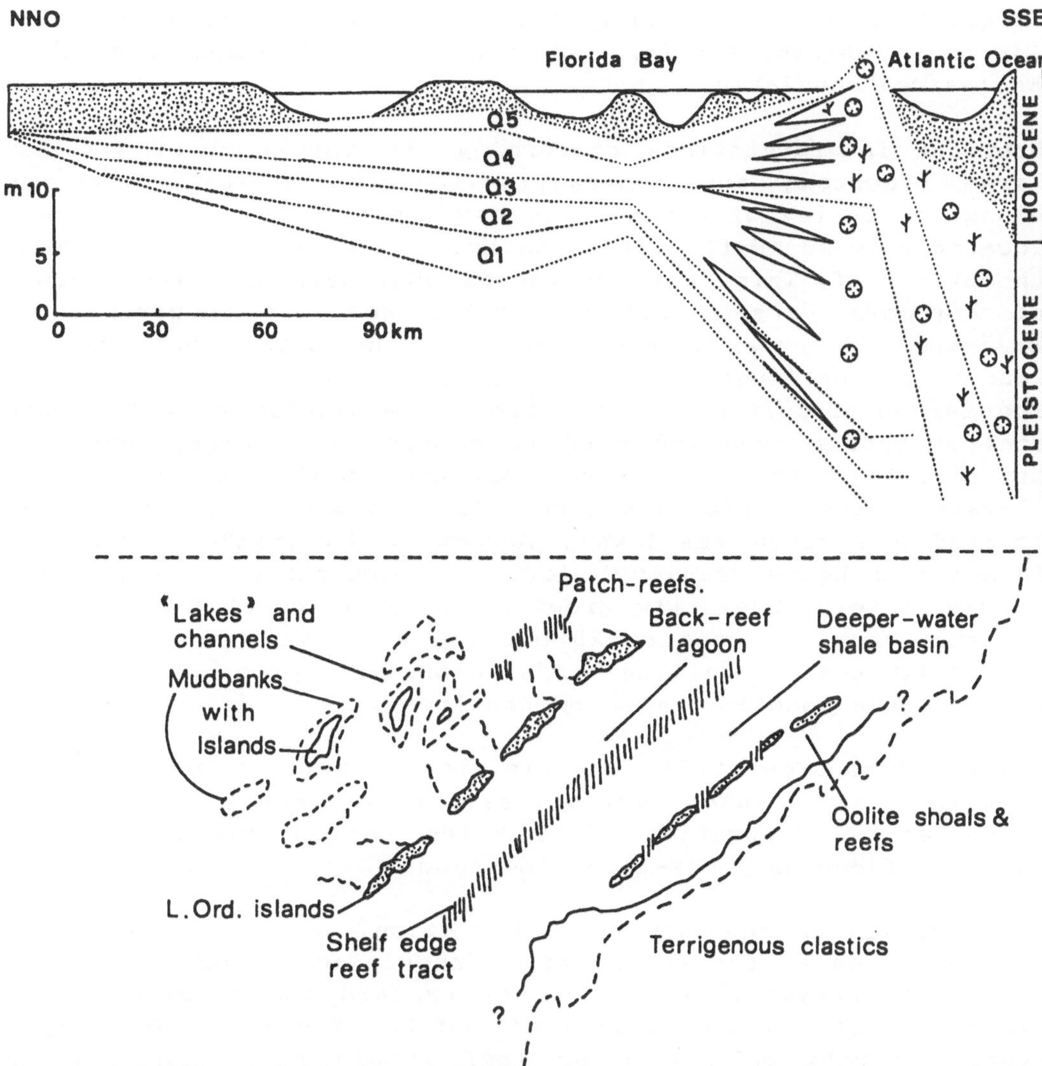

Fig.40 - Cross section of the Pleistocene of Florida platform (from Perkins,1977),core #34 by Perkins (1977) taken along the Florida Keys showing a shallowing-upward sequence developed above coral deposits, and environmental reconstruction of the Middle Ordovician of Tennessee (Walker,1974) showing several supposed sill areas oriented NE-SW.

The Key Largo Formation occurring in Florida platform constitutes a coral-bearing Formation. Maximum thicknesses are of about 60 m (20 on the average). The Formation constitutes slightly arcuate, elongated islands which extend for about 240 km along a ENE-WSW direction, 8 km inland from the outer border of the platform. The faunal composition is homogeneous. Coral zonations do not occur (Stanley,1966; Hoffmeister & Multer,

1968). Main corals are *Montastrea annularis*, *Diploria strigosa*, *Diploria labirintiformis*, *Porites*. The most common fossil is *Montastrea annularis*.

Previous interpretations concerning the formation of the Key Largo Limestone are contrasting.It was interpreted as an elongated series of patch-reefs developed behind a coral bank located seaward (Hoffmeister and Multer, 1968), or a remain of an outer reef. This last hypothesis contrasts with the absence of *Acropora palmata* and *Millepora*; and the occurrence of *Halimeda* plates which are absent in the outer reef, but an important constituent of the Key Largo Formation.
The main difficulties arising from an exclusion of a tectonic control result from the need to combine the former depth of deposition of the Key Largo Formation with its present-day elevated topographic location. The Formation is presently located 6 m above sea-level. According to Stanley (1966) it formed 9 m below sea-level (cf. the occurrence of *Montastrea annularis* which typically grows in quiet lagoonal areas, below the surf zone, as much as 25 m below sea-level, 9-10 m on the average).Coeval, coralline reefs along the border of the outer platform are located 8 m below sea-level.

According to Enos (1977) the Pleistocene surface of the outer reef is a depositional surface, as the buttresses maintain the patch-reef configuration. This rules out an erosion of the outer reef during a sea-level lowstand.

The location of the Key largo Formation, 6 m above sea-level was explained as the result of a deposition during a previous highstand. However, this does not explain the discrepancy in elevation between the outer reef and the Key Largo Formation, because corals on the outer reef should have caught-up to sea-level at the same rate as those located on the Key Largo Formation. The Pleistocene surface, along the outer platform, is slightly inclined seaward: the depth is of about 75 feet (1 feet= 33 cm) along the outer border, 50 feet in the middle and much less close to the Keys (Enos,1977:P.1947).

This inclination can not be explained as the result of a greater erosion along the outer border of the platform, due to the greater wave energy resulting in a higher sediment trasport of sediment, because sedimentary trends indicate a landward sediment transport. On the other hand, remains of *Acropora palmata* in the outer platform (here referred to as the deep ramp sector) suggests a former depth of deposition close to the surf zone.A gradual eastward dipping of the Keys is evident from topographic data: height values are 18 feet at Windley

Key,2-3 feet at Matecumbe Key, 0 feet at Big Pine Key, 6-8 feet below sea-level at Boca Chica.
These differences indicate an elevation difference of 8 m from Windley Key to the westward side of the Keys. Hoffmeister and Multer (1968), among the possible causes,suggested a downfaulting to explain these height differences.

It is suggested here that rotational subsidence led to the emergence of the keys and to a differential subsidence on the outer platform (deep ramp).

Some areas along the keys underwent relief inversions and emergence.The core shown in Fig.40 (core #34 by Perkins) shows the development of a shallowing-upward sequence above coral limestones. The sequence is similar to a sill sequence.The emergence of the Keys is documented in some cases by the development of vadose diagenetic features and caliche-like crusts above the Key Largo Formation and the eteropic Miami Limestone.

An alternative to a syndepositional tilting is represented by a filling of a previously downfaulted area by vertical sediment aggradation during a highstand. In the case of the prismatic structure of the Pleistocene Formation,the depth of deposition should have been -46 m during the deposition of the Q2 unit,then -17 (Q3),and -9m (Q4).The depth of the sedimentary interface would have diminished gradually by vertical aggradation and gradual infilling.

Nevertheless, the coral composition is incompatible with such a hypothesis, based on bathymetric and ecologic constraints.In fact, a depth of -46 m contrasts with the occurrence of *Porites* (which grows above -15m), *Chione cancellata* (which lives in shallow lagoons) and *Montastrea annularis*.
Conversely, corals living at greater depths (i.e. *Meandrina mussa*) do not occur.Analogous considerations can be made for the upper, Q3 and Q4 units.

A tilting antecedent to the infilling would have probably led to the development of offlap ramps (cf. Fig.75).

A situation partly similar to that found in the Key Largo Formation occurs in the intrashelf ramp from Belgium (Preat,1984: Fig.3) where the reef is 60 m thick. This anomalous thickness was related to a differential subsidence (Preat,1984).

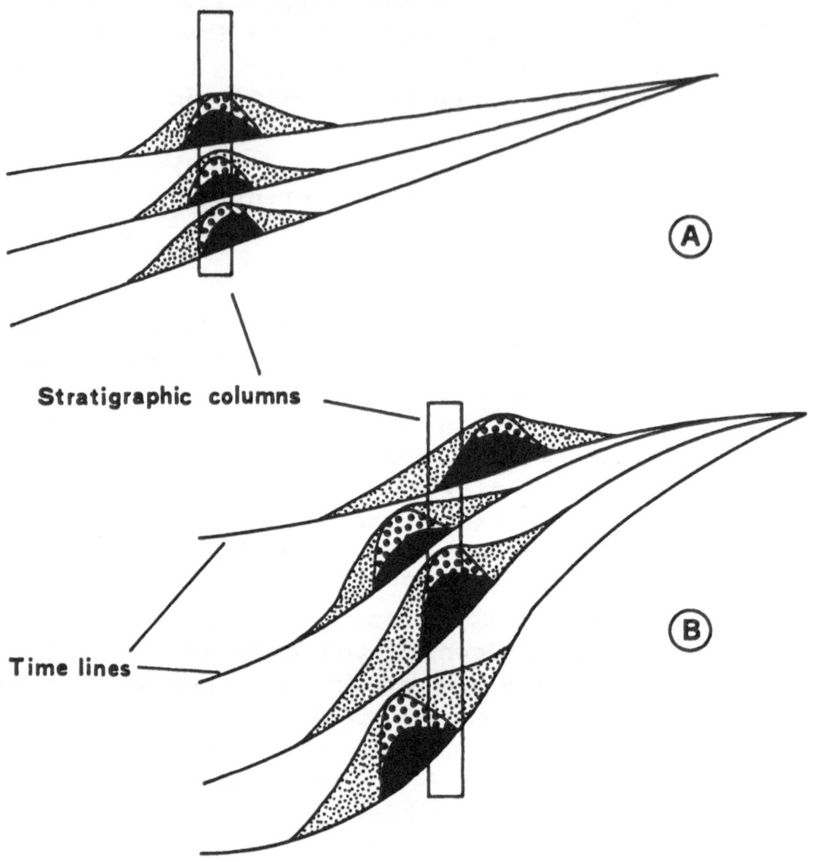

Fig.41 - Different controls played by the predepositional topography and related facies stacking in onlap ramps generated by rotational subsidence (A) and rotational subsidence + deformation (B).

Fort Thompson Formation, Pleistocene,Florida platform.

Introduction

The Fort Thompson Formation crops out in the south Florida plateau from Lake Okeechobee southward up to the Florida Keys (Fig.42).

The stratigraphy is outlined below.

The underlying Tamiami Formation (Miocene) represents several inner shelf environments ranging from nearshore to deltaic (Parker,et al.,1955). In the study area it consists of shelly sandstones, at places containing abundant *Ostrea* shells.
The Caloosahatchee marls were deposited in a paralic swamp above the Tamiami Formation as thin marly strings enriched in organic debris.The rapid vertical transition to the Fort Thompson Formation is characterized by the sudden disappearance of *Ostrea* and an increase in the carbonate content which reflect an environmental change from from paralic to a fully marine setting.
The Fort Thompson Formation consists of several alternations of marine - brackish - freshwater limestones which reflect high-frequency relative sealevel fluctuations which were correlated with interglacial highstands and glacial lowstands. The wedge-shape of the Formation,visible in Fig.42, corresponds to an onlap ramp geometry whose trough area is located in the south and southeast, and the hinge in the north and northwest.
In the east and northeast the Fort Thompson Formation interfingers with the Anastasia Formation which represents a complex of barrier beach, bar, dune and eolianite bodies (Perkins,1977) which controlled water circulation within the area of deposition of the Fort Thompson Formation during periods of relative sealevel lowstands.
In the south the Fort Thompson Formation is transitional to the patch-reef deposits of the Key Largo Limestone (Stanley, 1966).
The overlying Miami Oolite Formation represents a complex of oolite bars and sandwaves developed along a northeast - southwest strike (Evans and Ginsburg,1987).

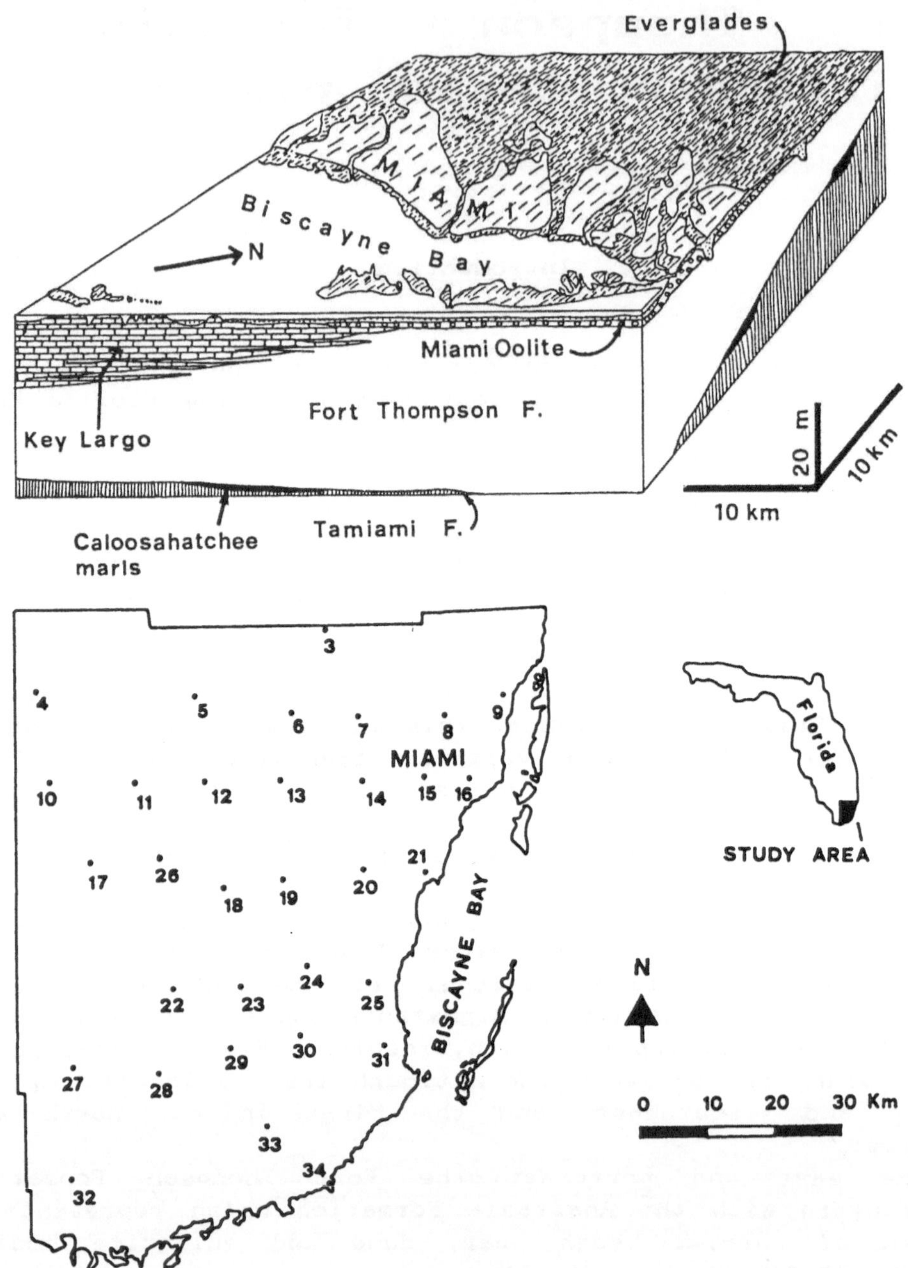

Fig.42 - Location of the Fort Thompson Formation, drill cores (Causaras,1987), and block-diagram (Parker, et al.,1955) showing the stratigraphy and the wedge-shape of the Formation.

Previous descriptions of the Fort Thompson Formation are those by Parker, et al.(1955), Cooke, et al.(1943), Brooks (1968), Perkins (1977),Mitterer (1975), Causaras (1987),Goldhammer, et al.(1990) and Galli (1991).

Outcrops are limited to rock pits and drainage canals; most of the actual surface is covered by a dense vegetation and a thin veneer of water.The study is based on the description of 34 cores (700 m of stratigraphic sections) drilled in Dade County area by air pumping techniques (Causaras,1987).

Lithofacies and environmental setting

The Fort Thompson Formation is composed of a complete spectrum of lithofacies ranging from marine bay (mollusk grainstones and packstones) to terrestrial carbonate subenvironments (*Helisoma* wackestones, root rock, phytogenic breccia).

Marine bay (Fig.43)

Description

This facies consists of poorly cemented, pale-orange to yellowish-gray, clean-washed calcarenite and calcirudite containing marine shells such as *Chione cancellata*, corals (*Archais angulatus*,*Montastrea annularis*, *Porites*), rare oysters and worm shells (*Vermicularia*). Mollusks are predominating: they are either disarticulated, concave upwards, or found in life position. They also form coquinae.Quartz sand composed of well sorted ,subangular grains constitutes approximately 15% of the totail grain percentage.

Two grain size trends occur in this lithofacies: 1) fining-upward trends accompanied by an upward increase in shell fragmentation and occurrence of *Callianassa* burrows; and 2) coarsening-upward grain-size trends consisting of an upward increase in size of shells,some of which occurring in life position.

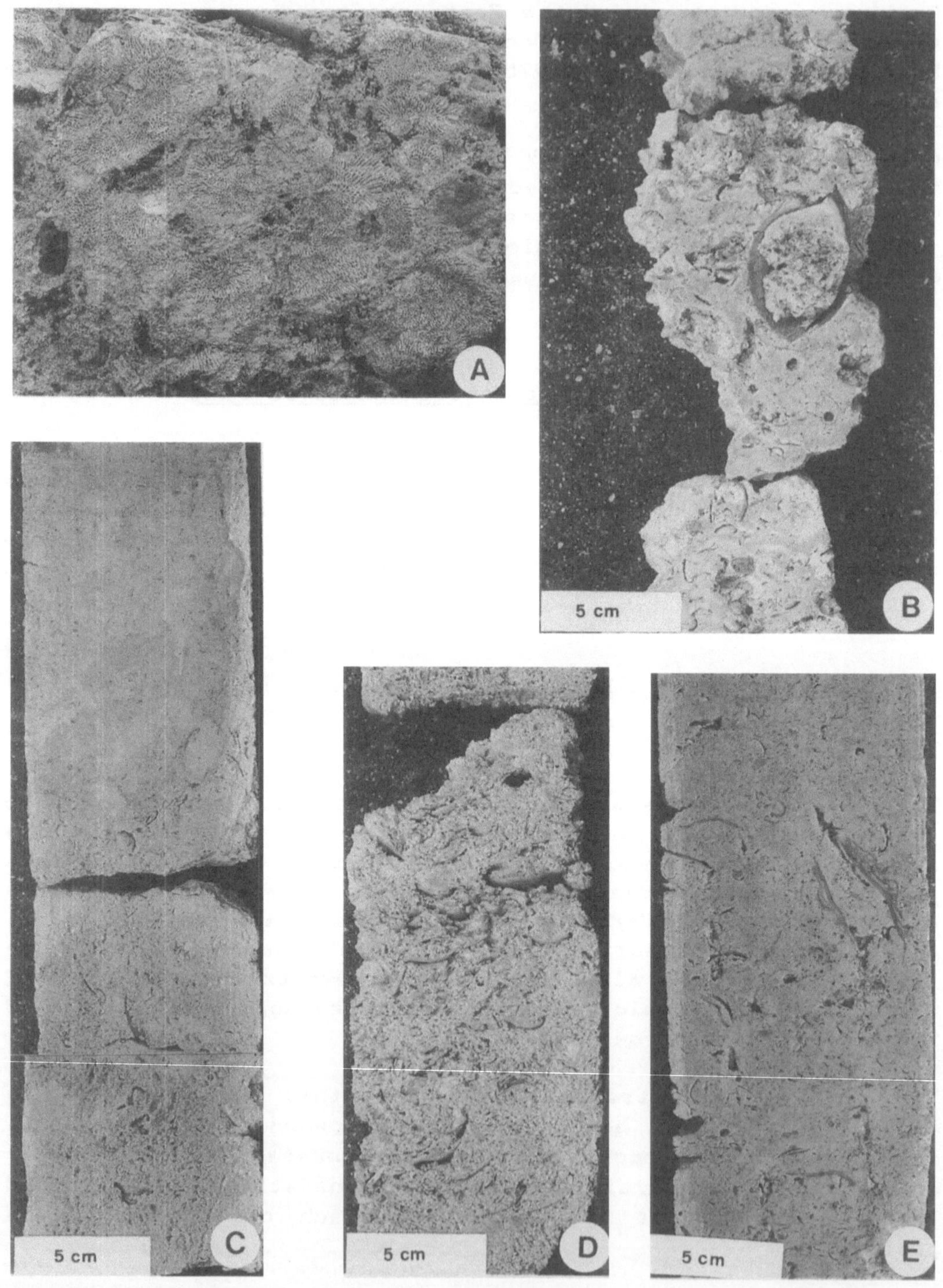

Fig.43 - Subfacies types of mollusk grainstones and packstones. A patch-reef. B marine bay. C marsh flat. D bar. E estuary.

Interpretation

The thanatocenosis is typical of present-day marine settings of Florida Bay and southwest Florida coast.This lithofacies includes several marine lagoonal subenvironments. Graded beds composed of fining-upward molluskan debris typify lagoons close to bars, beaches, or tidal deltas. The occurrence of *Ostrea* reflects a deposition in an estuary (Perkins,1977; Parkinson, 1989),more typical of the underlying Tamiami Formation. *Callianassa* mounds are found in lagoonal areas close to emergent and marsh flat environments. Coarsening-upward beds containing *Archais angulatus*, *Montastrea annularis* and *Porites* represent patch-reefs and bars developed in a marine lagoonal environment: they are common in the southeast where the Fort Thompson Formation grades laterally to the Key Largo Formation.

Freshwater swamp (Fig.44)

Description

This facies comprises root rock,phytogenic breccia and *Helisoma* wackestones.

Root rock, mainly developed within preexisting sediments, consists of an irregular network of circular, equidimensional rods whose rinds are lined with red-colored laminations and stained margins.

Phytogenic breccia consists of white-colored, dense masses of rootlets and traces of freshwater plants which impart to the rock a brecciated aspect.Casts of tree trunks and black pebbles are locally present. The matrix has a granular texture and consists of a mixture of roots and rhyzomes of freshwater swamp peat types.

Helisoma wackestones are constituted by dark-brown, well cemented mud containing abundant plane-spiraled and pulmonate gastropods (*Helisoma*) together with *Ameria* and ostracods. Organic matter and wood fragments are common components. Charcoal debris is not too frequent.Present-day cores of *Helisoma* wackestones show a fibrous dark-brown matrix with darker bands containing a greater amount of organic matter.

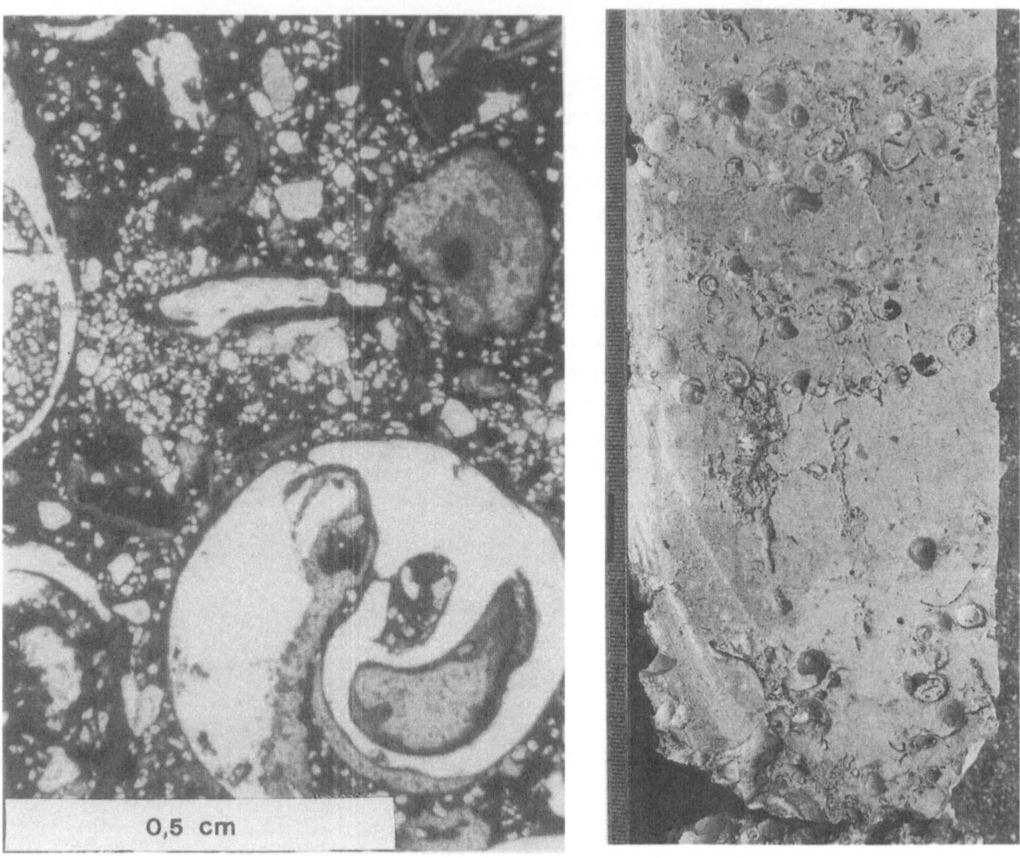

Fig.44 - Freshwater swamp as seen from core profile and thin section composed of mud, *Helisoma* snails, roots, vegetal debris and authigenic quartz grains (Galli,1991).

Interpretation

Root rock facies was described and interpreted by Bain and Teeter (1975) and Hoffmeister & Multer (1965) as black mangrove roots and mangrove reef rock respectively. The root rock facies represents the invasion and infilling by rootlets of *Avicennia nitida* of preexisting unlithified sediment. Vertically oriented root systems represent tap root systems.

Phytogenic breccias occur at present in the Everglades and are referred by Davies (1980) to as freshwater swamps.

Helisoma wackestones represent a deposition in a freshwater pond populated by *Helisoma* snails.The actual site of deposition of *Helisoma* wackestones is covered by few cm of water,saw grass and scattered red mangroves. The saw grass prairie is the site of accumulation of lime mud and marls and flocculation of a dense mass of rootlets, fragments and freshwater grass and sedges. The prairie is locally forested by buttonwood hummocks.The former existence of buttonwood hummocks is inferred from wood fragments observable in thin sections. Charcoal originated in hummock relieves which were the sites of aeration.The saw grass prairie is represented by lighter mud containing fewer shells and vertical root systems (*Distichlis spicata*).

Sedimentation took place in a low-energy, wide lagoonal environment, behind a complex of barrier island system represented by the Anastasia Formation and the Key Largo Formation.The lagoon, occupied by skeletal bars, patch-reefs and estuaries populated by oysters, graded landward, in the south and southwest,to muddy shorelines close to marshes and swamps, covered by mangroves; Further landward, the depositional surface was occupied by freshwater swamps.The depositional setting is analogous to the present-day environments bordering the Florida platform.

A paleoenvironmental profile of the low-energy shoreline is shown in Fig.45.

Facies associations

Lithofacies and alternations between freshwater and marine lithofacies are grouped into shallow ramp and deep ramp associations. The shallow ramp is located in the south; the deep ramp in the north. The intermediate ramp is absent and the transitions between the two sectors of the ramp are gradational.

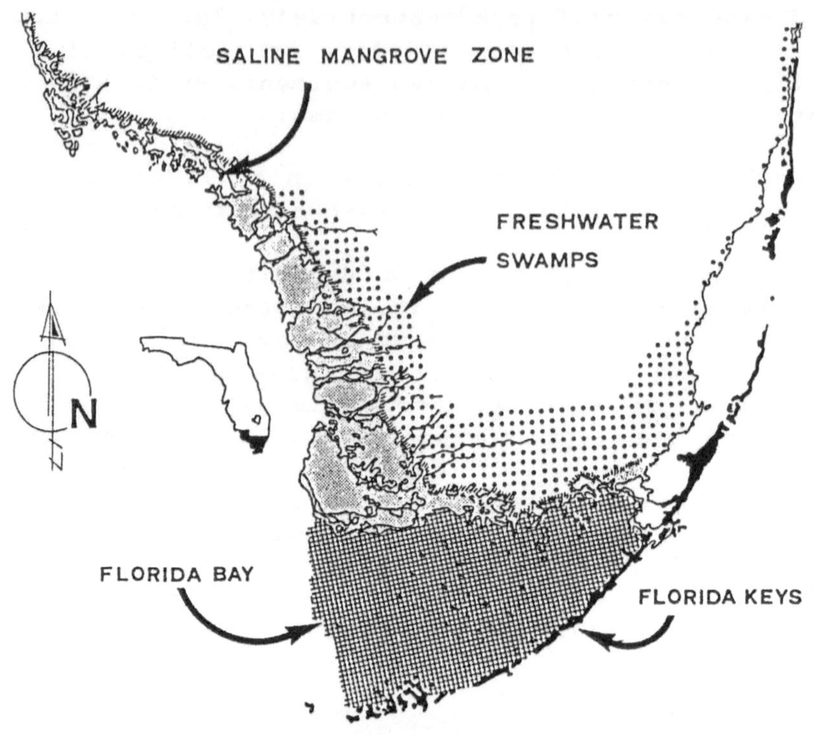

FRESHWATER SWAMPS SALINE MANGROVE ZONE MARINE BAY

BUTTONWOOD HUMMOCK | SAWGRASS PRAIRIE | FRESHWATER POND | BLACK MANGROVE PEAT | RED MANGROVE PEAT | MARSH FLAT | CALLIANASSA MOUNDS | BARS;PATCH PEEFS | (OYSTER BED)

(not to scale)

Fig.45 - Present-day environmental zones (from Craighead, 1969), and paleoenvironmental reconstruction of the Fort Thompson Formation.

Shallow ramp

The shallow ramp is represented by an alternation of marine - brackish - freshwater limestones.This sector consists of a

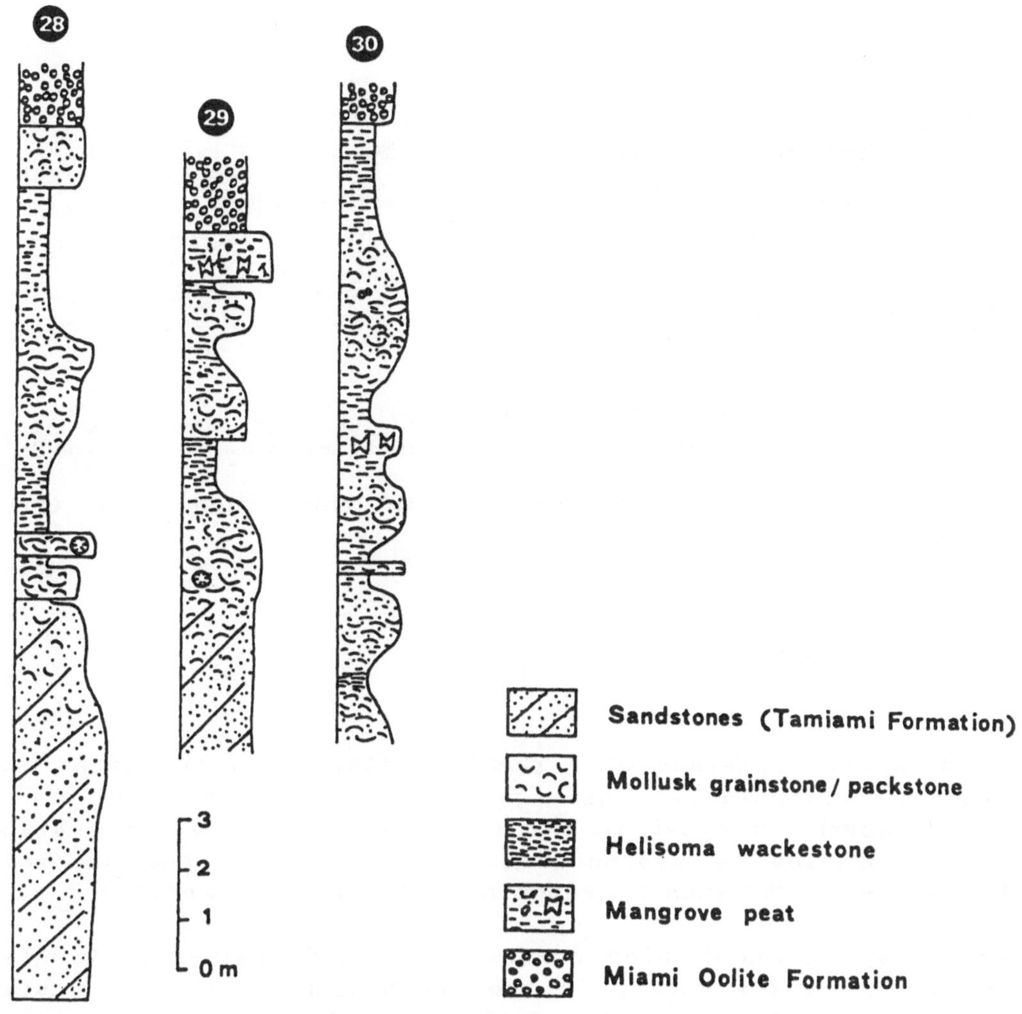

Fig.46 - Thickening-upward trends occurring in the shallow ramp.

maximum number of three alternations of facies (Fig.46) which are organized into symmetrical grain-size trends (coarsening-upward -->fining-upward).Remains of carbonized roots with black mangrove facies (root rock facies) are common. Transitions to freshwater swamp facies are more or less gradational.

The stacking of sequences results in a thickening-upward trend, also visible in Fig.47, which shows a stacking of marine lithofacies separated by discontinuity surfaces typical of hinge areas.

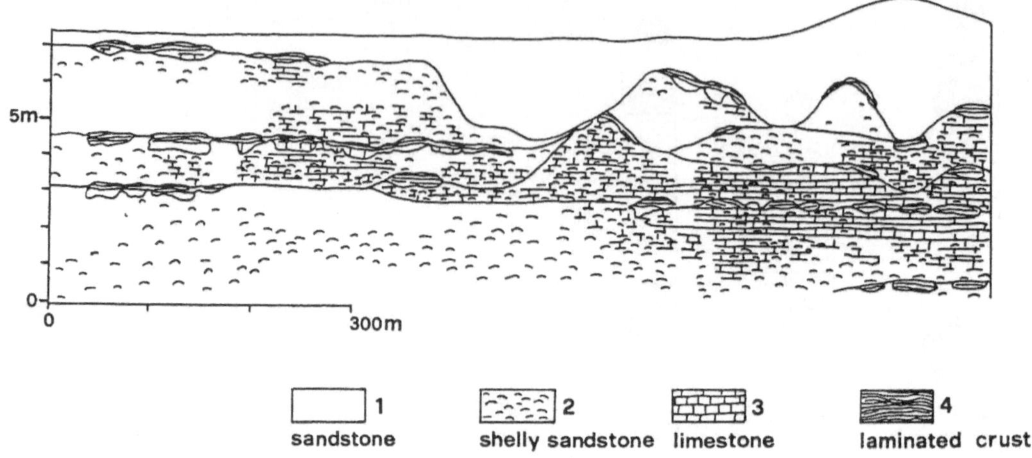

Fig.47 - Stacking of lithofacies in the hinge area (from Perkins,1977).

Deep ramp

The deep ramp consists of a regular alternation of marine and freshwater lithofacies that increase progressively in number towards north (from 4÷5 to 9 alternations: Fig.48). The marine bay lithofacies is preponderant. Close to the shallow ramp (cores #18 to 25) such a predominance is less apparent.

Marine bay facies display coarsening-upward grain-size trends, especially in the north; in the south,symmetrical grain size trends become more frequent. Carbonized plant remains and root rock facies are less common than in the shallow ramp.

The vertical stacking of alternations results in thinning-upward or asequential trends.

Depositional sequence

The Fort Thompson Formation represents a third order depositional sequence. The depositional theme is simple because the whole Formation consists of the vertical stacking of meter-scale alternations of freshwater and marine lithofacies (parasequences) which represent high-frequency, short-term phases of sedimentation.

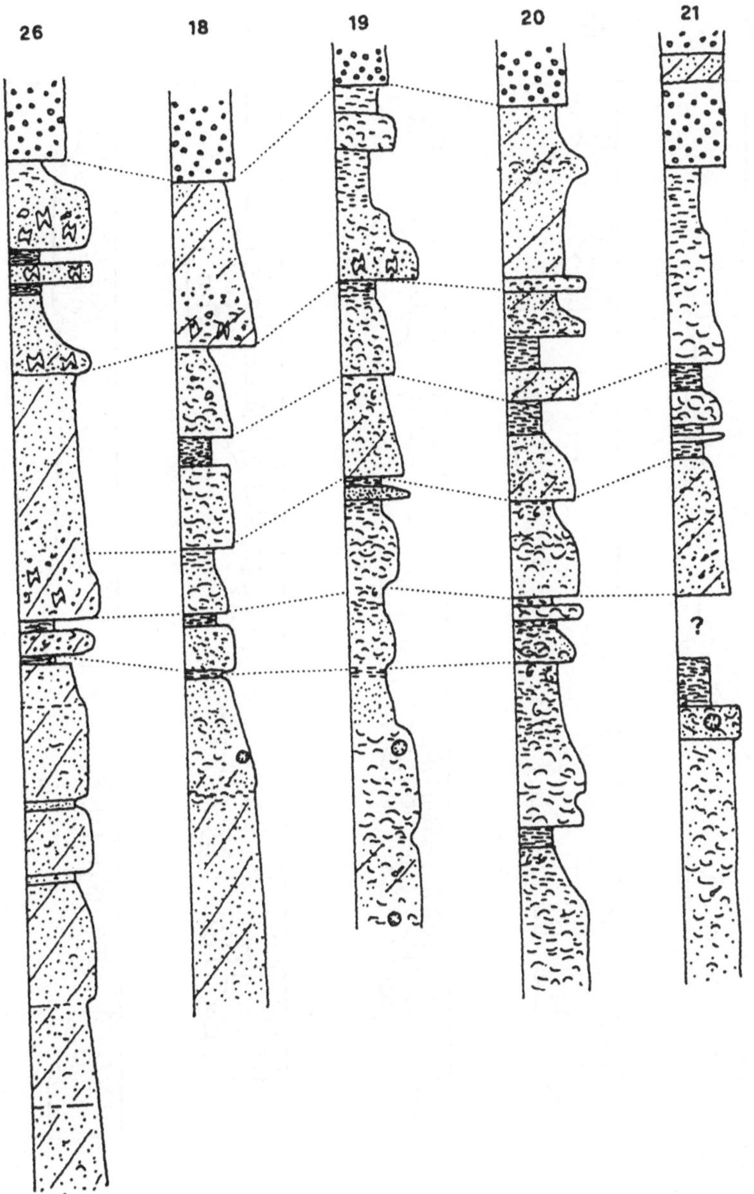
26
18
19
20
21
?

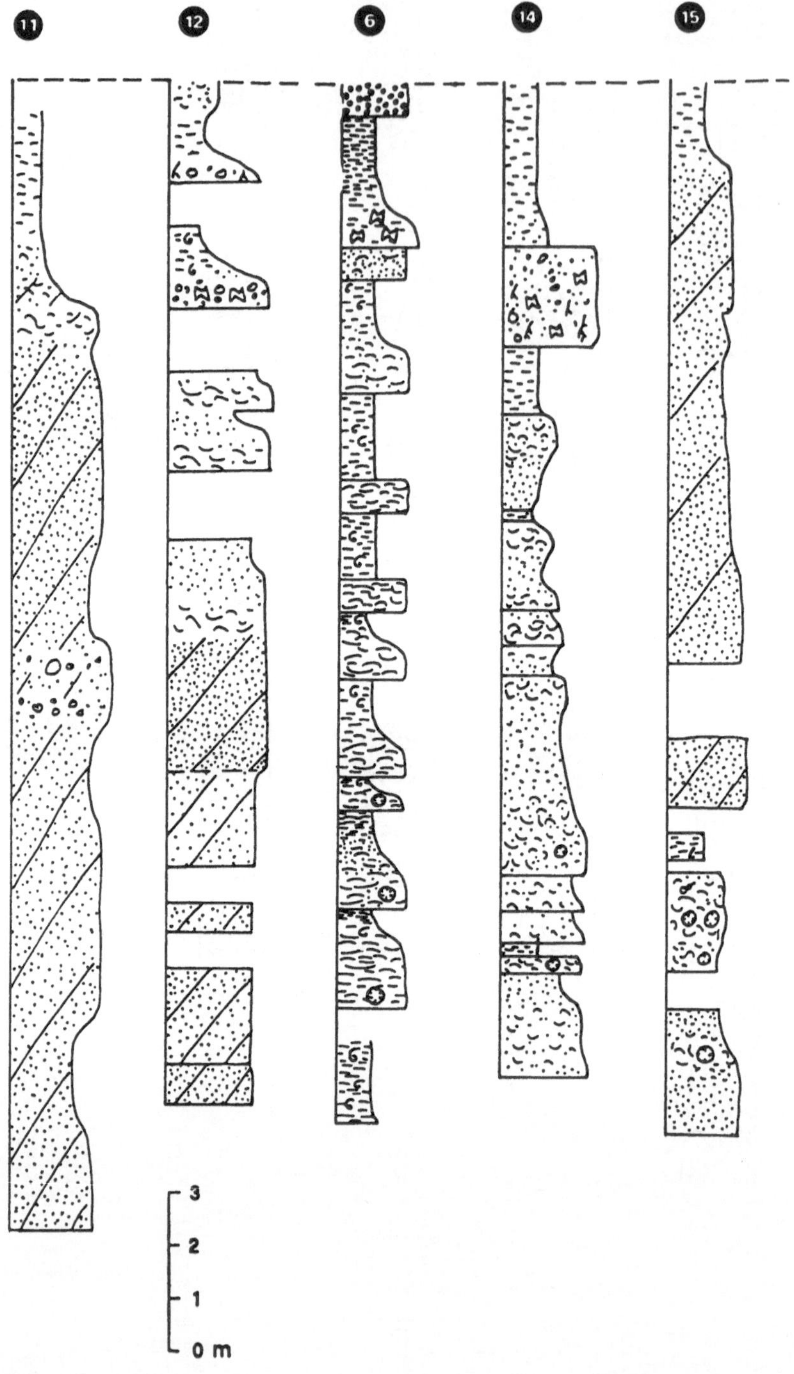
11
12
6
14
15
3
2
1
0 m

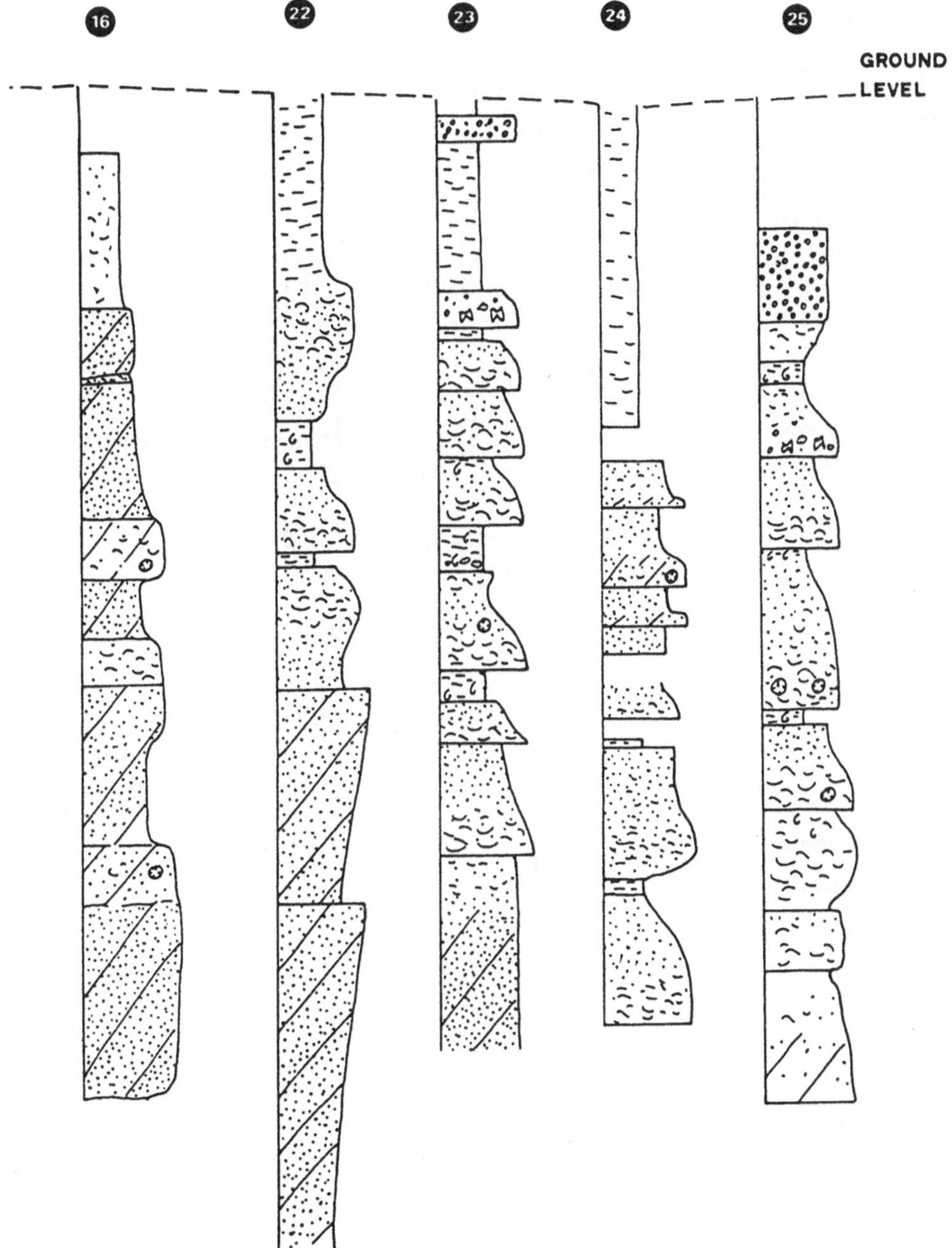

Fig.48 - Core profiles showing alternations between marine and freshwater lithofacies.

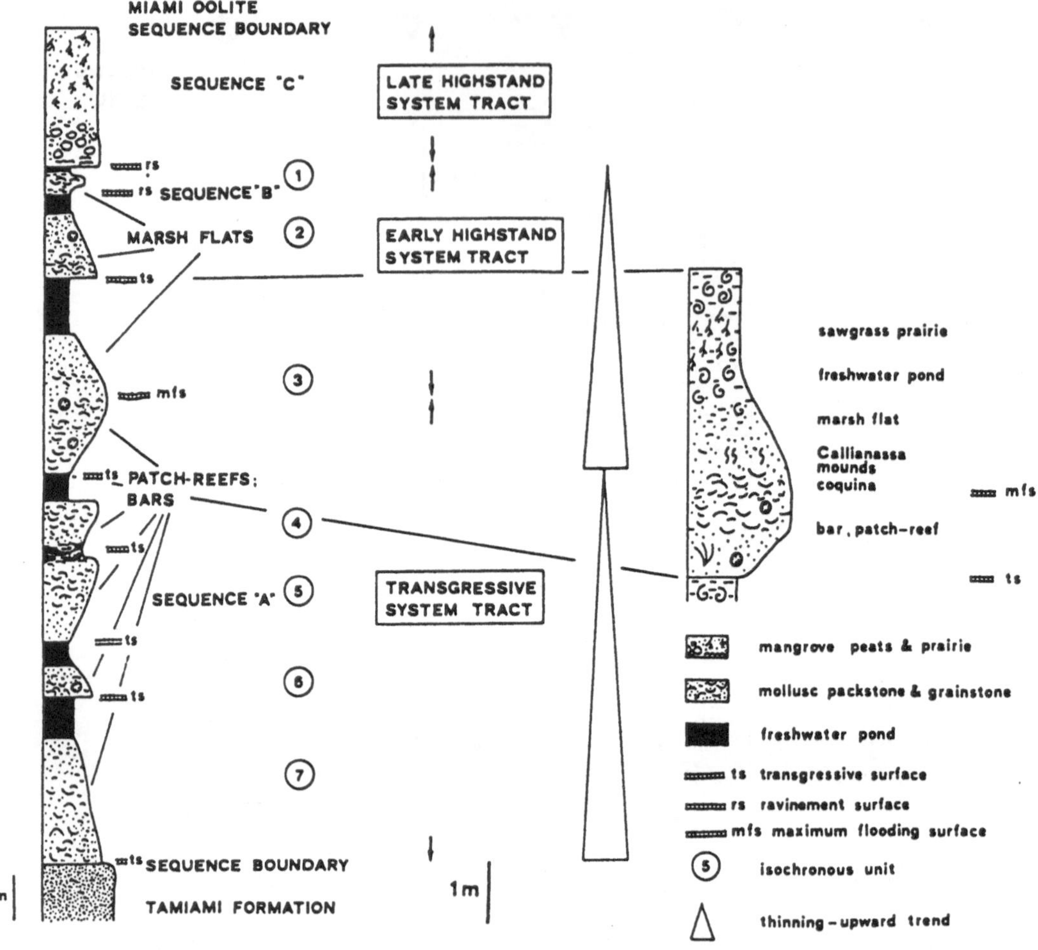

Fig.49 - Sequence stratigraphy depositional model of the Fort Thompson Formation.

The system tracts which compose the depositional sequence are the following, from bottom to top: 1) transgressive system tract; 2) early highstand system tract; and 3) late highstand system tract (Fig.49).

The discontinuity surfaces which separate parasequences are transgressive and ravinement surfaces. A maximum flooding surface separates the transgressive from the highstand system tract.

Variations in the internal characteristics of parasequences

reflect corresponding differences in the rates of relative sealevel rises.

Transgressive surface

Transgressive surfaces are represented by abrupt transitions from freshwater swamp to marine bay lithofacies occurring at the bases of lowermost parasequences (Fig.49). They reflect a rapid flooding of the lagoon located behind the barrier island complex of the Anastasia and Miami Oolite Formations, during sealevel highstands.Conversely, during sealevel lowstands the lagoon was a freshwater lake (Fig.50).

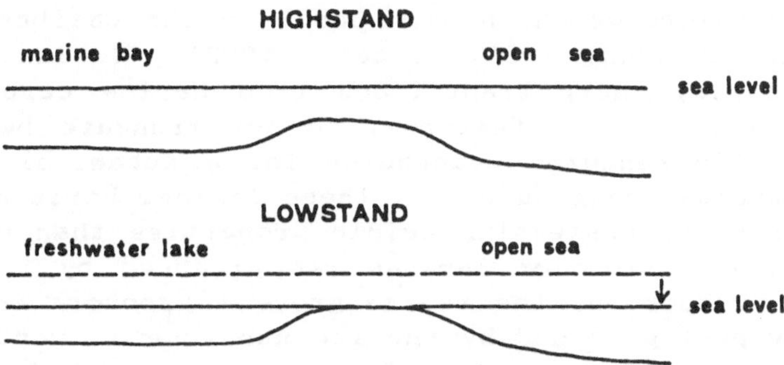

Fig.50 - Control of the sedimentary pattern of the Fort Thompson Formation by high-frequency relative sea-level changes.

Ravinement surface

This surface is a horizon comprising laminar micrites and pseudobreccias (Fig.51)

Laminar micrites consist of a few mm-thick, red colored laminations draping irregular subhorizontal surfaces developed mainly within the marine bay facies. These laminar micrites are composed of an alternation of darker and lighter laminae. Darker laminae are composed of a mottled, red colored, dense mass; lighter laminae consist of microspar filling irregular, sub-horizontal,contorted voids characterized by frequent bifurcations, labyrinthic structuress and pseudofenestral fabrics. fine vertical rods protrude downward from horizontal laminae, in some cases cutting through shells. As seen from outcrops parallel to the bedding plane, the lateral continuity of this thin horizon is interrupted by subcircular holes, averaging 8 cm in diameter. In some cases the holes are aligned along a circular perimeter.

Pseudobreccias generally overlying laminar micrites, are breccia-like features composed of red-colored mm-cm thick monogenic fragments surrounded by a matrix constituted by equidimensional quartz grains, and infested by the same quartz grains. Marine and freshwater shells occur within the matrix. Clasts look like pieces of a jigsaw puzzle; going downward, they become smaller and more numerous.

Profiles such as that shown in Fig.52,A are interpreted as the result of root penetration by mangrove root systems. Unlike caliches or calcretes which also occur in the Caribbean region (Beach and Ginsburg,1981; James, 1972), laminar micrite horizons display sharp transitions from marine carbonates to crusts,are accretionary features, do not truncate bedding and lack diagenetic textures evidencing for a subaerial exposure. The shell truncation produced by these laminar horizons is more typical for roots possessing acidic properties than of caliche crusts. The red color of laminae was produced by the tannine inclusions produced by the red mangrove *Rhyzophora mangle*. The present-day peat produced by the red mangrove is reddish brown to dark-brown and consists of a dense mass of rootlets. Likewise, water surrounding *Rhyzophora mangle* is red brown.

Circular holes visible on horizontal surfaces probably represent casts of former roots.White laminae inside the laminar micrite horizon represent the infilling by calcite of former thin horizontal root filaments.Laminar horizons are similar to those detailed and interpreted as root mats by Wright,et al.(1988).

Pseudobreccias are interpreted as the result of a dissolution produced by the acidic peat of the red mangrove root system. In several cases the transition between clasts and the matrix is gradational.The distinction between clasts and matrix is made possible by a greater abundance of equidimensional quartz grains within the matrix than within the clasts.It is possible that quartz grains originate from the siliceous material contained in the vascular tissues and periderm of roots, once they undergo peatification (cf. Hoffmeister and Multer,1965).

The development of red mangrove peat within marine bay facies

Fig.51 - Aerial views of some aras along the coasts of south Carolina and Georgia,thought to be representative of the different stages of flooding and shallowing up of the Fort Thompson Formation.

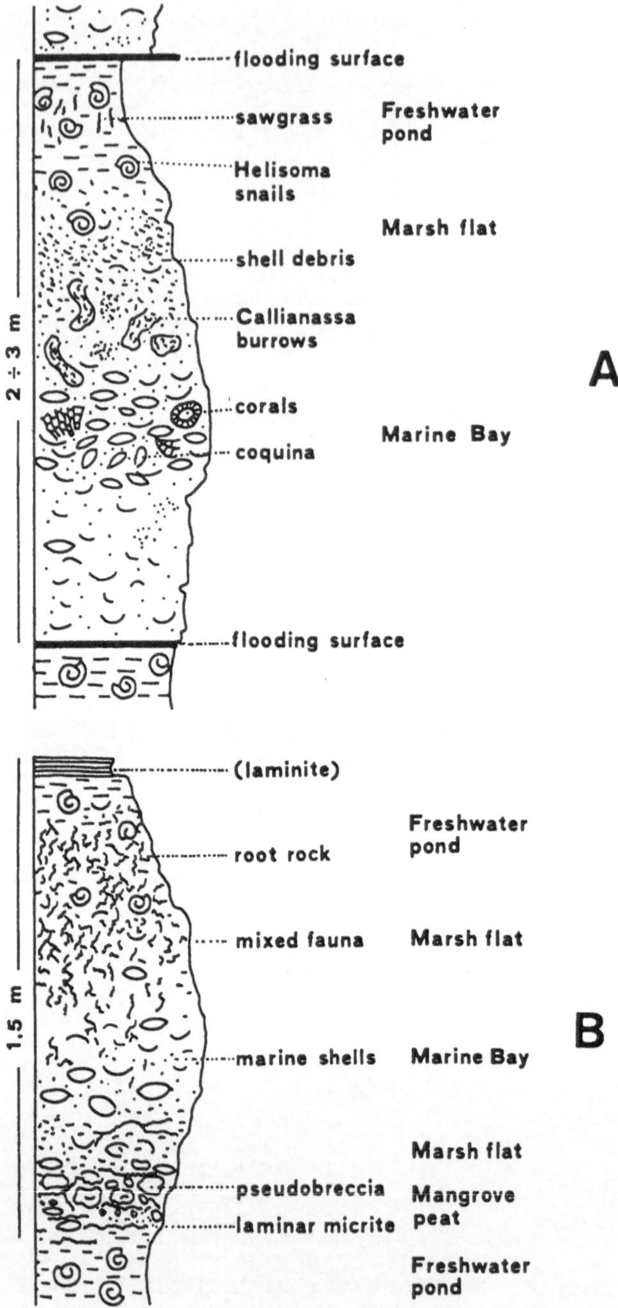

Fig.52 - Parasequences formed during a quick transgression (A) and a less rapid sealevel rise displaying laminar micrites and pseudobreccias (B).

Fig.53 - Ravinement surface. A Profile of the laminar micrite, pseudobreccia, root casts developed above marine bay facies. B Casts of prop roots and laminar micrites. C,D Pseudobreccia. E Root rock showing rhyzoturbated sediment. F,G Detail of laminar micrites (thin sections). The example visible in G shows a root lamina cutting through a shell.

is explained by the marine to brackish coastal, intertidal setting of *Rhyzophora mangle*.It forms during the initial phases of a relative sealevel rise.Quick sealevel rises prevent the development of the red mangrove because the limit of survival and colonization of the red mangrove seedling corresponds to the upper shoreface.The red mangrove peat transitional between freshwater and marine bay facies is interpreted as a ravinement surface because it represents a slow reworking of a previously exposed interface by an advancing sea. The pseudobreccia may be regarded as a biogenic transgressive conglomerate.

Parasequences formed during a quick transgression are different from those developed during a slow transgression (Fig.52).The first (Fig.52A) occur at the bottom of the Fort Thompson Formation; the second (Fig.52B) in its upper part and in the northern sector, in the deep ramp.

Maximum flooding surface

It consists of a coral-bearing horizon (Fig.53) containing *Montastrea annularis* and *Porites*. It is interpreted as a maximum flooding surface because records the time of maximum rate of accomodation increase: it records in fact a maximum deepening as shown by the occurrence of fossils such as *Montastrea annularis* and *Porites* which actually inhabit more open and deeper lagoonal areas (from -9 to -24 m below sealevel) than *Chione cancellata* and the other biota contained in the marine bay facies.

Transgressive facies tract

The transgressive facies tract, developed above the Tamiami Formation, consists of the stacking of transgressive parasequences (Fig.53,A).Their thickness range from 3 - 4 m to about 1.2 m.

The basal part is developed above a transgression surface. Lithofacies are mollusk grainstones and packstones (marine bay facies) in some cases with *Porites* and *Montastrea annularis*. The proportion and sizes of mollusks may increase upwards up to the half of the marine lithofacies, with the production of a coarsening-upward grain size trend.

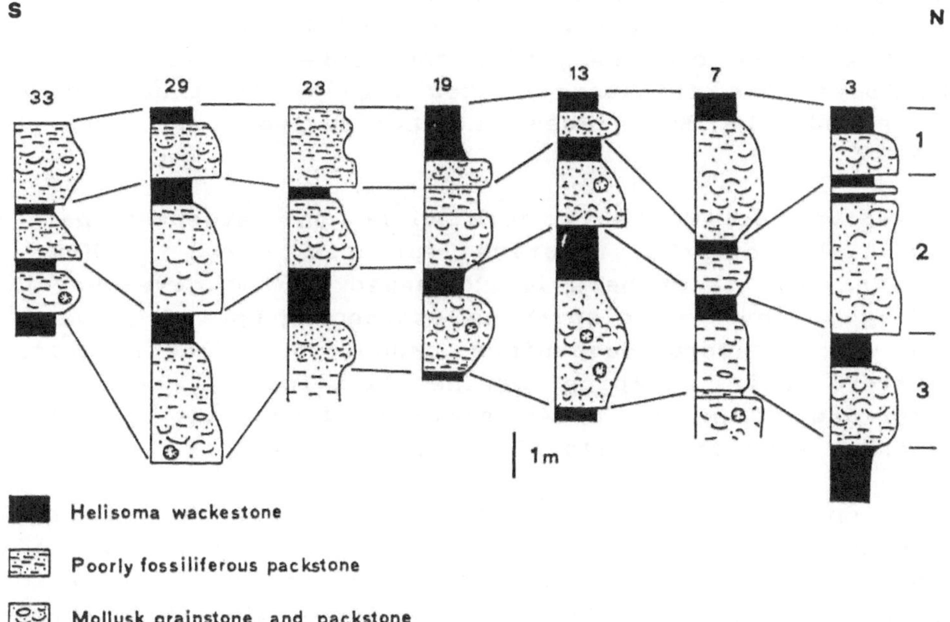

Fig.54 - Field aspect and lateral variations of the maximum flooding surface (parasequence #3).

The lower part of the parasequence grades upwards into a strongly bioturbated packstone, as is evidenced by sinuous galleries and shell debris-filled irregular patches. Going upwards, shell sizes decrease. This part in turn grades to a freshwater swamp lithofacies (*Helisoma* wackestone).This transition is evidenced by a mixed fauna containing marine mollusks and freshwater gastropods indicative for a brackish, transitional marsh flat environment.

Vertical transitions from marine bay facies to freshwater lithofacies indicate an upward shallowing in a low-energy shoreline. The basal coarsening-upward grain size trend in the sequence may by interpreted as a catch-up phase of sedimentation (Kendall and Schlager,1981) during which carbonate sediment production increase takes place by vertical growth as bars or patch-reefs in order to keep pace with the increased rate of sealevel rise. The successive fining-upward trend may correspond to the keep-up phase (Kendall and Schlager,1981) that evidences for a lateral accretion or outbuilding of islands during a period characterized by a reduced rate of relative sealevel rise, or a sealevel stillstand.

The transition to the freshwater unit may have been a consequence of an accentuation of the rate of sealevel fall during glacial periods. Barrier island faciess in the east during these periods became emergent and interrupted the water exchange with the open sea, with the consequent transformation of the bays into freshwater lakes similar to those actually existing inh the Everglades and nearby coastal areas (cf. Fig.51).

Trangressive facies tracts form along the steepest part of rising limbs of the relative eustatic curve (Haq, et al.,1987). This tract here is expressed by the development of aggrading patch-reefs displaying coarsening-upward grain size trends which record deepenings caught-up by sedimentation. Freshwater facies capping marine lithofacies document the interference of higher order sealevel fluctuations with the lower order frequency sealevel rise.

Correlations between cores mark a series of northward thickening wedges with irregular outlines formed by bulges and depressions that reflect the control played by the entecedent topography in the deep and shallow ramp. The Fort Thompson Formation is a northward and eastward thickening wedge (Fig.42,Fig.55).The reconstructions made by Parker,et al. (1955), Perkins (1977) and Causaras (1987) clearly show a

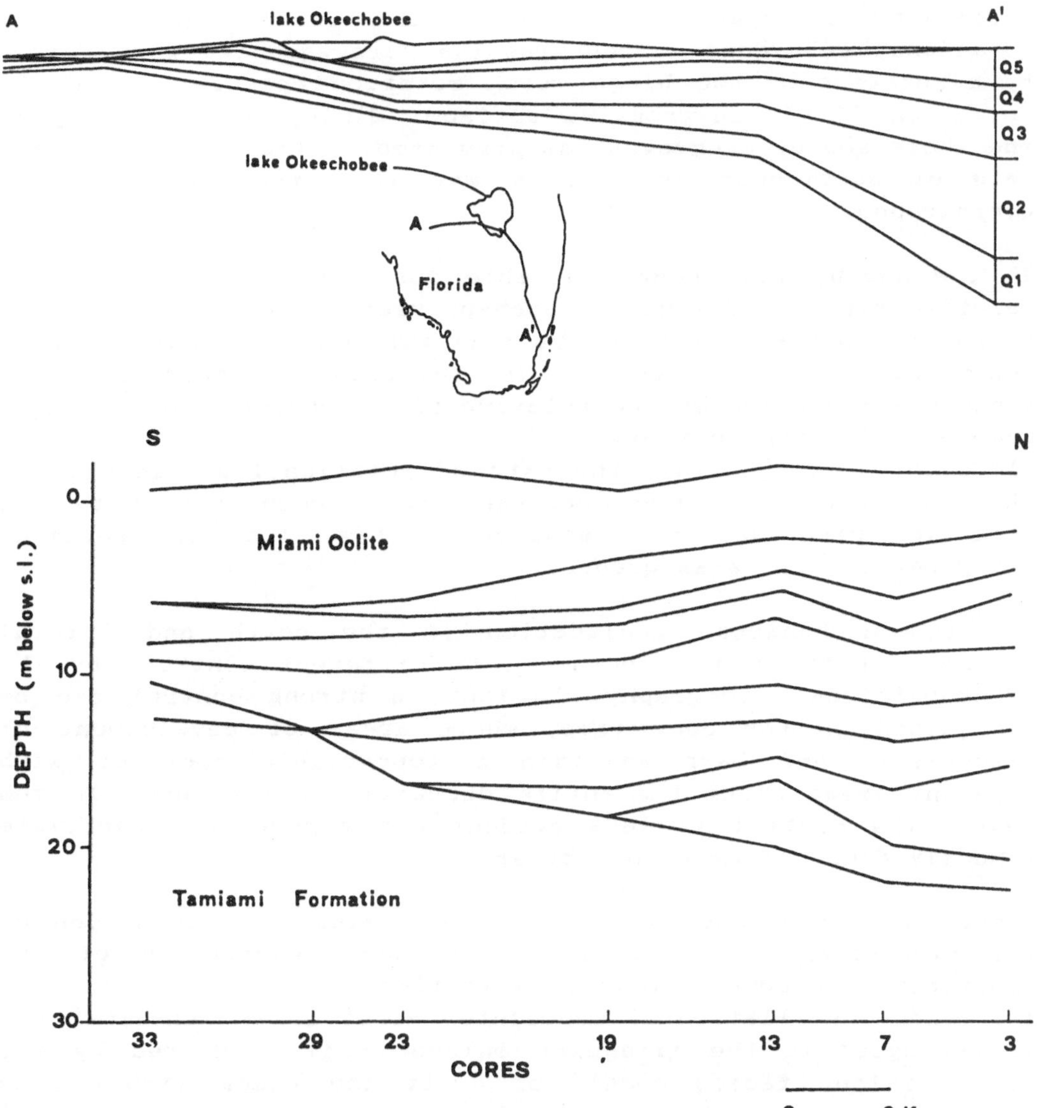

Fig.55 - Onlap ramp geometries of the Fort Thompson Formation (above: from Perkins,1977) showing southward and eastward thickening wedges and the control played by the antecedent topography in the trough and hinge areas.

gradual thinning-out and disappearing of parasequences towards a hinge located in the west and south that implies a rotational subsidence on a westerly striking hinge.

The rate of relative sealevel change varied not only temporally, but also with the position on the platform, due to variations in space and time of the amount and rate of rotational subsidence. Transgressive parasequences were formed by a rotation on the hinge: each rotation episode led to the formation of a transgressive surface. Biological reworking of the shoreface by mangroves was prevented by the rapidity of the rate of subsidence in such a way that the shoreface was overstepped.

In the north, the space available for sedimentation during a relative rise was filled by catch-up reefs.
Sites intermediate between the northern and the southern area underwent a lesser amount of rotational subsidence, and sedimentation outpaced the relative rise and migrated laterally over coeval catch-up reefs.
Close to the hinge line, the rate of rotational subsidence was at a mimimum: therefore, sediment could keep pace with the relative rise of the sealevel. This led to repeated amalgamations of parasequences.

The two mechanisms: aggradation in the north and lateral sediment shift in the south,led to different controls by the predepositional topography. In fact, a strong control can be recognized in the deep ramp, where areas of development of patch-reefs and bars maintain a topographic contrast with adjacent areas along the entire depositional sequence. In the south, the control is less evident and depocenters oscillate randomly from one area to another.

These two situations are also evident from an examination of the block-diagrams of Fig.56 which show variations of the uppermost lithosome geometries with time.
The random oscillation of depocenter areas in the shallow ramp is analogous to the irregular island shift predicted by the "tidal island facies model" of Pratt and James (1986). The resulting pattern along an isochronous stratigraphic interval is a lateral transition from coarsening-upward to fining-upward beds that is also demonstrated by symmetrical sequences developed in intermediate positions by a migration of the fining-upward beds over coarsening-upward beds (according to Walther's law).

Early highstand facies tract

This tract, developed above the maximum flooding surface, consists of three early highstand parasequences (Fig.53,B) organized into a vertical thinning-upward trend (Fig.49).In the basal part of these parasequences mangrove peat facies (ravinement surfaces) are interposed between underlying freshwater facies and marine lithofacies. The marine bay facies contains scattered shells of *Chione cancellata*, fine shell debris and several root traces. This lithofacies grades upwards into the root rock facies and to a lithofacies containing a mixed marine and freshwater fauna, indicative for a brackish environment; less commonly, the sequence is topped by supratidal laminites.

In contrast to the transgressive parasequences, these record gradual flooding rates (ravinement surfaces) and gradual sealevel falls (transitions from catch-up reefs --> marsh flats--> freshwater swamp). The occurrence of a marsh flat indicates that sedimentation could outpace the relative sealevel rise as a result of its slowing down.

The lower parts of these parasequences (mangrove peat-->marsh flat-->marine bay) are analogous to the sequence recently described by Parkinson (1989) from the southwest Florida platform which consists of 2 to 6 m thick transgressive - regressive couplets (paralic swamp--> restricted marine-->open marine) bounded by mangrove peat facies. It is also analogous to the cycle of Florida Bay (Enos & Perkins,1979).

As stressed by Goldhammer,et al.(1990), these cycles result to be controlled by 5th order sealevel fluctuations (rapid rise followed by a decelerating sealevel rise). The similar lithofacies and patterns of changes recorded in the early highstand parasequences suggest an evolution similar to that of Florida Bay and southwest Florida platform.

This part of the depositional sequence contains evidence for a eustatic control on sedimentation.Based on calculations by Mitterer (1975), the uppermost two parasequences result to have formed in 25.000 years (Galli,1991) which approximately fit Milankovitch precession cycles. Subsidence gradually diminished during the deposition of this facies tract, so that marsh flats could develop and eustatic effects, superimposed on tectonic subsidence, became more evident higher up the Formation.

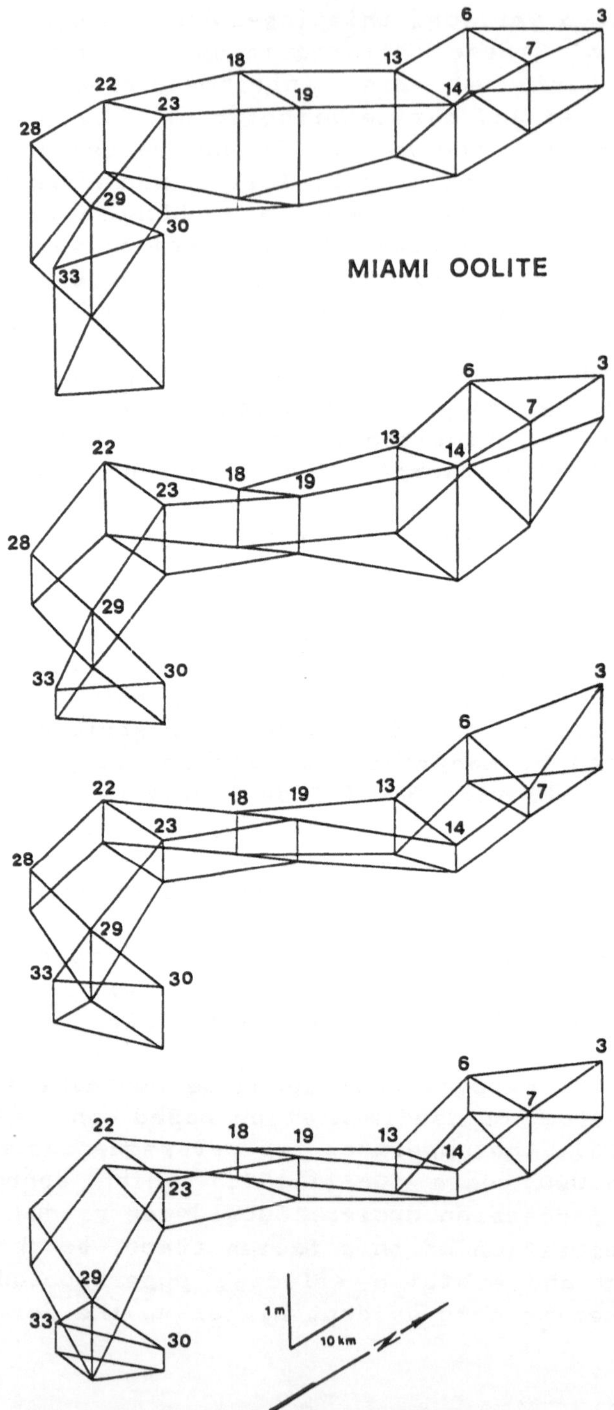

Fig.56 - Block-diagrams showing variations in volume and morphology of parasequences.A lack of progradation is apparent in the hinge, in the south, whereas a southward shift of depocenters (retrogradation) characterizes the trough.

Late highstand facies tract

This facies tract is constituted by a parasequence which varies in thickness from a dm to 4 m. (Fig.57).
The vertical succession of lithofacies with few changes is the same at all sites. The sequence passes from a laminar micrite horizon to a pseudobreccia, to the root rock facies and eventually to the freshwater phytogenic breccia which is commonly the thickest lithofacies. The upper part of this parasequence may contain oolites or mudstones typical of the overlying Miami Oolite Formation (Evans and Ginsburg,1987) which may correspond to a shelf margin facies tract.

A similar, modern sequence of sediments passing from mud banks to freshwater swamp facies was described by Craighead (1969), from the Everglades National Park (Fig.16),who showed that the Everglades has prograded seaward more than 8 Km, during the Holocene sealevel rise.

The late highstand parasequence is interpreted as a shallowing-upward sequence formed by a depositional regression in response to a sedimentation rate in excess of the relative sealevel rise. Lateral variations in thicknesses of the parasequence depended on local, former depths of the lagoonal floor: in fact, thicknesses are greater in the deep ramp and limited to a few dm in the shallow ramp.

By analogy with the modern cycle described by Craighead (1969) the parasequence records a seaward shoreline progradation of a mainland produced during a declining sealevel rise, as is also supported by the vertical transitions between mangrove and freshwater lithofacies. The late highstand facies tract records the maximum rate of accomodation decrease and a continuous shallowing-up. It vertically grades to the Miami Oolite which records a change from a eustatic fall to a slow rise in the sealevel. The progressive decrease in subsidence led to a progressive decrease in the control played by the predepositional topography: eventually, the deposition of the parasequence formed uniquely under eustatic controls which led to a levelling of the topographic relief, with the formation of a compensation cycle (cf. Enos & Perkins,1979).

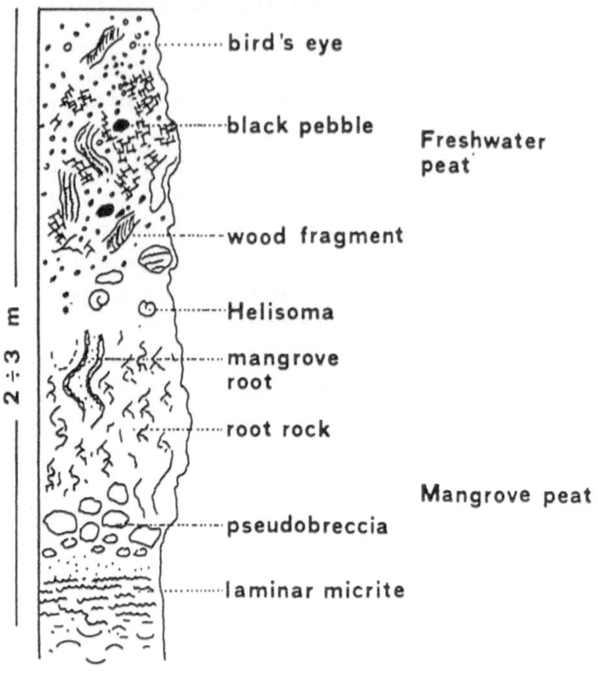

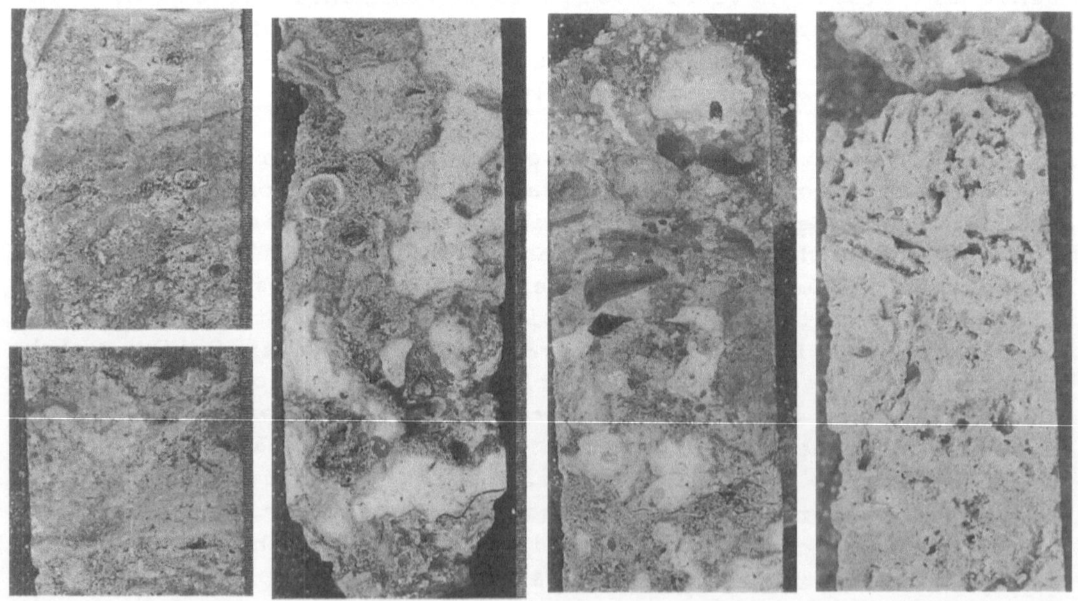

Fig.57 - Late highstand facies tract.

"Calcari Grigi" Formation, Jurassic, Venetian Alps

Introduction

The Jurassic Calcari Grigi Formation crops out in the Venetian Alps, Italy, at the top of the Trento platform (Fig.58), a rimmed, isolated platform, 8000 square Km wide, which is a portion of an Atlantic-type margin subjected to drowning as a result of a complex interplay of eustatic phases (Bernoulli and Jenkins,1974; Hallam,1978, 1981; Vail and Todd,1981), superimposed on synsedimentary tectonics (Castellarin, 1972).

The eustatic curves by Hallam (1978) and Vail, et al.(1977) show a continuous rise in the Jurassic. Major rises took place in the Hettangian, late Sinemurian, Pliensbachian, and at the beginning of the Toarcian. Conversely, sea-level falls were of less importance (Hallam,1981).

It is difficult to ascertain the role played by tectonics in the sea-level rise. Major facies changes were produced by a platform collapse in the european margin (Bernoulli & Jenkins,1974). The destruction of the carbonate platform begun in the lower Liassic (Lemoine, et al.,1978) but continued diachronously in the southern Alps. The destruction of the Trento platform probably initiated at a later time (Winterer & Bosellini,1981) , at the end of the middle Liassic (179 m.a.).

Listric synsedimentary faults oriented NNE-SSW producing half-graben structures, were mapped along the western margin of the Trento platform (Castellarin, 1972).
A non-isostatic subsidence antecedent to the drowning of the platform is evidenced in the study area by numerous features (Fig.59): synsedimentary faults,slumping features,crumpled beds,and tsunamites (Galli,1990).

Synsedimentary faults which were mapped in the area, projected on a Schmidt net (Fig.60) intersect in the SW quadrant and individuate a potential failure wedge oriented SSW. The NE quadrant conversely may represent a tectonically relieved area. This structural frame controlled bathymetry and facies distribution.

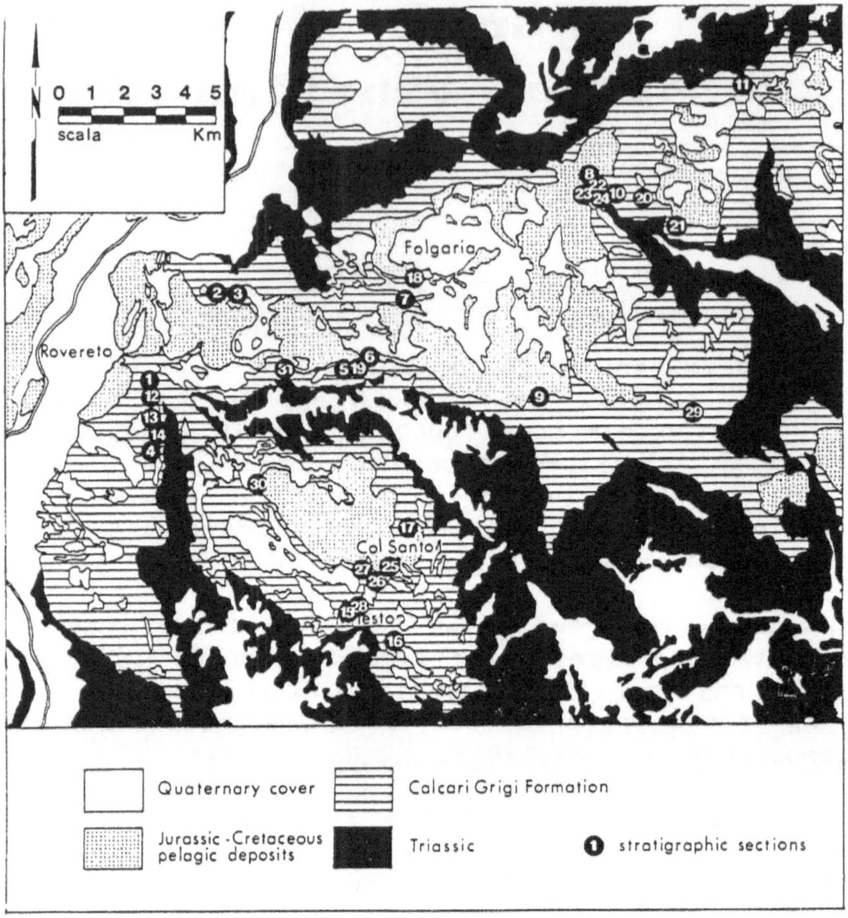

0 1 2 3 4 5
scala Km

Folgaria
Rovereto
Col Santo
Tiesto

Quaternary cover
Calcari Grigi Formation
Jurassic-Cretaceous pelagic deposits
Triassic
stratigraphic sections

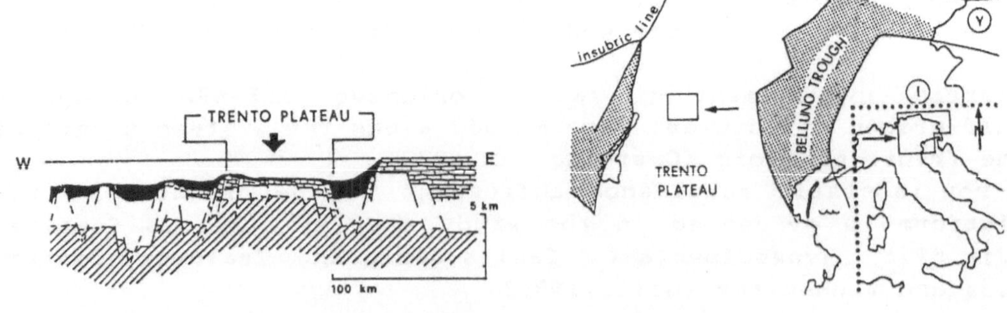

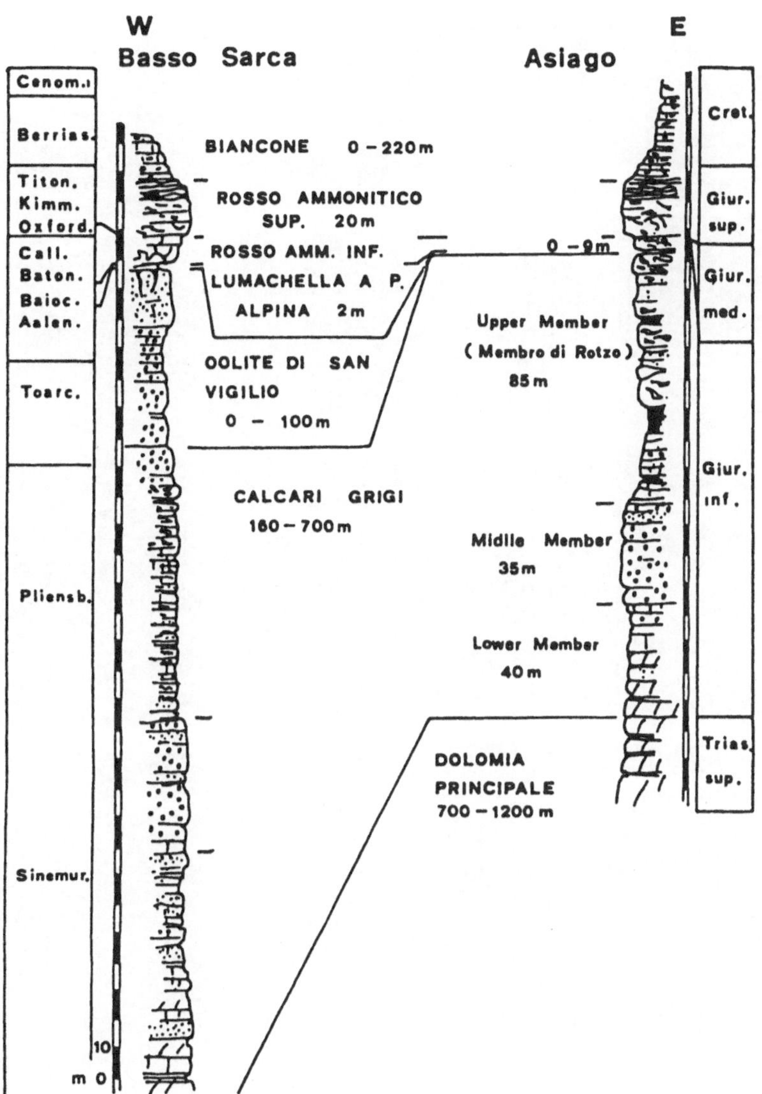

Fig.58 - Location of the study area. Above: from Göhner (1981). Left: from Bernoulli and Jenkins (1974) and Bosellini, et al. (1981).

Study area

The study area is 20 x 20 Km wide. It is situated approximately in the center of the platform. The thickness of the

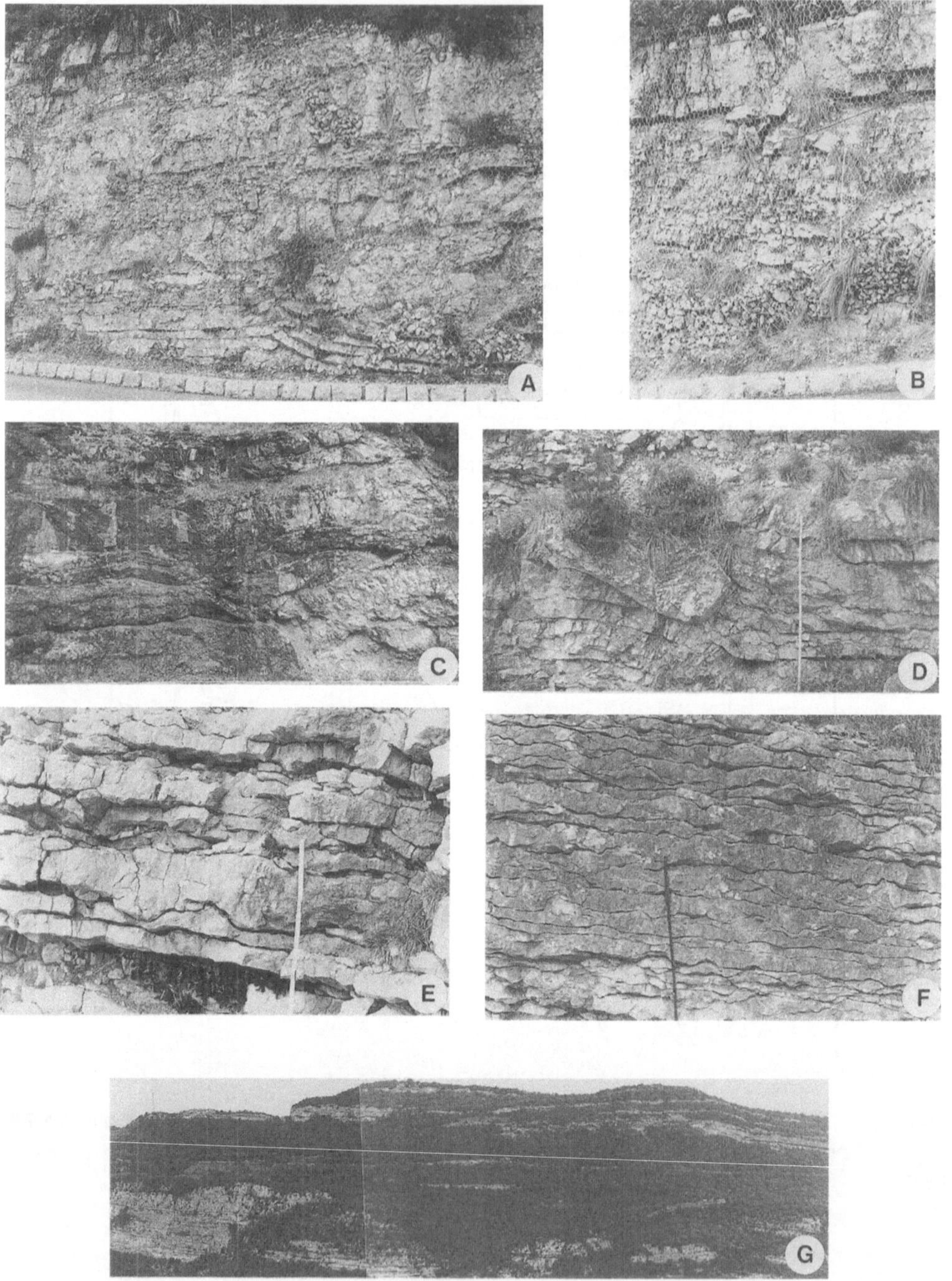

Fig.59 - Features indicative for tectonic instability in the Trento platform. A,D,G,E,F Crumpled beds and deformation features. B Intraformational discordance. C Small-scale synsedimentary fault. G Panoramic view (locality section #4-14) showing a deformed stratigraphic horizon.

stratigraphic interval ranges from about 20 to 60 m.The investigation was conducted on the upper part of the 'Calcari Grigi' Formation (upper part of the Rotzo Member according to the local stratigraphic terminology), within the *Orbitopsella praecursor* zone,by means of facies analysis of 32 stratigraphic sections amounting to about 800 m.
Most of stratigraphic sections were correlated by means of 'event correlation' by using physical surfaces as time lines (for example, a triple disconformity: Riding & Wright,1981, a dm-thick level containing radial oolites; tsunami-generated horizons).These surfaces provided a few transects which allowed for the subdivision of the stratigraphic columns into 4 isochronous units, successively related to different system tracts.

Previous studies

Various aspects of the 'Calcari Grigi' Formation were investigated by a number of authors.

The stratigraphy and regional geology reconstructions were made by Venzo (1963),Auboin,et al.(1965), Castellarin (1972), Bosellini (1973a,b), Bernoulli & Jenkins (1974), Winterer & Bosellini (1981) and Barbujani, et al.(1986).

Most of the work on the 'Calcari Grigi' Formation has been concerned with paleontology (Parona,1924; Fabiani & Trevisan, 1939; Wesley, 1956; Venzo,1963).Several of these papers are concerned with the description and interpretation of *Lithiotis* shells, a huge mollusk which is particularly abundant in the study area: Berti Cavicchi, et al.(1971), Bosellini (1972), Benini & Broglio Loriga (1974), Broglio Loriga & Neri (1976), Accorsi Benini & Broglio Loriga (1977), Geyer (1977), Accorsi Benini (1979).

The sedimentology was investigated by Venzo (1963), Fuganti (1964), Fuganti & Mosna (1966), Bosellini & Broglio Loriga (1971), Castellarin (1972), Castellarin and Sartori (1973a,b), Clari (1975), Göhner (1980,1981) and Galli (1990).

Facies associations

The lithofacies distribution is far more complicated than in the previous case history.A puzzling feature of the 'Calcari

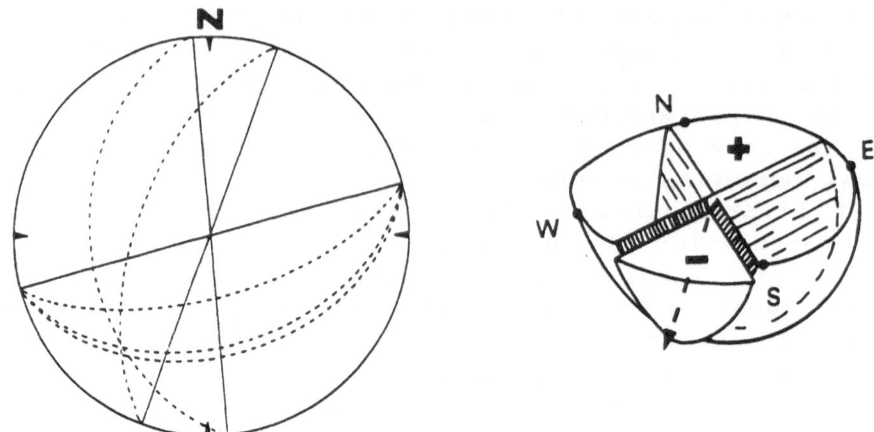

Fig.60 - Equiareal projection of planes of dip-slip synsedimentary faults in the study area.

Grigi' Formation is the complex alternation of several lithofacies, grouped here into the following facies associations:

shallow ramp (oncolite packstone, wackestones and grainstones);

intermediate ramp (oolite packstones and grainstones; bioclast wackestones, grainstones and packstones); and

deep ramp (thick *Lithiotis* banks).

Lithiotis shells are ubiquitous and occur in all sectors of the ramp; thick banks as much as 9 m thick are thought to have formed in the deep ramp.

The different bedding styles of the three sectors of the ramp (alternations of medium and thin beds in the shallow ramp; medium and thick beds in the intermediate ramp and thick beds in the deep ramp) can be appreciated on outcrop and panoramic views (Fig.61).

Shallow ramp

Oncolite grainstones and packstones

Description

This lithofacies consists of m-thick grainstone - packstone

beds containing abundant coated grains floating in a matrix constituted by a poorly sorted admixture of bioclasts and lithoclasts.Sedimentary structures are traces of cross bedding, rare dessication and subaerial features and channel-fills. Coated grains occur as surficial and large oncoids ('macroids' according to Peryt,1982). Surficial oncoids average 3 mm in diameter, have an intraclast/bioclast, monomictic core, a clastic texture and elliptical to ovoidal shapes. 'Macroids' reach 2 cm in diameter; often they have a polimictic core, a clastic to botryoidal texture and a crustose to oval shape.

This lithofacies commonly forms 0.5 m thick, coarsening-upward cycles, developed above wackestones. These are characterized by the following vertical trends: a gradual increase in grain size, grain abundance and percentage of surficial oncolites (10-25%), and sparry cement. Scours, infilled with coated grains and bioclasts (brachiopods, bivalves, gastropods and *Lithiotis* fragments) occur in the uppermost part. At few localities the tops of oncolite-rich beds are mud-cracked, or constituted by thin, reddish horizons enriched in lithoclasts, or also by parallel laminated, dm-thick, yellow, silt-size sands.

Interpretation

In the study area oncolite beds are often associated in space with *Lithiotis* banks and oolite beds. They represent a deposition in interbank, interbar areas or ponds located in the outermost part of the shallow ramp. Traces of scouring reflect episodic mechanical reworking.

Oncolite deposition is an indicator of breaks or slowing down in the rate of sedimentation, in a nearshore or very shallow environment (Weiss,1969). The occurrence of large quantities of large oncoids within a mudstone lithofacies was taken by Catalov (1983) as an evidence for low rates of subsidence.

Coarsening-upward cycles are interpreted as shallowing-upward sequences related to an upward, gradual decrease in sedimentation rate following a relative fall of the sealevel. Macroids may reflect the onset of hypersaline conditions and local exposures, when associated with mudcracks, reddish horizons and keystone vugs.

Bioclast-lithoclast grainstones and packstones

Description

This lithofacies is represented by massive to thin bedded, intraclast-bioclast grainstones and packstones containing in addition to lithoclasts variable percentages of coated grains, peloids and bioclasts (bivalves, foraminifers, *Lithiotis*, algae, crinoids, etc.), often enveloped by algal coatings. Oolites are rare. Lithoclasts are both rounded and angular, light-gray to reddish in color.

Examples of bedding styles, thick-bedded and thin-bedded alternations are shown in Fig.14,15 and 17. The most typical sedimentary structure is given by scours ranging in amplitude from a few dm to 10 m. Some beds may result from merging and stacking of channel structures which do not evidence for any facies vertical trend (Fig.17).

A common sequence encountered in this lithofacies ranges between 0.5 and 1.5 m in thickness and consists of three units.The basal part has an erosional base, a disorganized massive bed containing fossils which in some cases are disposed in parting lineations,traces of hummocky cross-bedding and undulations of uncertain origin. This basal part appears structureless when the composition and grain size are homogeneous.The basal part passes up gradationally to a better sorted unit characterized by intrastratal, scalloped undulations and various types of wave-generated structures such as climbing-wave ripple lamination (Kreisa,1981), or very thin, slightly undulated, flattened plane lamiantion. Traces of ripples are also present with amplitudes of about 8 cm and heights of 4 cm.The upper unit consists of well sorted intraspatites orgainzed into laminae, interbedded in some case with lime mudstones. This cycle records an overall grain size fining-upward trend.

Interpretation

This lithofacies reflects a deposition in a shallow lagoonal environment.
The cycle is analogous to interpreted nearshore storm deposits described by Kreisa (1981),Brenchley & Newall (1982), Kumar & Sanders (1976), Mount (1982), and others. They are interpreted as thick-bedded alternations (see above). The tops of these cycles characterized by thin storm-generated beds alternating with fairweather muds are interpreted as thin-bedded alternations .

Lime mudstones

Description

This lithofacies occurs as thin interbeds with other facies. It overlies subaerial surfaces or marine unconformities. It is also frequently sandwiched between marine lithofacies. It consists of black, calcareous, clayely deposits containing at places abundant plant debris and more rarely traces of sulphate minerals. Thicknesses average 10-20 cm. They form thin-bedded alternations with homogeneous, dm-thick mudstones containing rare ostracods. These interbeds are completely lacking of macroskeletal constituents. Similar lithofacies investigated by Castellarin and Sartori (1973) in a nearby location revealed that the mud contains traces of illite and hematite, and/or goethite ,quartz and chlorite.

Interpretation

This lithofacies was interpreted as a mud flat or marsh environment (Bosellini & Broglio Loriga,1971). Marl deposits form in a number of distinct subenvironments, which may be colonized by a dense vegetation.

They were described from shallow ponds ,coastal marshes, and freshwater lakes. In the ponds and lakes, such as the Everglades, marls occur at the base of a transgressive sequence as the sealevel rises and as a result of a progressive rise of the freshwater lens (Monty and Hardie,1976). It may also form at the top of a regressive sequence when the marsh progrades over a retreating shoreline, as occurs on the eastern half of Andros Island,Bahamas. Marls also form over exposed surfaces during prolonged periods of lowstand. Coastal marshes can represent the transitional zone between freshwater marls and marine calcareous mud. In these cases, intrusions by marine sediments during storms produce interbedded freshwater and marine sequences (thin bedded alternations).

Intermediate ramp

Oolite grainstones and packstones

Description

On outcrop this lithofacies is a massive bedded, light-gray to creamy, homogeneous packstone and grainstone containing concentrical, tangential ooids (amounting to about 30-40%). Wackestones are much less common.

This lithofacies is most widespread in the eastern area of the Trento platform and in its topmost part.

Two lithofacies can be distinguished (Fig.22). The first, poorly sorted, is composed of surficial oolites and lumps and a mixture of intraclasts, peloids and coated grains, other than several types of skeletal grains such as foraminifers,ostracods and algae which constitute the nuclei of ooids. The second lithofacies,less common,consists of well sorted oolite grainstones.

Sedimentary structures are large-scale hummocky cross-bedding, rare tabular cross bedding, symmetrical megaripples and horizontal lamination. Smaller scale structures include flute casts, scours infilled with mud and coated grains, and dubious load casts. Half-cm thick, lenticular coquinites composed of densely packed, imbricated, both articulated and disarticulated shells of bivalves are found frequently intercalated with some of the thicker beds. Graded beds are common. Some outcrops show upward transitions from the poorly sorted, to the well sorted oolite lithofacies.

In the study area this lithofacies has a patchy distribution: it mostly occurs on top of shoals.

Interpretation

These massive bedded, oolite grainstones and packstones are interpreted as shoals, banks and sandwaves situated in a storm-dominated area, at a shallow-water depth.Storms were mainly responsible for the ooid migration.

Oolitic sand shoals are found actually along the edges of several Bahama areas (i.e. Cat Cay, Joulters Cay, Berry Islands, etc.). They occur as 1) narrow, active ooid shoals marginal to the open sea; and 2) as stabilized ooid-aggregate grains-pelletal sands flats forming widespread blanket sheets behind active sand shoals and grading to other platform sediments (Multer, 1977). Skeletal admixtures are greatest in the deeper sites.

These two modern sediment types correspond respectively to the well sorted and poorly sorted oolite lithofacies. Transitions from the poorly sorted to the well sorted lithofacies indicate a shallowing of the sedimentary interface consequent to a depositional regression (cf. Van Steenwinkel,1990).

Skeletal wackestones

Description

These thin-medium bedded, light-gray wackestones contain at places abundant, thin shells of bivalves (*Pholadomia*, *Gresslya*, *pectinidae*) and minor quantities of gastropods. Other skeletal constituents are thin-shelled brachiopods, crinoids (*Isochrinus*), foraminifers (*Paleodasycladus*, *Orbitopsella praecursor*), occasional *Lithiotis* and other undetermined microfossils. Small, reddish intraclasts are found near the base of some of the less fossiliferous beds. Coated grains occur infrequently throughout this lithofacies.

A variety of this lithofacies is constituted by thin beds (10 - 20 cm thick) of poorly fossiliferous, dark-colored mudstones with interspersed sand-size grains. This lithology is typified by nodularity which gives way to pseudobudins, and lensoid nodules with laminated clay seams.Bioturbation is dominated by *Thalassinoides* burrows which may have contributed together with pressure solution to the formation of the nodularity.

Sedimentary structures consist of irregular, erosional scours (a few cm to some dm wide), symmetrical megaripples, gutter casts and hummocky cross-bedding. Fenestrae are rare. Coquinites are composed of 1) gastropod streaks forming pebble-cluster alignments parallel or draping the symmetrical undulations, and, more commonly, of 2) lenses of bivalves that are essentially thin-shelled, of the same size, convex-side up and draping the topsets of symmetrical megaripples. Few of these lenses are organized into 20 cm thick fining-upward cycles composed of: 1) a lower grainstone-packstone unit consisting of randomly oriented bivalves; 2) a thinner unit with convex-up shells; and 3) an upper mud rippled top or an argillaceous, yellow cm-thick horizon. Multistored, complex lenses containing coquinites are volumetrically less represented than wackestone layers.

Interpretation

This lithofacies was deposited at a deeper depth than the oolite grainstone and packstone lithofacies, as is suggested by

Fig.61 - Bedding styles of deep ramp (A:thick banks), intermediate ramp (B:thick and thin beds) and shallow ramp (C: thin beds).The panoramic view of M.Testo (D; locality sections #15 and 28) shows a transition from shallow to deep ramp evidenced by an upward increase in bed thickness and declivity (a higher erosion in the shallow ramp is favoured by frequent intercalations of lime mudstones).

the transition and lateral changes to *Lithiotis* wackestones. Similar bathymetric relationships between skeletal and oolite beds occur in modern areas, for example at Lily Bank (see also Hine,1977: Fig.24) where skeletal wackestones occur in deeper, seaward sites (-5 to -10 m below sealevel).

This facies formed as lime mud thickets. Intense bioturbation by *Thalassinoides* occurred in more protected, less populated areas. Truncated *Thalassinoides* burrows indicate however episodic erosion, as is supported by the occurrence of other storm-generated structures, such as hummocky cross-bedding, wave megaripples, coquinites, etc.

Deep ramp

Lithiotis wackestones

Description

Lithiotis wackestones are widespread in the Trento platform. They are typified by thick beds (1 to 9 m) of tightly packed accumulations of huge pelecypods (*Lithiotis problematica, Cochlearites loppianus,Lithopedalium, Gervilleioperna*, etc.) as much as 40 cm long and embedded in a wackestone matrix. Other fossil types such as crinoids, corals, brachiopods, algae, foraminifers (*Orbitopsella praecursor, Glomospira,Textularia*) and sponge spicules are only accessory components, as this biofacies represents a suspension feeder, olygotypic association (Broglio Loriga & Neri,1976).

Shells of *Lithiotis* display various fabrics: vertical, fanning-upward clustering, imbricated, wave - knitted (bidirectional shell orientations). The upward decreases in shell sizes result in a fining-upward trend.

Sedimentary structures are various types of scours and undulated bedforms (Galli,1990). In this sector of the ramp thick beds composed of *Lithiotis* alternate with skeletal wackestones (trough sequences: Fig.23D;Fig.61A).

Interpretation

Thick banks represent a deposition in deeper areas of the ramp

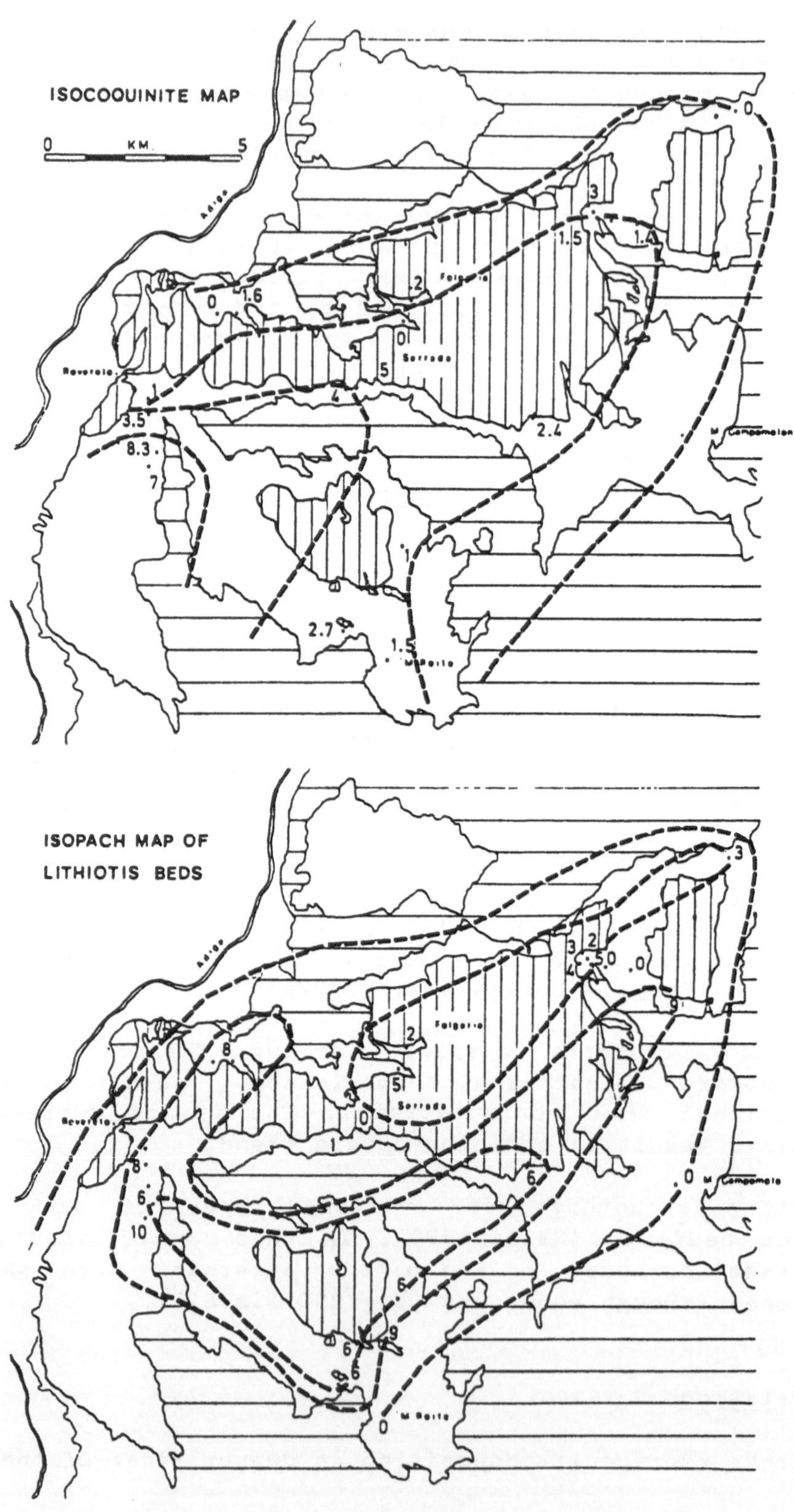
ISOCOQUINITE MAP
0 KM. 5
Adige
Rovereto
Folgaria
Serrada
M. Campomolon
M. Boite
ISOPACH MAP OF
LITHIOTIS BEDS
Adige
Rovereto
Folgaria
Serrada
M. Campomolo
M. Boite

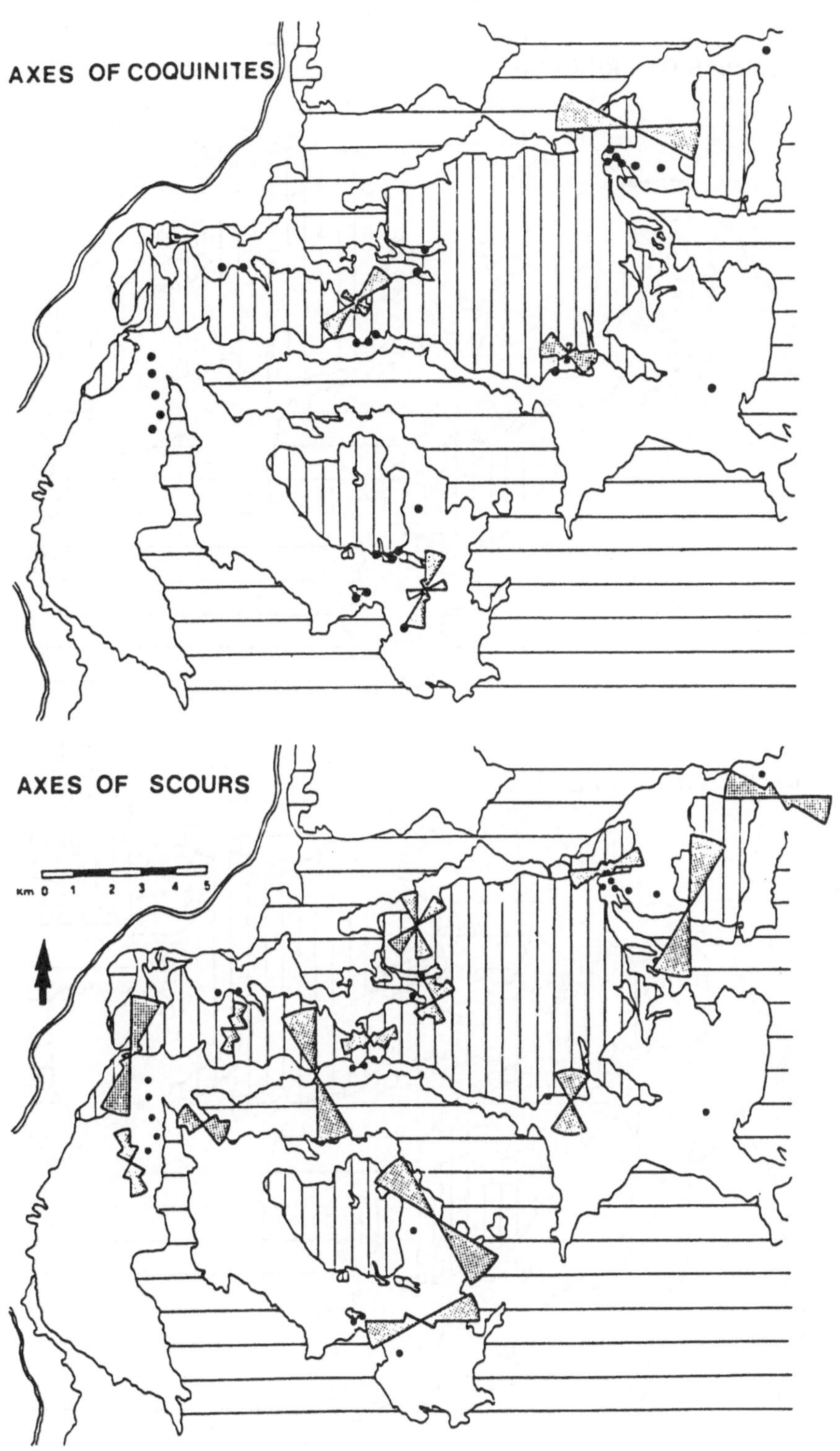
AXES OF COQUINITES
AXES OF SCOURS
Km 0 1 2 3 4 5

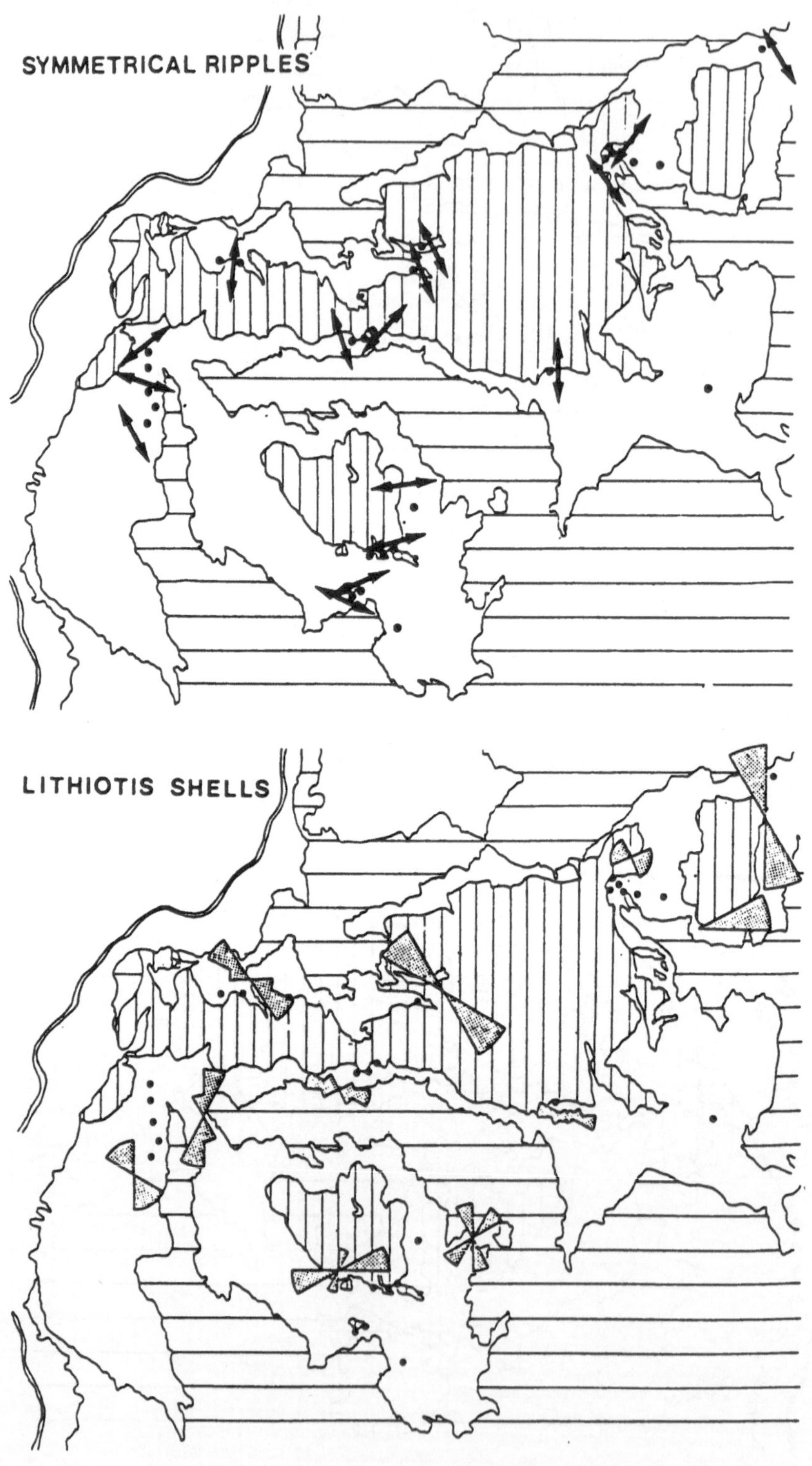
SYMMETRICAL RIPPLES
LITHIOTIS SHELLS

Fig.62 - Contour maps and paleocurrent data of the study area.The isocoquinite map results from contouring sites characterized by the same numbers of coquinite lenses; the values were obtained by dividing the number of coquinite lenses within skeletal wackestones by the thickness of skeletal wackestones occurring in stratigraphic sections.The frequency of coquinite lenses decreases towards northeast, which is the shallowest sector of the ramp (cf. the proximality - distality concepts shown schematically in Fig.15).The isopac map of *Lithiotis* banks also reflects changing water depth because thick beds of tightly packed accumulations of big shells of *Lithiotis problematica* Gümbel took place in the deep ramp.An examination of the contour maps reveals the existence of an elongated lagoonal depression oriented NE-SW. The lagoonal floor was uneven due to the development of an array of shoals.Directional data indicate a dominant rotary high-energy path oriented SW-NE driven by the lagoonal corridor configuration and a minor mode, oriented SE-NW which indicates currents flowing oblique to the lagoonal corridor.Measurements from shell imbrications and parting lineations suggest that currents moved from SW to NE.A refracted wave pattern which displays a counterclockwise sense of rotation probably resulted from the impingement of the SW-NE oriented current upon shoals located in the east. Probably most of currents and waves were produced by tsunamis (cf. Galli,1990).As suggested elsewhere (Galli,1990),the storm system,rather than actively transporting sediment, determined near-bottom oscillating currents acting through strongly pulsating bursts of energy.Strong pressure pulses on and below the lagoonal floor and strong shear stress produced an 'in situ' reorientation od shells. Large-scale bedforms,not described in this work, were probably generated by tsunamis, as sugegsted by their formation by the action of surface waves, a great lateral extent of exposures and their restriction to the same stratigraphic horizons (for discussion see Galli,1990). Earthquakes within the platform may have produced sudden oscillations of water which incorporated the whole water column.

Fig.63 - North-south cross section showing the wedge-shaped geometry of the studied part of the 'Calcari Grigi' Formation.A lack of parallelism between time lines (base of the early highstand system tract and radial oolite horizon interpreted as a type-2 unconformity) is taken as an evidence for synsedimentary tectonics.

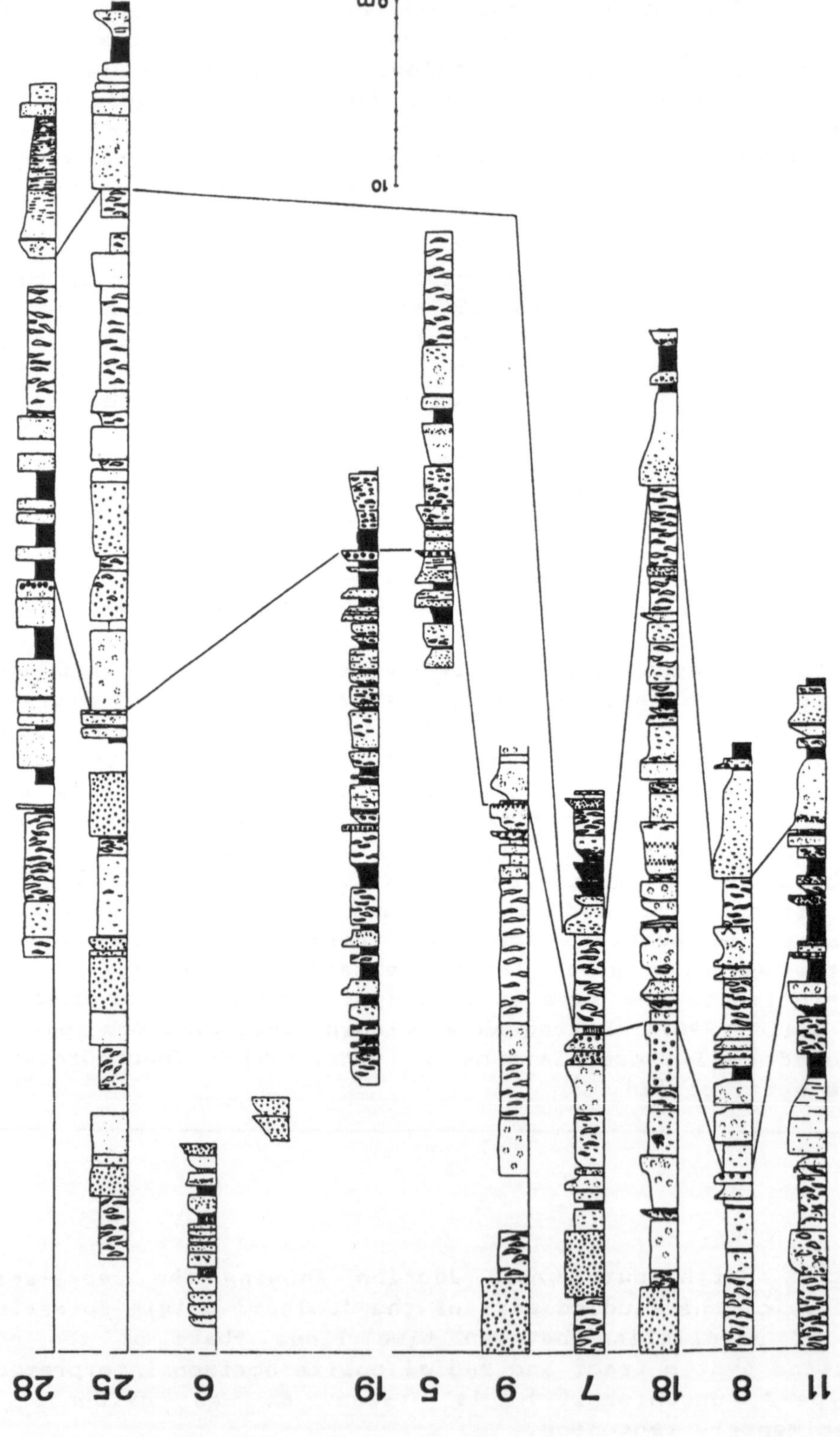

as is suggested by the great thicknesses (as much as 7 - 9 m), lack of erosional structures, lack of supply of intraclast sand and/ or skeletal debris. Water circulation was limited as the faunal diversity is low.

Thin *Lithiotis* beds, 1 to 2 m thick ,were probably deposited in shallower areas, close to the intermediate ramp. These thinner beds contain in fact thin-shelled brachiopods, *Opisoma*, and other skeletal fragments typical of the intermediate ramp.

Hummocky structures and other mechanical sedimentary structures evidence for strong episodic disturbances by waves responsible for reworking of *Lithiotis* shells.

The occurrence of trough sequences suggests some bathymetric fluctuations possibly related to relative sealevel fluctuations.

Paleobathymetry

An estimate of bathymetry of the study area was carried out by constructing the isopach map of the maximum thicknesses of *Lithiotis* banks. The existence of a lagoonal bucket oriented NE-SW is revealed by the isopach map. Deeper lagoonal areas (deep ramp) are located in the west and south. Directional data, summarized in Fig.62 and 26, collected from orientations of symmetrical wave ripple crests, gutter casts, axes of coquinite lenses, channels and from the longest axes of *Lithiotis* shells, indicate that, whatever the origin (storms, tsunamis,etc.) currents paths were controlled by the paleotopography and bucket configuration.

In cross section, the studied upper part of the 'Calcari Grigi' Formation is wedge-shaped (Fig.63; cf. also Fig.58). The northeastern side, 20 m thick, is mainly composed of shallow ramp lithofacies. It formed at a shallower depth than the southern side where the sedimentary prism, 60 m thick, was the site of accumulation of the thickest *Lithiotis* banks.

The sedimentary wedge corresponds to an intrashelf onlap ramp whose hinge is located in the north and northeast; the flexure area is oriented north-south. The lagoonal trough strikes northeast - southwest. A graphic simulation obtained by interactive modelling shows a hypothetical representation of the paleobathymetry of the ramp surface which was inclined

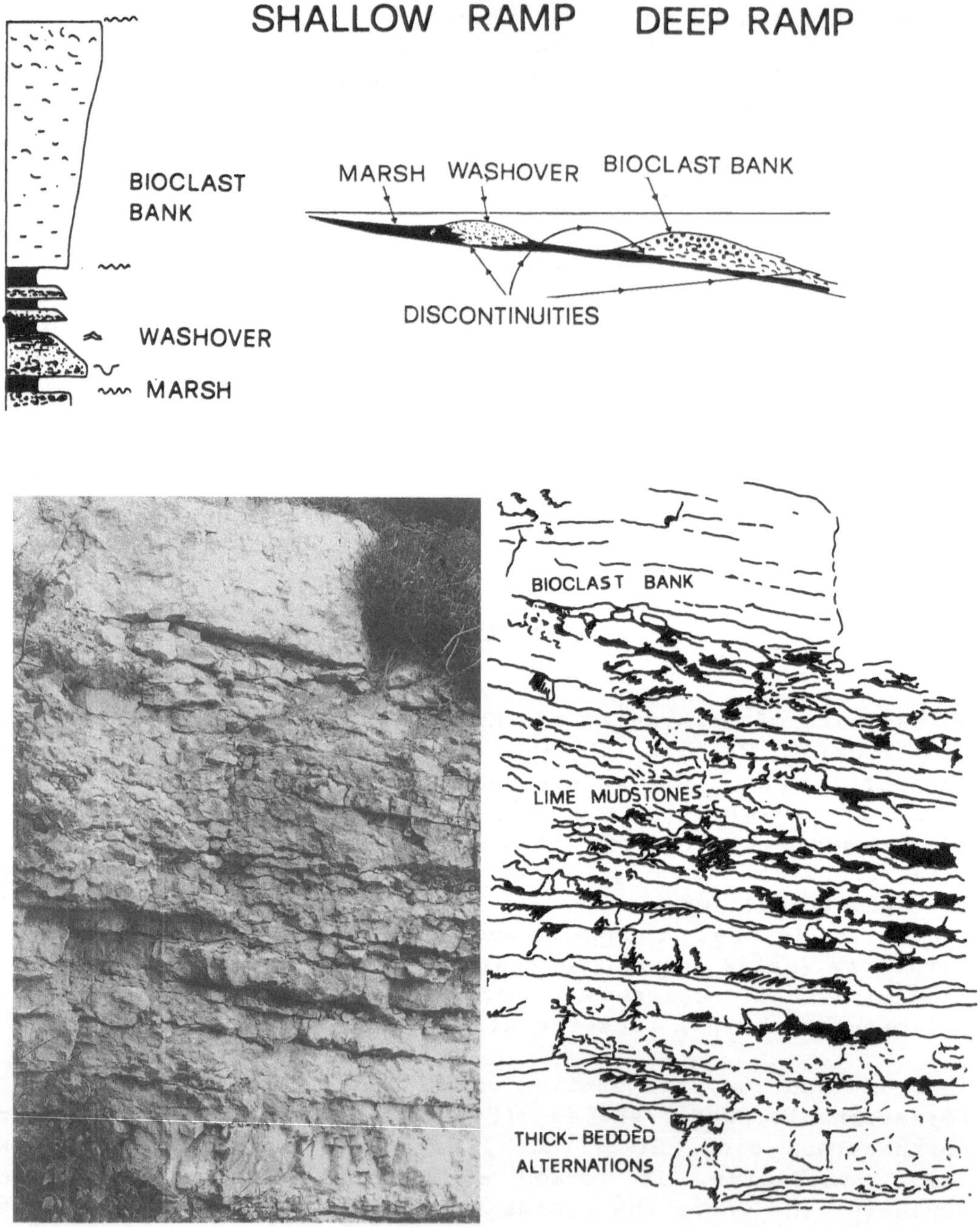

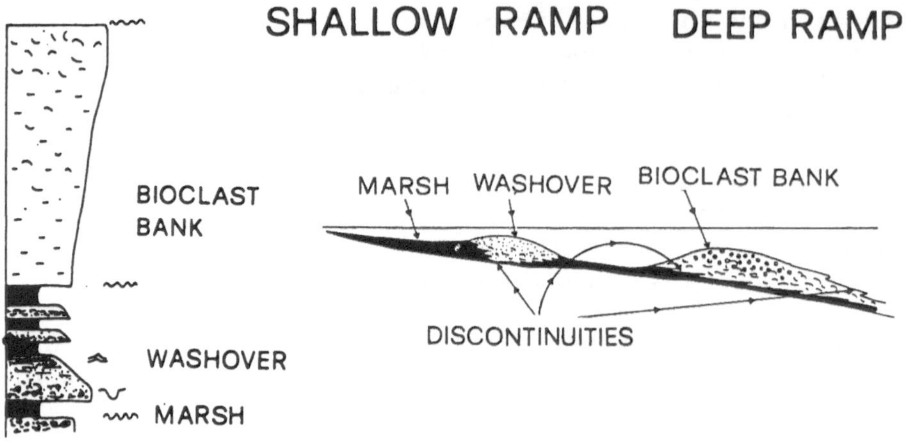

SHALLOW RAMP
DEEP RAMP
BIOCLAST BANK
WASHOVER
MARSH
MARSH
WASHOVER
BIOCLAST BANK
DISCONTINUITIES
BIOCLAST BANK
LIME MUDSTONES
THICK-BEDDED ALTERNATIONS

Fig.64 - Transgressive system tract.

towards west and southwest and punctuated by relieves corresponding to sandwaves and banks (Fig.26).

Depositional sequence

The stratigraphic interval represents a third order depositional sequence which is subdivided from bottom to top into a transgressive facies tract, an early and a late highstand facies tract and a shelf margin facies tract.

Transgressive facies tract

Description

This facies tract is constituted by the following alternation of lithofacies, summarized below from bottom to top (Fig.64).

1) Scoured packstone with reddish lithoclasts passing to undulating, nodular black lime mudstones (shallow ramp);

2) Graded bioclast layers, each characterized by an upward increase in the proportion of lithoclasts and coated grains (washover deposit - thick-bedded alternation: shallow ramp);

3) Lime mudstones (shallow ramp);

4) Stacked coarsening- and thickening upward layers recording a progressive increase in the bioclast percentage. Grain sizes display some bimodality; peloidal grains are well sorted and micritized (intermediate ramp).

Interpretation

The sequence is interpreted as a hinge sequence. It is analogous to the littoral barrier described by Riding and Wright (1981) and to the Hampole beds occurring in the English Zechstein.

Transitions recorded by the sequence from shallow ramp ponds and mud flats to an intermediate ramp submarine bioclastic bar point to a deepening-upward trend.

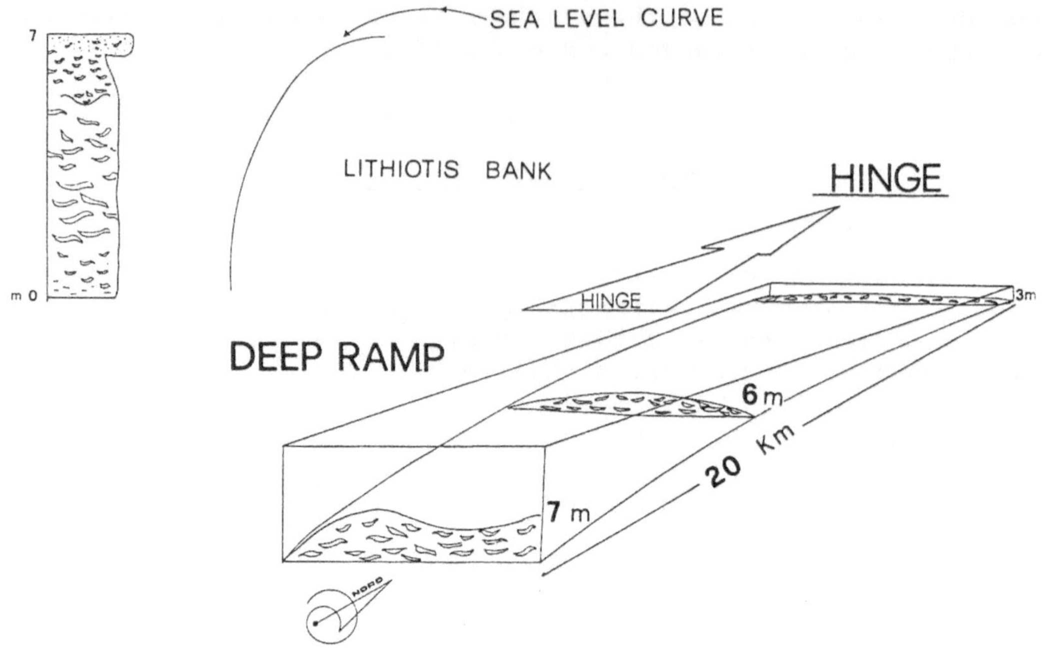

Fig.65 - Early highstand system tract.

Changes in the relative sealevel were discontinuous as individual assemblages of lithofacies are separated by discontinuity surfaces.
The lowermost surface is scoured. It formed above a horizon of likely subaerial origin and is interpreted as a transgressive surface. It corresponds to the lower sequence boundary of the depositional sequence.The former pedogenetic horizon above which the sequence was deposited is rarely preserved. It is mostly inferred from abundant red-colored lithoclasts occurring at the very bottom. The scarce micritization of grains contained within the lowermost graded beds (washover deposits in the shallow ramp) and the preservation of washover deposits point to an increase in the rate of the relative sealevel rise.A successive slowing in the speed of the relative sealevel rise si indicated by the occurrence of intraclasts and micritized grains on top of the shallow ramp lithofacies.

The transition to the bioclast bank marks an increase in the deepening-upward trend. The surface which separates the two sectors of the ramp corresponds to a retrogradational line (see chapter #3 for the definition of retrogradational line).

The sequence represents a retrogradational transgressive facies tract. Sediment was transported hingeward and formed a string-like body which may correspond to a thin-sheet lithosome as defined by Burchette, et al.(1990). The low sedimentation rate was barely sufficient to couple with the rising sealevel.

Early highstand facies tract

Description

The early highstand facies tract is a wackestone bank composed of thick shells of *Lithiotis* (Fig.65). The thickness decreases from the trough area in the southwest (7 m: section# 25) to the hinge area in the NE (2.7 m: section #8). Faunal diversity is higher in the trough area.

The thickest beds in the southwest (trough area) record vertical changes in fabric and composition of *Lithiotis* and sedimentary structures. Fabrics vary from parallel (in some instances found in physiological position) to wave-knitted, to random towards the top.*Lithiotis* comprise *Lithiotis sp. ss.* and *Cochlearites* in the lower and middle part of the bank, and *Gervilleioperna* in its upper part. Shell sizes also decrease from 5cm - 20÷30 cm at the bottom, to 1-3 cm towards the top. Traces of scours filled with randomly oriented *Lithiotis* shells and traces of cross bedding occur at the top of the bank.This *Lithiotis* bank grades upwards to grainstones containing small-size dispersed *Lithiotis* and other bioclasts such as foraminifers, algae and gastropods. The top is also enriched in oncolites and intraclasts. Allochems are micritized. The upper grainstone unit records also coarsening-upward as well fining-upward grain size trends.

Interpretation

This bank formed in the deep ramp.The lithosome geometry is that of a wedge. Decreases in thickness from NE to SW indicate that the sedimentary interface was slightly inclined towards south.

The lower part of the bank containing thick shells of *Lithiotis* formed when the accomodation potential was highest: the high sedimentation rate favoured the development of *Lithiotis*; this part of the bank corresponds to a maximum flooding surface. The bank is comparable to other catch-up reefs described in the literature (Fig.28) which record a transition from a quiet, deep water stage to a shallower water stage typified by detritus and lithoclasts associated with fossils.

This early highstand facies tract records an aggradational trend during which the space created by the rising sealevel was infilled by vertical sediment growth ('catch-up phase' by Kendall and Schlager,1981). The progressive relative sealevel fall of the sedimentary interface produced by the piling-up of shells led to the deposition at a shallower water depth which favoured the formation of *Gervilleioperna* thickets.Then, the decreasing space available to sedimentation favoured a lateral facies shift and progradation which is documented by the set of compositional and fabric features occurring at the top of the bank.These changes reflect the initiation of the late highstand facies tract phase of sedimentation.The vertical transition from *Lithiotis* to *Gervilleioperna* appears to be a primary function of the decreasing sedimentation rate, in keeping with the results obtained by Rey et al.(1990) from south Spain, rather than a main function of different depths of deposition of the two bivalves,as suggested by Broglio Loriga & Neri (1974); interpretations which assign different depths to fossilized organisms within a carbonate platform are frequently based on some sort of circular reasoning; conversely, their inferred dependance of the relative rise in the sea level may be confronted with independent data.

Late highstand facies tract

Description

This phase is approximately 7 m thick and consists of a number of grain-supported, thinly bedded oncolite - bioclast - intraclast- bearing lithofacies interbedded with lime mudstones

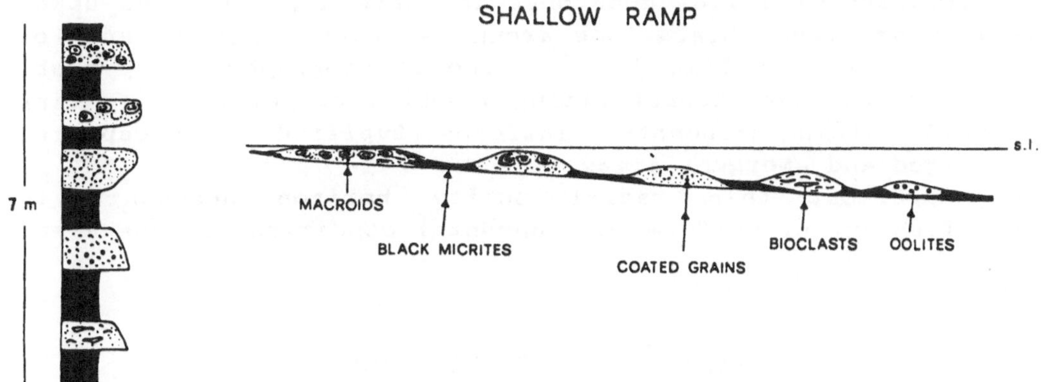

Fig.66 - Late highstand facies tract.

Individual thin beds are organized into fining- and coarsening-upward grain size trends. In the northern area, close to the hinge (sections # 8,11 and 18: Fig.63), coarsening-upward cycles are predominating. Fining-upward cycles are most common towards the top of this tract and in the south. In the most part of the stratigraphic sections this tract records a recurring change in the composition of grain-supported beds intercalated with lime mudstones from bioclast --> oolite---> to oncolite. This tract is topped by a cm-thin horizon of radial oolites.

Interpretation

Deposition took place in a shallow ramp.The frequency of oncolites within micrites and pisolite-oncolite grainstone beds point to a reduced sedimentation rate. Southward changes in the thickness, lithofacies and types of beds show that the southern zone was deeper. The occurrence of 6-7 m thick oolite banks in the south (section # 25: Fig. 63) and oolite storm deposits (spillover and washover deposits) in the north along the same stratigraphic horizon suggests northward,hingeward directed storm processes. Grain-supported beds are interpreted as thick-bedded alternations which formed shallowing-upward cycles as is documented by emersion features on tops resulting from high-frequency sealevel changes.

The overall lithologic transitions of grainstone-packstone beds intercalated with lime mudstones indicate a progressive upward shallowing trend. Migrations around scattered depocenters took place during this time interval.The physiography was probably analogous to that characterizing a number of peritidal settings located within carbonate platforms typified by prospicient submerged and emergent areas.
The uppermost thin radial oolite horizon documents the establishment of uniform environmental conditions in the area.

Shelf margin facies tract

Description

This tract overlies the highstand facies tract and is overlain by pelagic lithofacies. It is characterized by the predominance of lithofacies typical of the intermediate ramp.

The uppermost part of the 'Calcari Grigi' Formation consists in fact of several repetitions of the following type of alternation of lithofacies (Fig.67): oolite grainstones --> *Lithiotis* bank--> skeletal wackestones (intraclast - bioclast packstones and grainstones). This sequence averages 7 m in thickness. Its time interval of formation, calculated by using the method described by Grotzinger (1986),is of about 190.000

years.

This tract is characterized by a lack of thin beds.Facies transitions are sharp. Thin, lime mudstone horizons occur at the bottom and top of the oolite lithofacies. Bioclasts are scarcely micritized.

Interpretation

These 7 m thick alternations are the product of a hingeward, stepwise migration of intermediate lithofacies,as is shown by

the distribution of oolite bodies (Fig.63).
The relative sealevel rise initially led to the formation of oolite sediments; successively,the space created by the sealevel rise was colonized by *Lithiotis* which formed banks thinner than those occurring in the deep ramp. These banks are in turn overlain by skeletal wackestones and/or bioclast and intraclast grainstones which point to a shallowing-up.

These sequences are interpreted as deepening-upward sequences capped by shallower water facies which were deposited following a slowing down of the rate of sealevel rise.The deepening-upward trend is also documented by the scarcity of micritized grains and intraclast lithofacies.

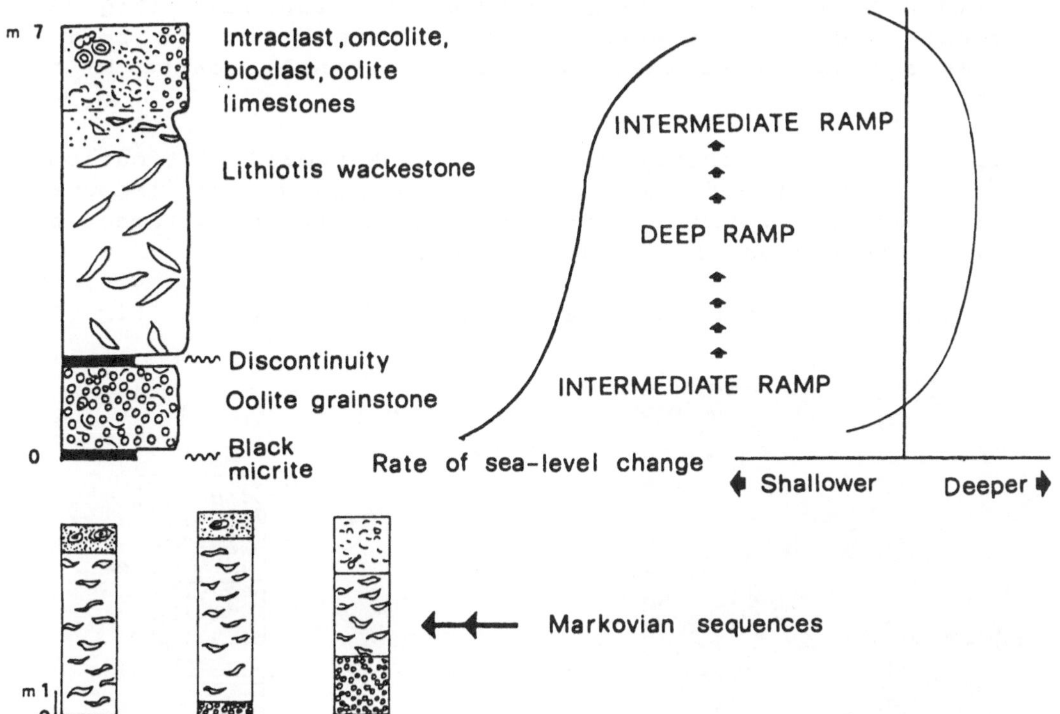

Fig.67 - Modal cycle of the shelf margin facies tract.

In the study area the SMST is overlain by different pelagic facies of various ages (Rosso Ammonitico; Oolite di San Vigilio; Tenno Formation). It follows that the diachronous top of the Calcari Grigi cannot be considered as a flooding surface as stated by Barbujani, et al.(1986).

Model of deposition

The TST corresponds to a barrier-lagoon littoral facies; the early HST to a catch-up reef; the late HST to a prograding mud flat; the SMST represents the landward migration of an oolite barrier facies. The stacking of these different depositional faciess is typical of mature, long-lived platforms.

Sedimentation took place mainly by aggradation and retrogradational mechanisms. A limited progradation occurred during periods of slowing-down of the rate of sealevel rise (late HST and top of the SMST).

A general transgressive trend is demonstrated by the following:

1) facies belts are dipping towards the deep ramp (Fig.63). Depocenters of the intermediate ramp, namely oolite bodies, display a progressive shift towards the hinge located in the north.

2) Spillover and washover deposits (thick-bedded alternations) indicate a hingeward, onshore storm transport of lagoonal-barrier sediments.Along the vertical, thick-bedded alternations overlie thin-bedded alternations (Fig.68).

3) Deep ramp associations overlie shallow ramp facies (Fig.69).

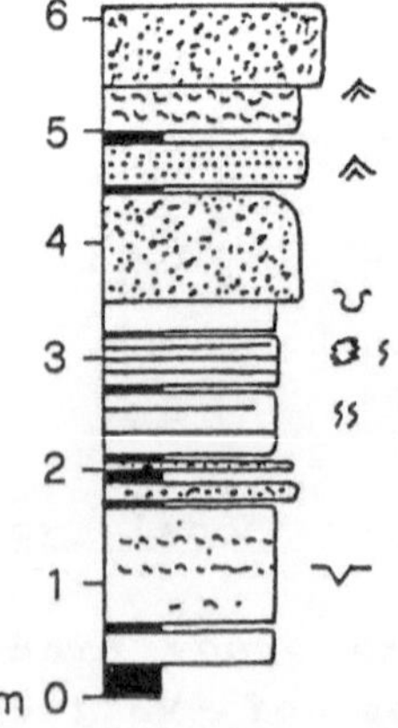

Fig.68 - Overposition of thick-bedded alternations over thin-bedded alternations. Right:section #7;left:section #20.

Fig.69 - A,B Stratigraphic section #15 showing gradual supplantation of intermediate and shallow ramp by thick *Lithiotis* banks. C Transition from shallow ramp to intermediate-deep ramp (section #5: oncolite packstones and grainstones --> skeletal wackestones --> lime mudstones --> *Lithiotis* bank.

The transgressive trend was marked by stepwise phases which led to 7 m thick lithofacies assemblages which differ depending of the position on the surface of the ramp.

The sedimentary wedge and the transgressive trend resulted from a retrogradation of the ramp towards the hinge, namely during the TST and SMST formation.

Differential,rotational subsidence, played an important role which is suggested, other than by local tectonic synsedimentary features which may have been local in extent, by the lack of

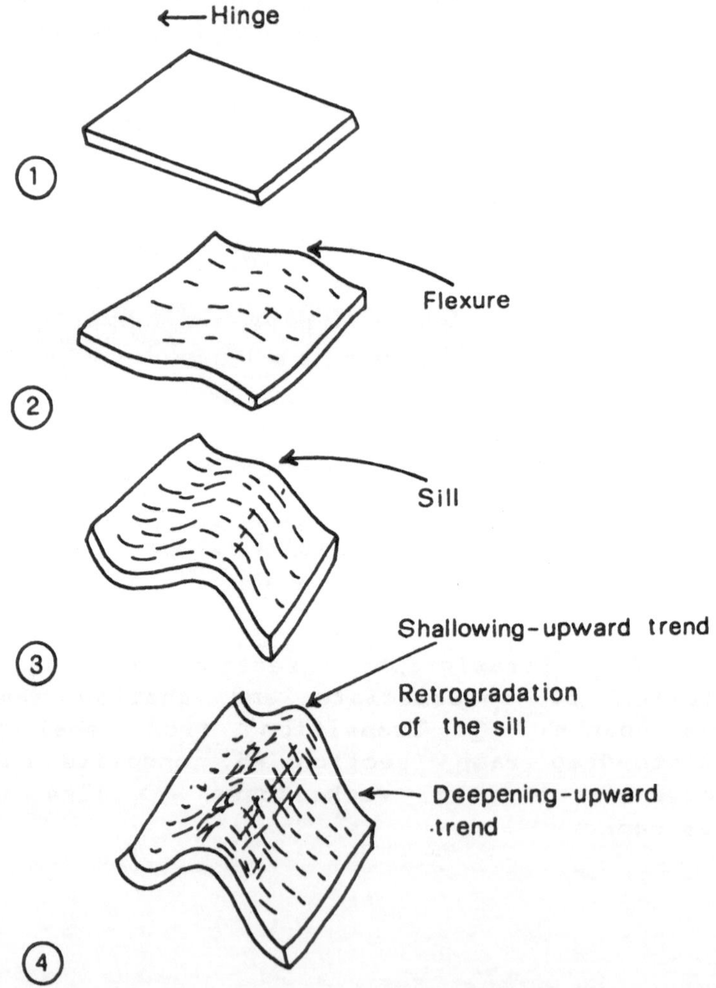

Fig.70 - Progressive deformation responsible for the formation of sills,secondary trough areas and shift of the flexure towards the hinge.

parallelism between time lines (Fig.63).The discontinuous migration of the oolite barrier followed the direction of migration of the flexure which offered the optimal bathymetric conditions for tangential oolite formation.The tectonic tilting determined a progressive folding and shift of the flexure towards the hinge (Fig.70).

This determined a differential deformation and downwarping between flexure and hinge responsible for the formation of 1) a

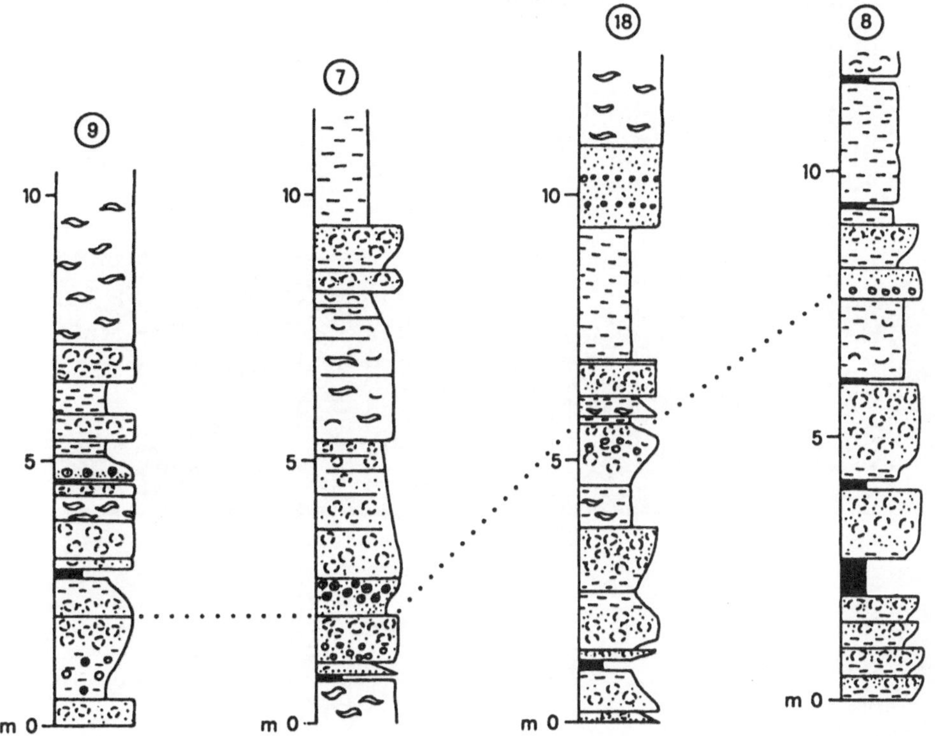

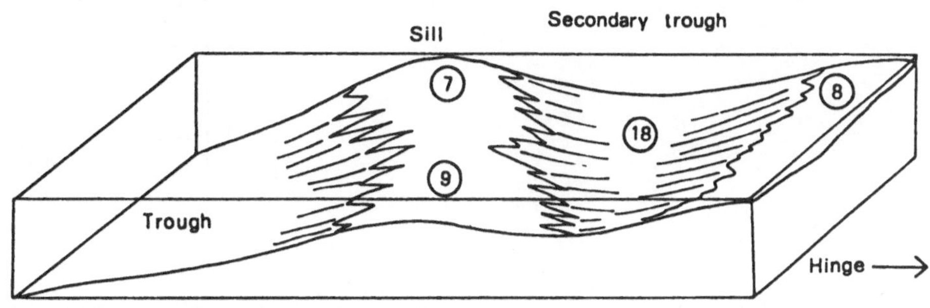

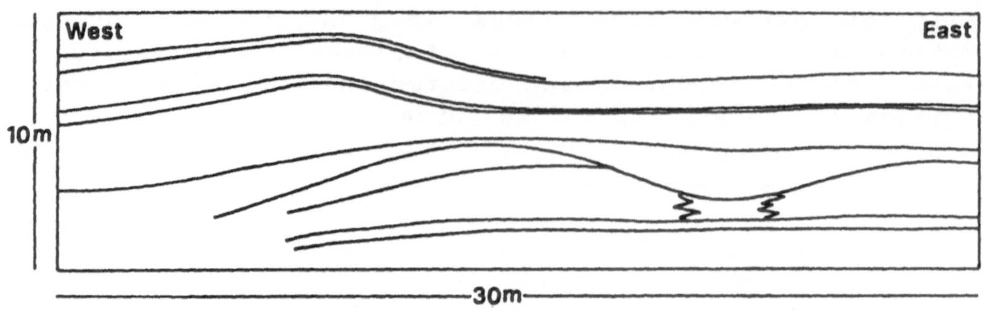

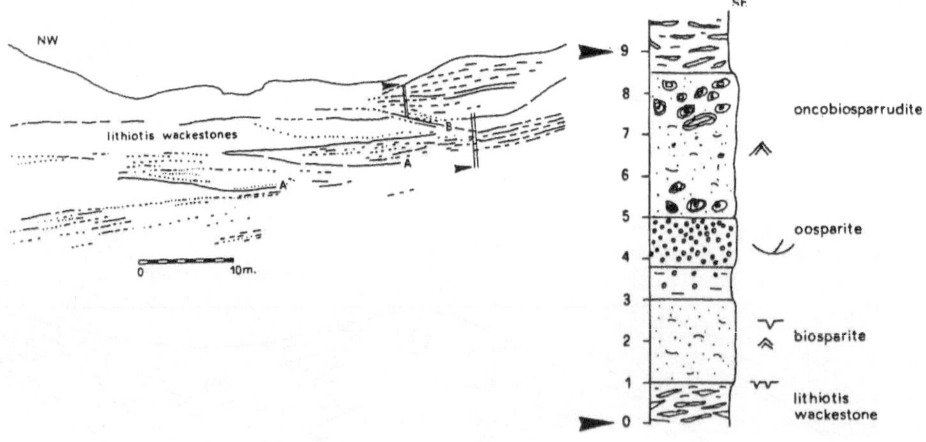

Fig.71 - Examples of sill sequences and bedding styles of the sill area.In the example below the accentuation of channel traces (transitions from flattened to semicircular scours) is seen as the result of a progressive uplift and transformation of a former deep ramp floor into a sill.

secondary trough in the north that complicated facies distribution; and 2) sill sequences.

Some examples of sill sequences are given in Fig.71. The formation of a secondary trough and sill sequences are seen in terms of progressive increments and decreases of vertical space consequent to the differential tectonics (Fig.70).This mechanism is explained by the model of relief inversion described in chapter #4.

The interbedding of lime mudstones with deeper ramp lithofacies, especially recorded in the shelf margin facies tract, may have been produced by uplifts of the platform which caused the emersion of some areas. Fig.72 shows for example a tilted substrate sutured by lime mudstones.

Fig.72 - Tilted beds overlain by dm-thick lime mudstones. The tilting produced emersion of the interface which became a swamp. This is a small-scale example of relief inversion (chapter #4).

Devonian carbonate platform, Carnic Alps, Italy

Introduction

Devonian limestones in the Carnic Alps occur along a 20 Km east-west trend as faulted tectonic thrusts (Fig.73) resulting from the complex Hercynian and Alpine tectonic phases (Vai,

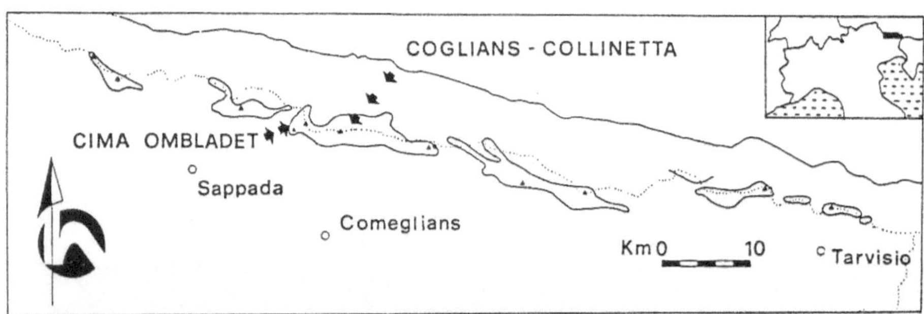

Fig.73 - Location of the study area and outlines of the hercynian tectonic sheets (Venerandi Pirri,1977).The panoramic view shows the Cima Ombladet tectonic sheet in the center and the Coglians-Collinetta sheet in the right embedded within the turbiditic Hochwipfel Formation.

1980).They are part of an epioceanic shallow-water carbonate complex included in a continuous sequence from the Caradoc to the Westphalian.The stratigraphy, tectonics and paleontology of the Paleozoic of the Carnic Alps were extensively studied. A reference list is found in Vai (1980).

A hypothetical cross section of the Coglians- Collinetta reef complex, lower to Middle Devonian, was constructued by Vai (1980) and is shown in Fig.74. However wrong (bioherm areas are overrated),such a cross section is useful as it emphasizes the well pronounced progradational trend that characterized the platform growth during that time interval.

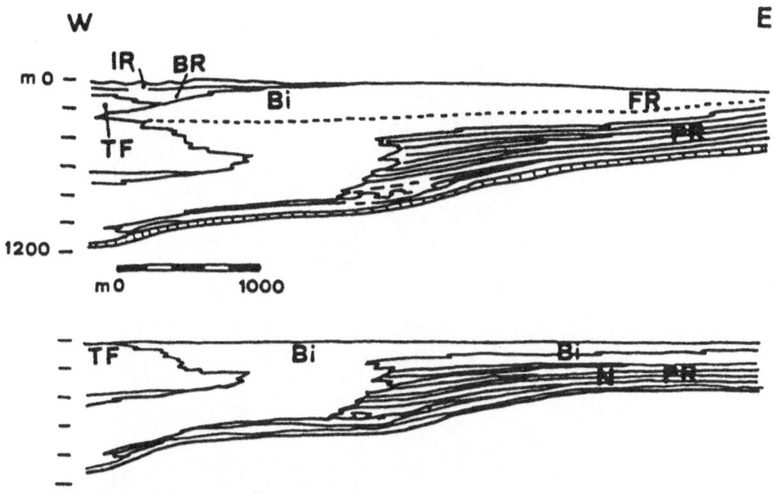

Fig.74 - Cross section of the Coglians-Collinetta reef complex (from Vai,1980).TF:tidal flat; BR: Back-reef area;IR: inter-reef area; BI: Bioherms; FR: Fore-reef area; PR: Peri-reef area.

The stratigraphic sequence described here,Givetian to Upper Frasnian in age, forms the uppermost part of the shallow - water complex.

The paleoenvironmental situation of adjacent areas of the Coglians-Collinetta shallow-water complex is simple, as is exemplified by the stratigraphic sections measured in the Volaia-Coglians (Fig.75).These five stratigraphic sections indicate a deposition passing from a reef flat in the east to a tidal or storm flat (Wanless,et al.,1989) in the west.The general vertical and westward trend is fining-upward and shallowing-upward. Grain sizes show a westward decrease from 0 phi to 4 phi.The fossil composition shows a change from open, agitated environments (corals, algae, brachiopods,foraminifers, etc.) to semirestricted, protected conditions (*Amphipora*, calcispheres,ostracods).

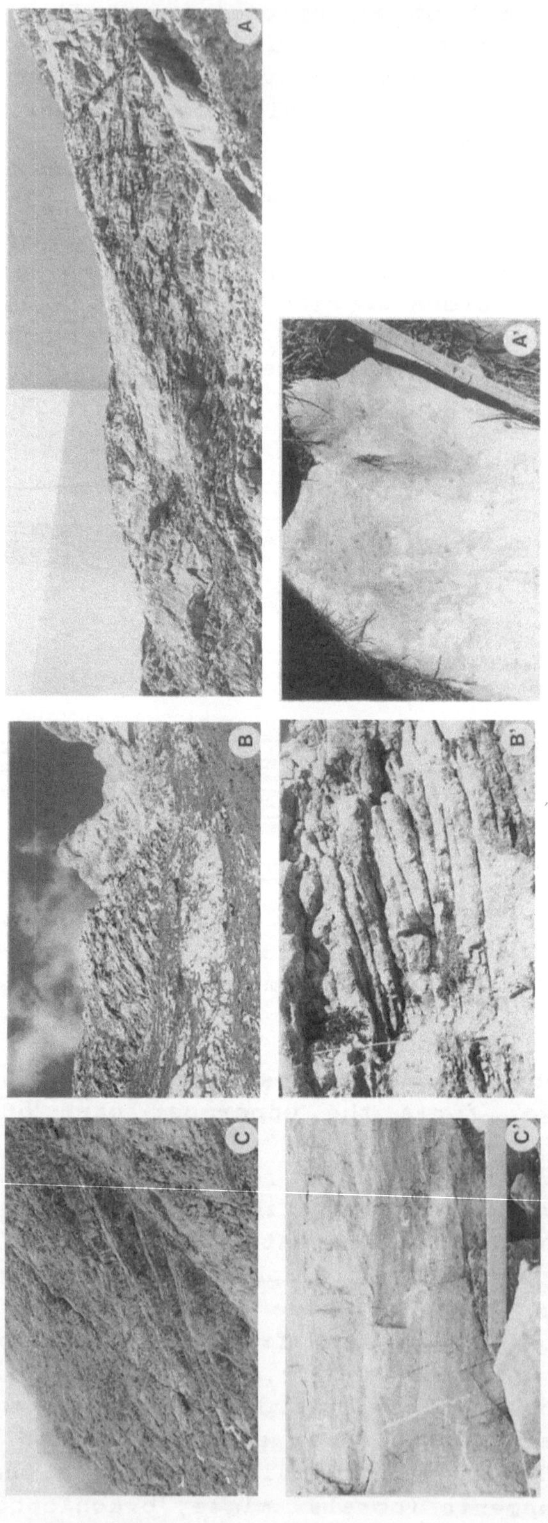

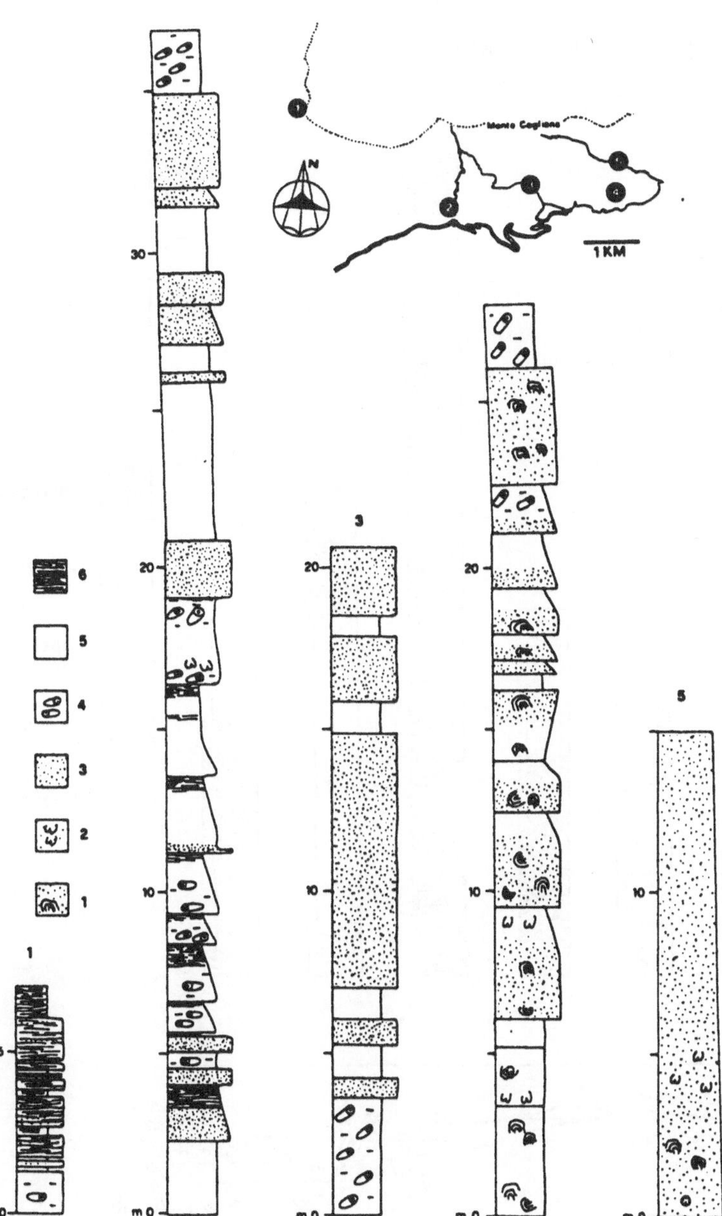

Fig.75 - Cross section of M.Volaia-Coglians shallow-water complex showing vertical and landward shallowing- and thinning-upward trends.(Section#4 measured by A.Argnani). A,A': massive beds composed of coral rubble (outer reef flat; section #5).B,B':Sigmoidal calcarenite beds probably developed as sandwaves (section #3).C,C': inner lagoon and tidal-storm flat (sections #1,2).

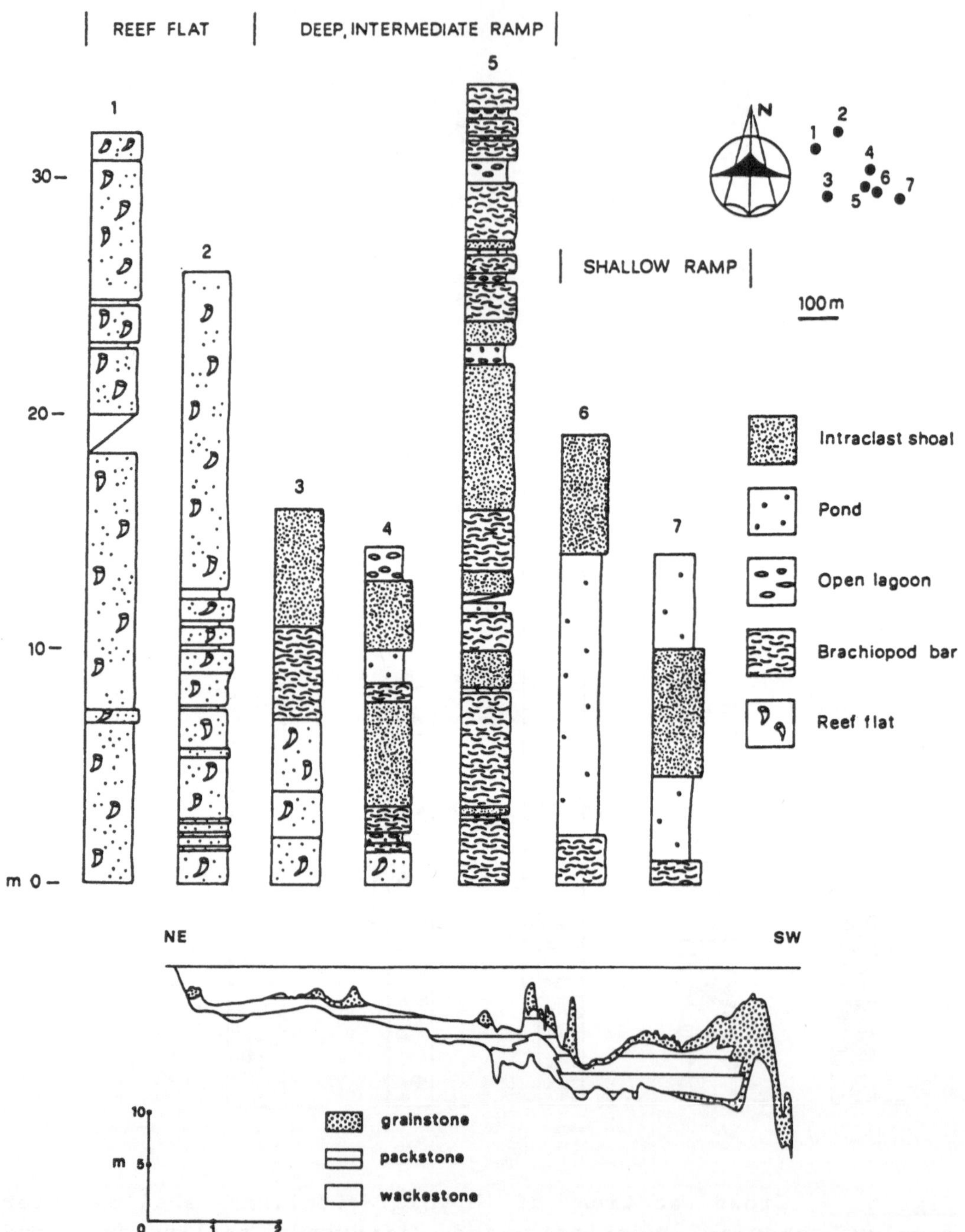

Fig.76 - Stratigraphic sections of the Cima Ombladet succession and (below) cros section of the Florida platform (Enos,1977) which shows analogous vertical and lateral sediment trends.

These changes which also occur along the stratigraphic sections are related to a predominating progradational trend. The sedimentary interface was inclined towards east and subject to flooding by storm currents.The physiography and sediment distribution are comparable to a 'Motu-Hoa' configuration (Bourroulh-le Jan and Talandier,1985) where onshore directed storm floods loose gradually energy and competence towards inner areas, with the formation of a shoreward fining-upward grain size trend.

The Cima Ombladet carbonate succession displays remarkable differences in facies organization with respect to those occurring in coeval, adjacent parts of the shallow-water limestone complex. These dissimilarities were superficially explained (Galli,1984,1985) as a result of a deposition within an isolated atoll within a major carbonate comlplex.

Such differences may be better explained by considering the Cima Ombladet succession as having formed as an onlap, intrashelf ramp structure.

In the Cima Ombladet carbonate succession a transition from a reef flat to semirestricted, inner lagoons is recorded (Galli,1984; 1985a,b,c; 1986). The paleobathymetry (Fig.77), reconstructed by means of integration of statistical and facies analyses, consists of a series of lagoons separated by islands and banks which complicate the environmental trends.

Environmental gradients were studied by means of quantitative modal analyses of paleontologic and lithologic components of 76 thin sections.

From the left to the right side of the diagram of Fig.77A the increase in ostracods, calcispheres and *Amphipora* percentages corresponds to a transition from the outer reef-deep ramp to shallow ramp sectors. The decrease in sedimentary influence of the outer-inner reef flat towards the shallow ramp is gradational.For each lithofacies the average fossil abundance was correlated with the corresponding detritus:matrix ratio (biopeloidal + intraclast detritus : matrix + cement). This ratio, being a measure of the packing, gives an estimate of the environmental energy (Fig.77C).With the exception of crinoids and some stromatoporoids, inverse correlations between fossils and detritus:matrix ratio indicate that organisms lived in muddy habitats. Brachiopods (*Stringocephalus burtini*, *Pentamerus*) lived in the deep ramp, probably in a series of

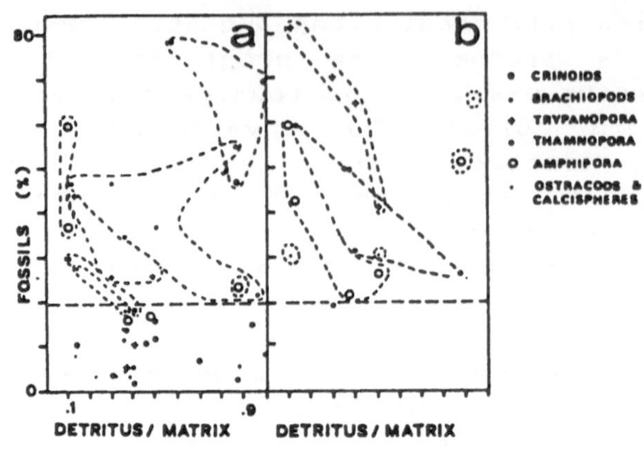
a
b
80
0
FOSSILS (%)
.1
.9
DETRITUS / MATRIX
DETRITUS / MATRIX
CRINOIDS
BRACHIOPODS
TRYPANOPORA
THAMNOPORA
AMPHIPORA
OSTRACODS & CALCISPHERES

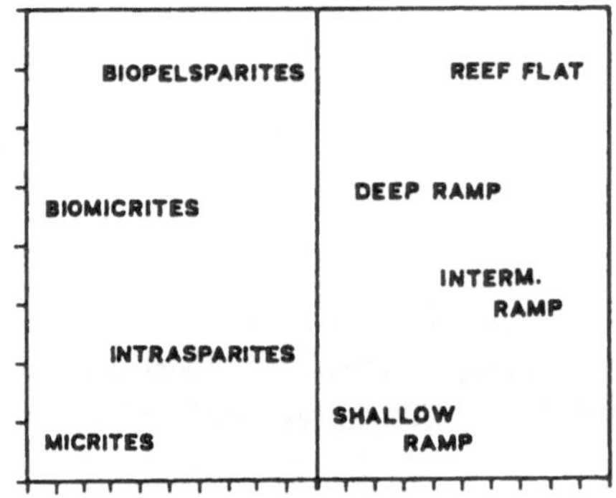
BIOPELSPARITES
REEF FLAT
BIOMICRITES
DEEP RAMP
INTERM. RAMP
INTRASPARITES
SHALLOW RAMP
MICRITES

SIMILARITY

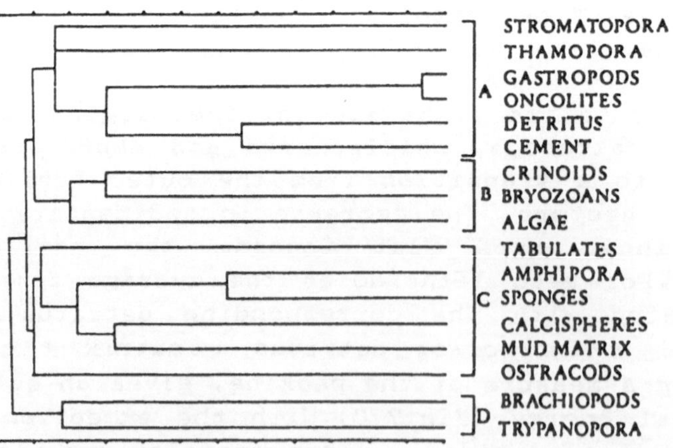
STROMATOPORA
THAMOPORA
GASTROPODS
ONCOLITES
DETRITUS
CEMENT
A
CRINOIDS
BRYOZOANS
ALGAE
B
TABULATES
AMPHIPORA
SPONGES
CALCISPHERES
MUD MATRIX
OSTRACODS
C
BRACHIOPODS
TRYPANOPORA
D

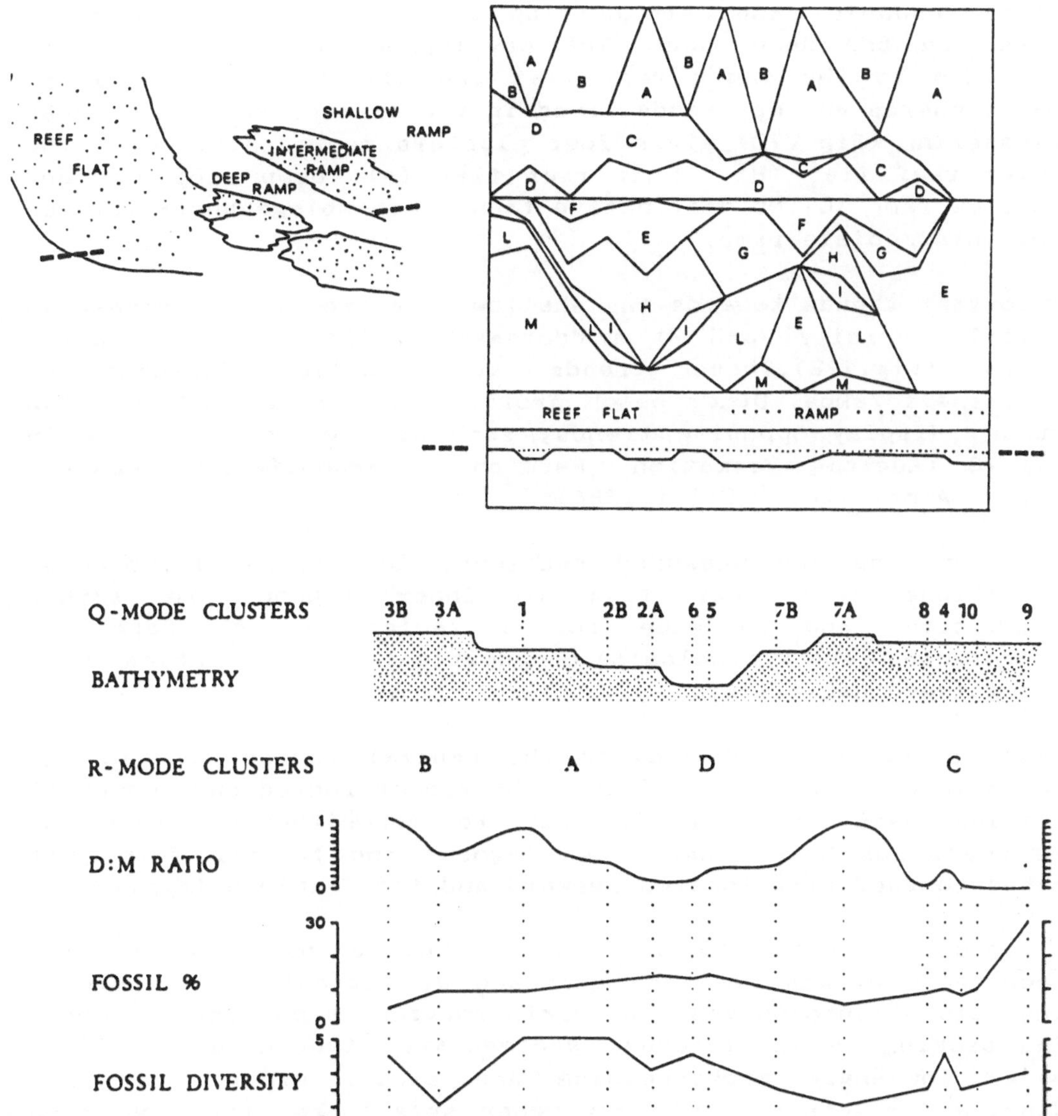

Fig.77 - Paleoenvironmental reconstruction and facies distribution of the Cima Ombladet carbonate succession (for more details see Galli,1985), paleobathymetry, petrographic trends and Q,R-mode clusters.

tidal channels and bars cutting through inlets. *Trypanopora* lived in the deep ramp. The development of *Thamnopora* was confined to the inner reef flat and the intermediate ramp. Calcispheres and ostracods lived in the shallow ramp.The R-mode clustering (Fig.77D) gives four clusters representative of the outer reef flat (B), inner reef flat (A), deep ramp (D) and shallow ramp (C).No distinctive faunal assemblage characterizes the intermediate ramp.

Important trends towards the shallow ramp are: 1) a decrease in fossil diversity; and 2) a decrease in the detritus: matrix ratio (Fig.77B).These trends are peculiar features of intrashelf ramps. Other beach profiles, not developed in onlap ramps, display opposite trends, such as the example shown in Fig.78 (Auernig Formation, Permian - Carboniferous, eastern Carnic Alps, Italy: Galli,1986).

As seen from the measured sections, faunal and lithofacies variations form reef flat to inner lagoon are rather complicated and evidence for a facies mosaic. Based on microfacies and lithologic composition, twelve facies were recognized (Galli,1985).

The hypothetical map showing the general facies zonation is shown in Fig.77A (Galli,1985b). The reconstructed environmental setting bears a general similarity to some situations occurring in the Exuma Cays, where open lagoons and beaches form just close to reef flat located leeward and tidal inlets (Fig.79).

A peculiar feature occurring in the study area is the occurrence of massive beds composed of intraclast grainstones and dolointramicrudites. The grain sorting is moderate to poor. The packing is low.Internal sedimentary structures are first order low-angle cross-bedding and second order high-angle dipping foresets within first order sets.These beds have been regarded as the lagoonward terminations of the reef flat occurring as linear sand ridges.These bodies have some analogy with rampart deposits occurring in the reefs inside the Great Barrier Reef (Scoffin,1977).

Brachiopods are the most abundant fossils in the succession as occurr in a wide range of subenvironments. As is shown schematically in Fig.80,different biostratonomic data of these fossils characterize the three sectors of the ramp.

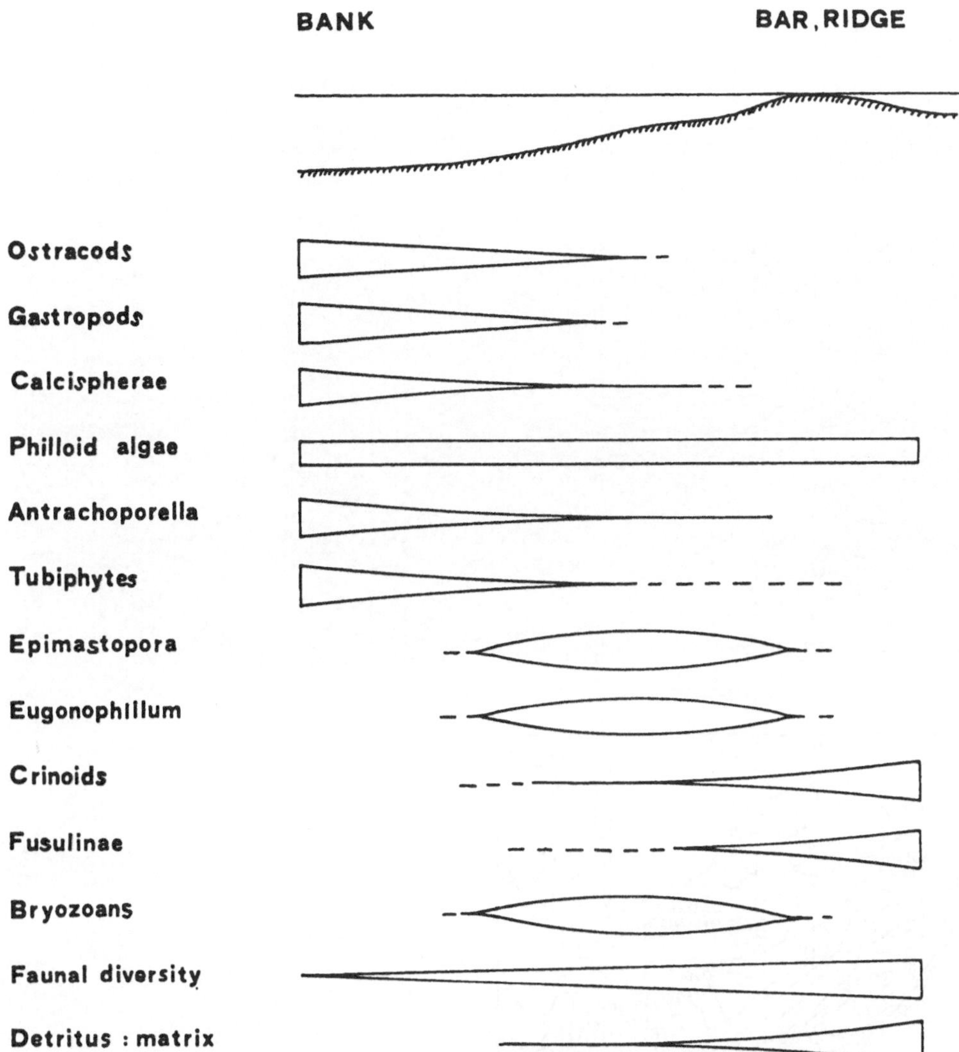

Fig.78 - Allochem distribution in a beach profile in the Auernig Formation ('Permo-Carbonifero Pontebbano, Carnic Alps, Italy: Galli,1986)

Facies associations

Shallow ramp

The shallow ramp consists of pond facies and thin- and thick-bedded alternations.

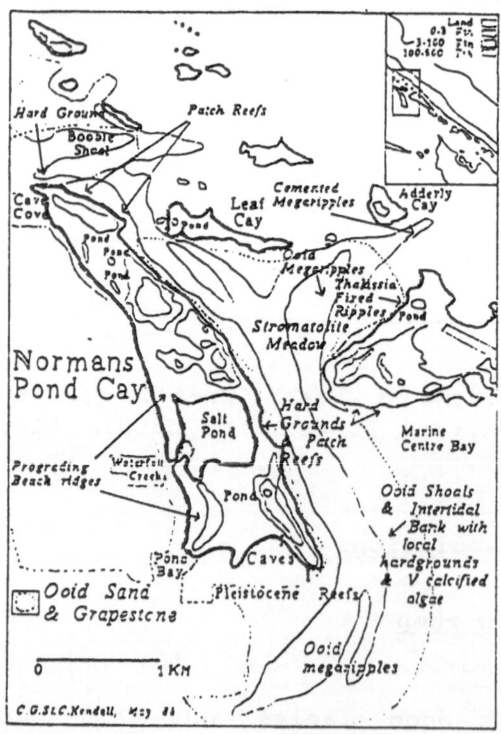
Land
0-3
3-100
100-500
Hard Ground
Patch Reefs
Boobie Shoal
Cave Cove
Leaf Cay
Pond
Cemented Megaripples
Adderly Cay
Ooid Megaripples
Thalassia Fixed Ripples
Stromatolite Meadow
Normans Pond Cay
Hard Grounds
Patch Reefs
Salt Pond
Marine Centre Bay
Prograding Beach ridges
Waterfall Creeks
Ooid Shoals & Intertidal Bank with local hardgrounds & V calcified algae
Caves
Pond Bay
Pleistocene Reefs
Ooid Sand & Grapestone
Ooid megaripples
0
1 KM
C.G.St.C.Kendall, May 88

	DEEP RAMP	(BAR)	INTERMEDIATE RAMP	SHALLOW RAMP
FAUNAL COMPOSITION	HOMOGENEOUS	HOMOGENEOUS	HOMOGENEOUS	HETEROGENEOUS
DISSOCIATION OF SKELETAL PARTS	VALVES ARTICULATED (DISARTICULATED FOR BURROWERS)	VALVES BOTH ARTICULATED & DISARTICULATED	VALVES BOTH ARTICULATED & DISARTICULATED	VALVES NEARLY ALL DISARTICULATED
ORIENTATION	LIFE POSITION, PARALLEL	IMBRICATED, INCLINED, PARALLEL	RANDOM	RANDOM
PACKING	LOW	HIGH	VARIABLE	HIGH
GEOPETAL INFILLING	INTRAMICRITIC	INTRAMICRITIC	INTRACLASTIC, NONE	BIOPELOIDAL
BRACHIOPOD %	66	78	90	29
ASSEMBLAGE	"IN SITU"	LOCALLY TRANSPORTED	TRANSPORTED	TRANSPORTED

Fig.80 - Biostratinomy of brachiopods. Brachiopods found in the deep ramp represent an 'in situ' assemblage, as is shown by shells found in life position, articulated and floating in a micrite matrix with a micrite geopetal infilling. Brachiopods of sublittoral bars in the deep ramp underwent a local selective transport, by means of tidal-longshore currents (bipolar beddings). Brachiopods occurring in the intermediate and shallow ramp (within thin- and thick-bedded alternations) are smaller, display a higher degree of transport (random orientations,a greater fragmentation, and various geopetal infillings). Brachiopods piled up in the intermediate ramp underwent a mass transport, as is evidenced by random orientations, poor sorting and variable packing. Unlike brachiopods of the deep ramp, those of the intermediate and shallow ramp underwent transport and mixing with other lagoonal fossils (see Fig.13 for some additional details).

Pond facies

Description

This facies consists of poorly fossiliferous, well-bedded, thin, black micrites and barren dolomicrites, with scarce calcispheres, ostracods and *Trypanopora* (Galli,1985), and very thin shelled brachiopods. Pyrite, organic matter horizons and rare algal mat with mm-size laminations occur.Other sedimentary structures include vertical burrows, some wavy lamination, flat pebble conglomerate levels and pebble clusters.

Intercalated with this facies are thin-bedded and thick-bedded alternations. The last consist of bioclast interlayers of variable thickness. Rapid pinch-outs into micrite facies can be seen in some instances.The faunal content is heterogeneous. Intraclasts of variable shapes consist of black micrites. These beds are organized into a fining-upward sequence which is composed of two units: 1) a lower part, consisting of dark calcarenitic beds, composed of brachiopods,corals, crinoids, calcispheres, peloids and intraclasts; and 2) an upper part, composed of thinner beds with disarticulated both thin- and thick-shelled brachiopods. This upper part grades quickly into calcilutites with thin algal laminae.

Interpretation

A very shallow environment for this facies is indicated by vertical burrows ,paucity of fossils,algal layers and pyrite horizons.This facies represents intertidal pools and ponds whose extension and depth depended on the local intraclast shoal configuration, and other topographic barriers.

Thick- and thin-bedded alternations represent a deposition by storms, as washover deposited, as evidenced by the following: sharp basal contacts,couplets of shelly layers and laminated mud, grading, escape structures in the underlying mud and screening fabrics.The tripartite subdivision of the thick bedded alternations is analogous to that occurring in Florida Bay.

Intermediate ramp

Intraclast shoal

Description

This facies consists of poorly sorted, disorganized, coarsely grained intraclast grainstones and packstones, up to 3 m thick.Lithoclasts are of variable size,shape and composition. Fossils are all transported and of variable provenance (lagoons, reef flat, ponds).Common sedimentary structures are vadose silt,cut-and-fills, keystone vugs, gradations, cross-bedding and flat-pebble conglomerates.

This lithofacies is overlain by cryptalgal laminites, represented by light-gray, well-sorted, planar- bedded, laminated intrasparites, 60 cm to 1 m thick, with very thin micrite laminae, which are undulated, slightly inclined and stylolitic, with a horsetail filigree pattern.

Interpretation

The intraclast shoal facies represents a deposition in a beach environment,as is indicated by the occurrence of flat pebble conglomerates,fringing cement, keystone vugs, vadose silt, occurrence of various textures and absence of a grain size signature (Davis,et al.,1972). The occurrence of wide textural and grain-size ranges, disorganized beds and massive bedding indicate a deposition under complex hydraulic conditions, as a result of island shifting.

Cryptalgal laminites are interpreted as beach-ridge laminations. The sediment was deposited by storm floods and currents by-passing ridges and bars.

Deep ramp

Brachiopod grainstones and wackestones

Description

Brachiopod wackestones contain abundant monotypic asemblages of thick-shelled brachiopods (*Stringocephalus*, *Pentamerus*) occurring together with smaller amounts of other fossils such as crinoids, and occasional solitary corals.

Brachiopod grainstones consist of brachiopod shelly layers containing well-sorted shells. This unit ranges in thickness from 1.5 to as much as 3 m.Imbrications, geopetal structures, planar and bipolar cross-bedding were observed in some instance.Some of the thicker beds contain a greater percentage of intraclasts, a lesser percentage of brachiopods,and a lesser shell orientation.

Interpretation

Brachiopod wackestones represent a deposition in a marginal bay. The environment is subtidal,as shown by the occurrence of patchy and distinct,"in place" accumulations of brachiopods and other types of fossils. Zonations of endemic populations and textures indicate that this lagoon was rather large and deep.

Brachiopod grainstones may be referred to 'in situ' lagoonal bars and banks.Some reworking by storm agents is envisaged for the thicker beds, which may represent reworked lagoonal bars and banks close to the intermediate ramp which was heterogeneous in composition.

Depositional model

The measured stratigraphic section records complex and recurring facies transitions.The paleoenvironmental reconstruction is shown in Fig.21 (Galli,1986). The Markov chain analysis (Miall,1973),applied to the succession (Galli,1985) in order to discriminate deterministic from random facies transitions,was an aid in the identification of two main sequences corresponding to:

1) a storm bar sequence; and
2) a beach bar sequence.

These two sequences (Fig.81) are the result of two opposite,different depositional mechanisms which were operating in the area: 1) aggradation; and 2) progradation.

The complexity of facies transitions, shown by the rhythmogram of Fig.82,is a consequence of the interference between these two depositional mechanisms.

The progradational trend produced the deposition of shallow-water deposits over deep ramp facies. As indicated by the detailed microfacies analyses (Galli,1985) these shallow-water sediments prograded into the lagoon, from reef flats as linear

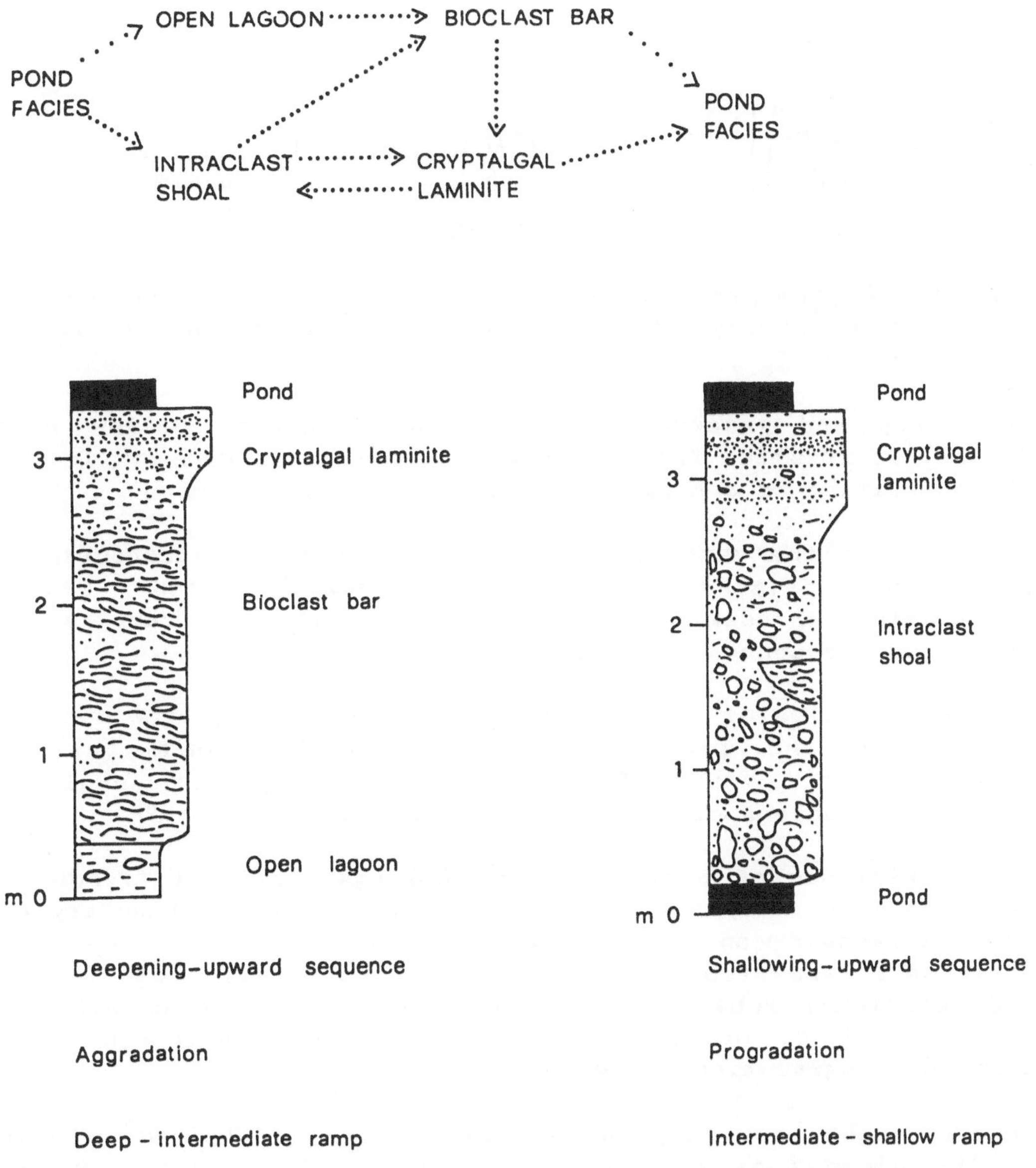

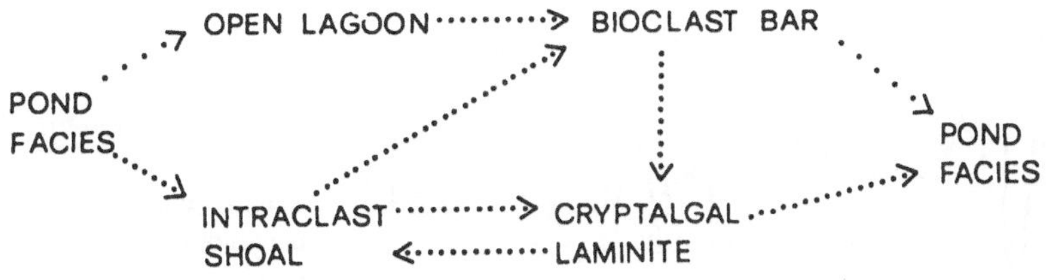

OPEN LAGOON
BIOCLAST BAR
POND FACIES
POND FACIES
INTRACLAST SHOAL
CRYPTALGAL LAMINITE
Pond
Cryptalgal laminite
Bioclast bar
Open lagoon
3
2
1
m 0
Deepening-upward sequence
Aggradation
Deep - intermediate ramp
Pond
Cryptalgal laminite
Intraclast shoal
Pond
3
2
1
m 0
Shallowing-upward sequence
Progradation
Intermediate - shallow ramp

Fig.81 - Facies flow chart showing more common than random facies transitions (Galli,1985) and modal sequences (storm bar sequence:left;beach bar: right).

sand shoals and ramparts. Topographic highs were the sites of formation of beaches.This progradational trend is a local response to the dominant progradational trend recorded in the adjacent areas of the shallow-water carbonate complex (Fig.74 and 75).

The aggradational trend led to the deposition of deeper,

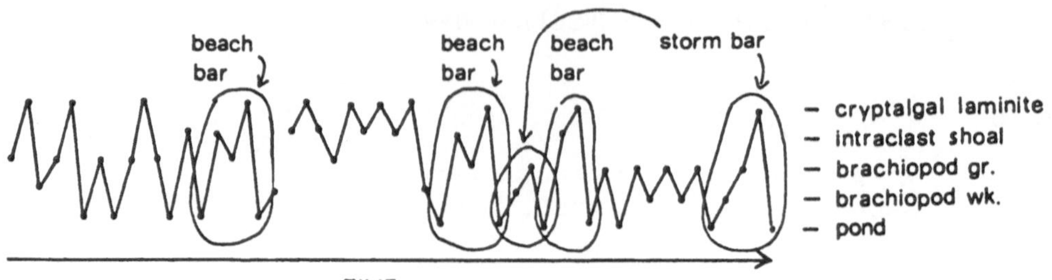

Fig.82 - Rhythmogram of facies transitions evidencing for an aperiodic cyclicity (Schwarzacher,1975), obtained by Markov chain analysis.

deep ramp and intermediate ramp facies above shallow ramp facies. It resulted from a deepening-upward trend, and led to the deposition of a hinge sequence (storm bar sequence).

As seen from the vertical changes throughout the stratigraphic succession (Fig.83), the transgressive trend became progressively predominating upwards, where the deep ramp facies and trough sequences overlie hinge sequences.

The complexities of facies transitions suggest a scarce control exerted by the predepositional topography. This may be attributed to a model of formation of onlap ramps involving complex mechanisms of rotations and tilting (cf. Fig.39).

The coexistence of progradational and aggradational trends in the same area is typical of areas located near the boundary of the intrashelf onlap ramp,close to the hinge area, in intrashelf ramp structures formed by a combination of tilting and rotational subsidence. In these sites the onlap ramp is transitional to other areas of the platform governed by a dominant progradational trend.

Interferences between progradational and aggradational trends in the hinge area are also recorded in the 'Calcari Grigi' Formation. Thickening- and coarsening- upward trends, such as that shown in Fig.84, occur in the eastern border of the ramp, where it is transitional to areas subjected by progradation. In these stratigraphic sections, the material supplied by the nearby prograding areas produced an equilibrium between subsidence and sedimentation rate.

These complex situations are in accordance with the result of the study conducted by Eberli & Ginsburg (1981) on the Bahama platform (Fig.5).Additional work is needed to confirm the possibility that some ancient carbonate platforms, rather than

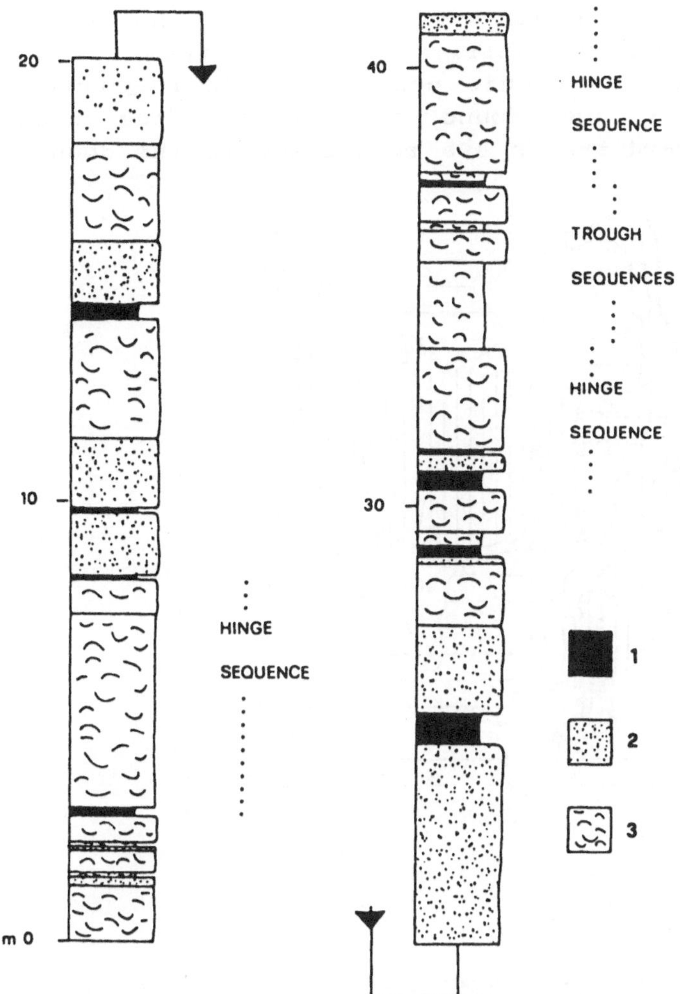

Fig.83 - Stratigraphic sequence showing a progressive increase in proportion of deep ramp facies. 1:pond facies (shallow ramp); 2: intraclast shoal (intermediate ramp); 3: brachiopod grainstones and wackestones (deep ramp).

monolithic structures, are constituted by complex 'offshooting' prisms, resulting from the juxtaposition and lateral transition between different types of intrashelf ramp geometries.

The same distribution of thickening- and thinning-upward cycles and megasequences occurs in other areas and situations. In a different situation this distribution, found in the Pleistocene

- Holocene fan delta system close to the city of Bologna (Italy), studied by Galli,et al.(1985),seems to be controlled by differential tectonic control: thickening-upward trends of gravel bodies are found in the uplifted areas, whereas thinning-upward trends are restricted to the downfaulted sites (Fig.84).

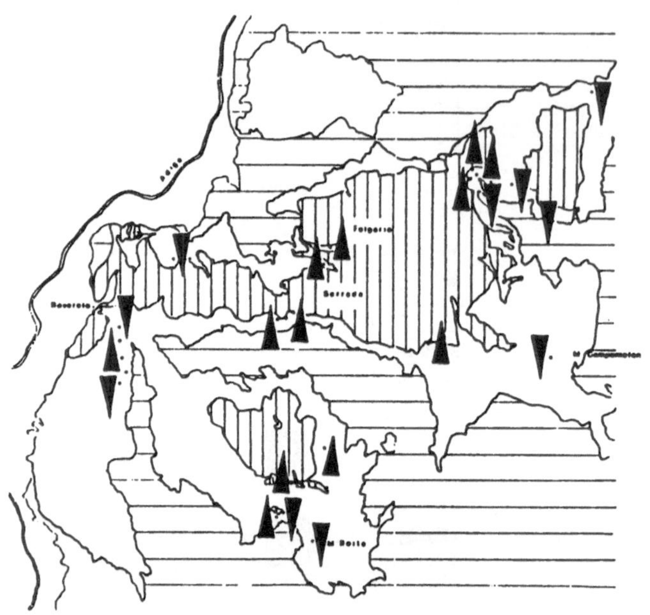

Fig.84 - Location of thickening- and coarsening-upward trends and thinning and fining-upward trends in the studied area of the 'Calcari grigi' Formation (cf. Fig.26 showing the paleotopography of the ramp);example of thickening-upward and coarsening-upward sequence in the 'Calcari Grigi' Formation (section #11);cross section showing the distribution of Holocene thickening- and thinning-upward trends of gravel bodies in the Po plain and the tectonic synsedimentary features (Galli, et al.,1985).

'Capo Rizzuto' shoreline sequence, Pleistocene

Introduction

Pleistocene marine deposits in southern Italy occur along the coastline of southern Italy and form the lowest ones of a series of marine terraces (Fig.85) deposited during the last climatic oscillation (Riss/Würm), 300.000 to 700.000 years ago.

Although the overall deposition of the sedimentary sequence described below records the onset of a climate change associated with glacial-interglacial fluctuations, shoreline

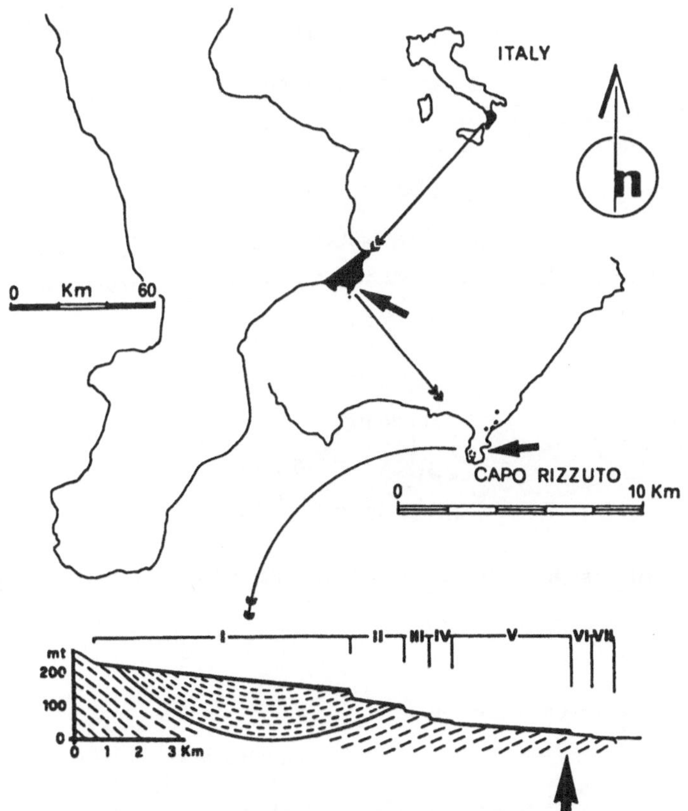

Fig.85 - Location of study area and cross section (from Selli,1962) showing marine terraces developed unconformably above older rocks.IV to VII represent Tyrrhenian terraces.

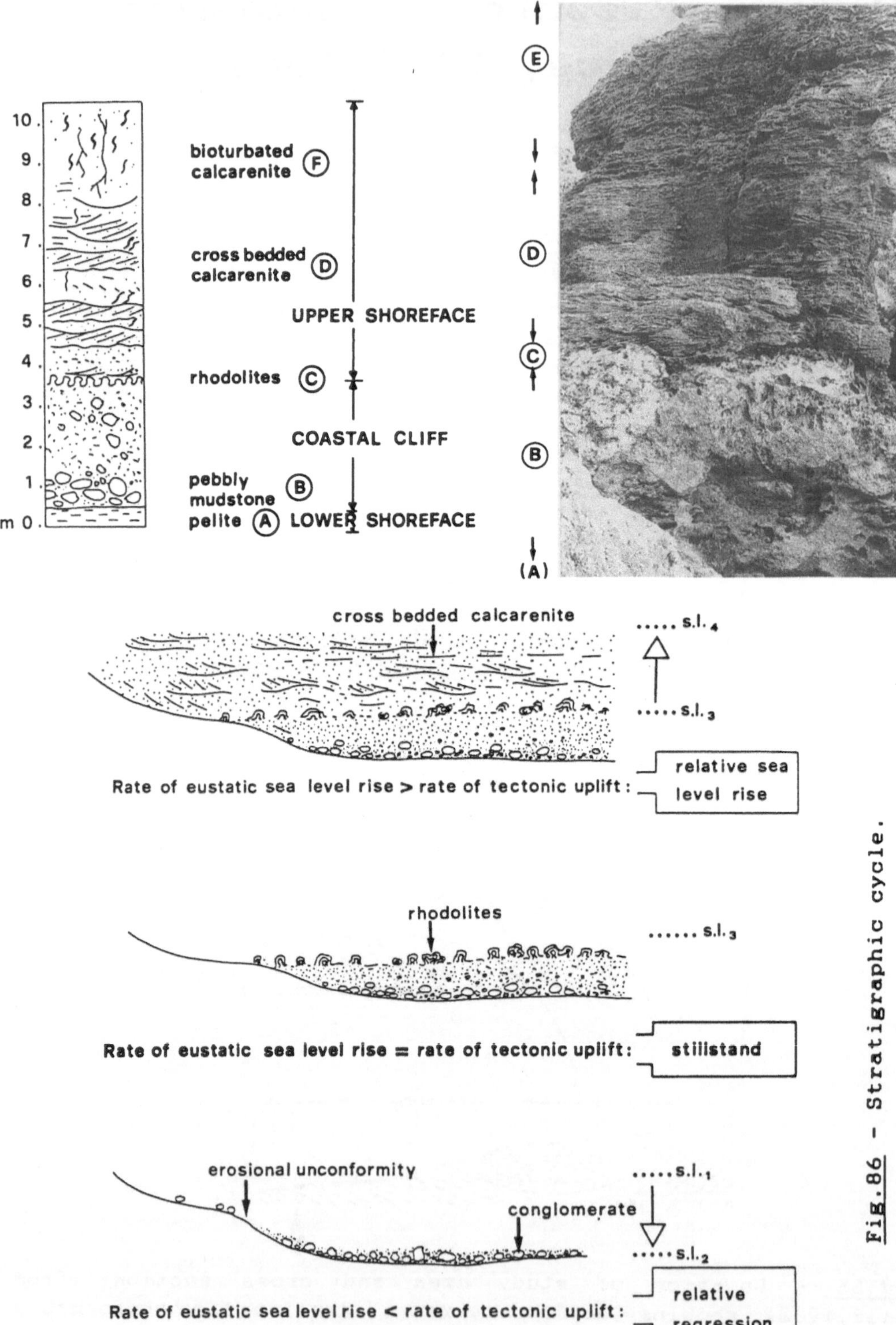

Fig.86 - Stratigraphic cycle.

erosion, deposition, and the location of cycles at variable altitudes were controlled by combined tectonics and eustasy (Selli,1962,1967; Desio,1973).

Depositional sequence

The measured stratigraphic sequence,10 m thick, consists from bottom to top of the following facies tracts (Fig.86):

1) Lowstand facies tract;
2) Transgressive facies tract;
3) Highstand facies tract.

Lowstand facies tract

Description

This facies tract overlies bioturbated siltstones with interbedded lenses of poorly sorted, coarse to fine-grained calcarenites containing some shells interbeds (Fig.86A).

It consists of a 2 m thick fining-upward sequence; the very base consists of a thick unit composed of unsorted, intraformational, rounded blocks and pebbles engulfed within a moderately sorted calcarenite to calcirudite matrix (Fig.86B) whose allochems are given by abraded skeletal fragments of sponges, *Strombus bubonius*,*Cardium*, *Glycymeris*, *Cerithium*, shark teeths and bryozoans.

A discontinuous horizon of branching rhodolites caps this tract.

Interpretation

Calcarenite interbeds underlying the depositional sequence are common in nearshore to lower shoreface environment. They reflect a fine-grained shelf or lagoonal environment subject to an episodic input of bioclast calcirudite sand.These deposits were laid down in the lower shoreface.

The overlying lithofacies (Fig.86B) indicates a deposition

under high-energy conditions, in a coastline environment, as is shown by the grain-supported texture and the type of fauna. The rounding of clasts indicates a very shallow environment subject to a constant water motion.The formation of relict lithoclast blocks may have been a consequence of bioerosion and wave-action. Indented lateral contacts with the upper lithofacies outline an articulated, rough submarine paleotopography which suggests the existence of submarine cliffs.

Branching rhodolites on top suggest a slow sedimentation rate within a very shallow environment.

A rapid relative sea-level fall produced repeated reworking of carbonate sediments with the formation of relict lithoclast beds which mark the contact between lower shoreface and cliff.Tectonic uplift, faster than eustatic sealevel rise, led to a relative sea level fall and fragmentation of previously exposed carbonates into blocks which subsequently underwent reworking and rounding in a shoreline environment (Fig.86).The lower erosional contact between shoreface and overlying conglomerates does not corresponds to the onset of the Tyrrhenian phase, but represents probably a minor relative sea level fluctuation.

The rate of tectonic uplift, initially geologically rapid, (cf. Cisne,1987), gradually decreased and at a certain time reached the same rate as the eustatic sea level rise. Consequently, the sea level remained stable for a short time interval; these conditions, favourable to the formation of oncolites, led to the development of the rhodolite horizon, at a depth less than 2 meters, under a low sedimentation rate.

Transgressive facies tract

Description

This consists of coarse-grained, moderately sorted, cross-bedded biocalcarenites (Fig.86D).
Some accumulations of *Ostrea* occur towards the top of this lithofacies which is 4 m thick. Sedimentary structures are given by trough cross-bedding and planar, sigmoidal and hummocky cross-bedding.

Introduction

The multistored channelized geometry is typical of the upper to lower high-energy shoreface environment (Galli,1986). The trough cross bedded sets and hummocky cross bedding were interpreted as the result of a deposition by storms, above fair-weather base (Galli,1989).

Tectonic uplift was unimportant and sedimentation caught up to sea level, mainly in response to the eustatic sea-level rise, by vertical aggradation (Fig.86).

Highstand facies tract

Description

This tract is represented by calcarenites (Fig.86E) whose sedimentary structures are nearly completely obliterated by root penetration and bioturbation by *Callianassa* and *Ophiomorpha*, well visible on bedding planes.

Interpretation

The record of episodic sedimentation is less apparent here due to bioturbation and rhizoturbation which point to an upward shallowing and decrease in sedimentation rate, in turn function of a decrease in the rate of sea level rise.

Concluding remarks

Although not directly related to a deposition in an intrashelf ramp, the sequence described above has been included in this work because of its similarities to a number of sequences described here, for example the Lofer cyclothem and the hinge sequence described above.

The succession of facies tracts and the relative sea-level curve is analogous,apart from differences in scale,to the second and third depositional sequence of the slope carbonates in the Gargano massif described in chapter #3. The sequence described in this section is also a small-scale example of relief inversion.

Part II

The computer simulation of prograding clastic wedges individuated two main patterns: 1) parallel-divergent; and 2) convergent obtained by different values of sedimentation rate, and complex interactions between subsidence and eustatic changes. In most cases divergent patterns form by a sea-level rise, a seaward tilting and a constant or increasing sedimentation rate. Convergent patterns form mainly by a negligible subsidence rate, a decreasing sedimentation rate and a stillstand or fall of the sea-level.
Divergent and convergent patterns occurring in the two case histories described below (Cretaceous slope deposits:Gargano Massif;Middle Triassic carbonate buildups: Dolomites) confirm the results of the computer simulations. Divergent patterns are overlain by convergent patterns.

Computer simulation of clastic wedges

Introduction

Most of siliciclastic depositional sequences developed on the outer edges of passive continental margins are progradational sedimentary wedges consisting of laterally stacked bodies developed above a progradational surface and delimited by chronostratigraphic surfaces.

Each of these surfaces can be divided into a nearly horizontal and a sloping portion which represent respectively the aggradational and the progradational component of the sedimentation.

Aggradation consists of a vertical build-up of sediment which takes place under an increase in the relative sea level rise. Stacking of units produces sigmoid or complex-oblique clinoform patterns (Mitchum,et al.,1977).

Progradation refers to seaward building-up of sediment and results in a net decrease or stillstand of the sea-level. Progradation produces oblique (tangential and parallel) patterns (Mitchum,et al.,1977).

The plane which ties all the lines of inflexion (platform break hinge) between horizontal platform-top sediments and inclined slope sediments is the <u>platform break plane</u> (PBP) (Doglioni and Bosellini,1989).The <u>downlap plane</u> (DP) (Doglioni and Bosellini,1989) is the diachronous plane intersecting all the lines where the toe of prograding clinoforms flatten out into the horizontal basin facies (downlap hinge).
The downlap plane may be horizontal, climbing and descending (Bosellini, 1984).

The inclination of the PBP may also vary from nearly horizontal to nearly vertical, with the formation of offlap relationships (Mitchum,et al.,1977) when lateral progradation takes place contemporaneously with aggradation, and toplap relationships (Mitchum,et al.,1977) when inclined strata terminate on top against the upper horizontal boundary. Also variable is the length and thickness of the clinoforms.

These parameters are shown in Fig.87. They change in response to: 1) eustatic changes; 2) tectonic subsidence; 3) sediment volume; 4) sediment penetration into the basin; 5) width of the margins; and 6) hydrodynamic factors (Burton,et al.,1987).

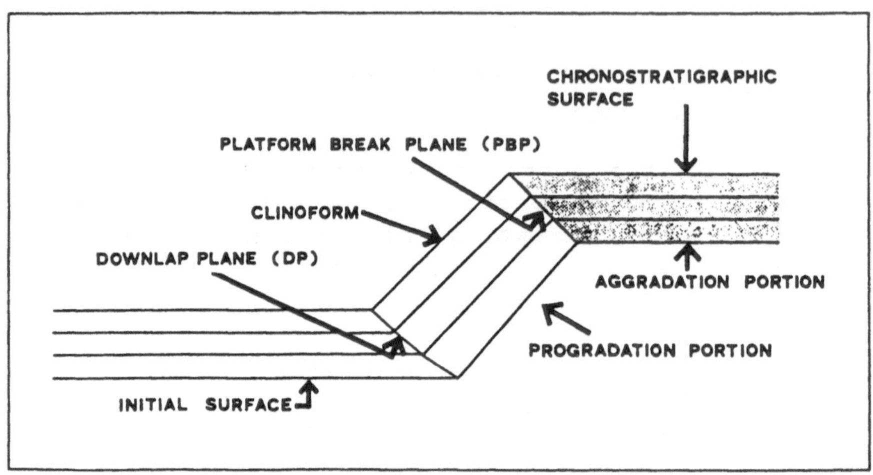

Fig.87 - Key elements of clastic prograding wedges.

This chapter presents the results of a computer simulation of stratigraphic models of accumulation of prograding clastic wedges, in response to sea-level curves, tectonic models and sediment inputs. For each pattern, the forward mathematical modelling generated a 'family' of solutions. They are a simplification of reality; nevertheless, they provide a tool in basin analysis because they allow for a distinction between the components of the relative sea-level rise, which are difficult to separate in stratigraphic interpretations (Burton,et al., 1988).

The mathematical modelling of siliciclastic wedges used in this simulation tracks the evolving geometry produced by the infilling of two-dimensional basin by a step-by-step addition of sediment increments, for user-defined, changing values of subsidence and eustasy. The simulation does not reproduces physical processes of deposition. It simulates the sediment geometries, based on the concept that important features of depositional geometries are controlled by macroprocesses such as changes of eustasy, subsidence, deposition and compaction of sediment. Each depositional surface is defined by an array of evenly spaced points. This gives the sediment surface a staircase-like appearance which is not seen, due to the large number of horizontal points. The sediment deposition is simulated as right triangles of specific length and thickness. The height of triangles is a measure of the volume of sediment deposited; the length is the distance of sediment penetration into the basin.

The sequence of processes at each time step is the following.
1) calculation of the sea-level;
2) tectonic subsidence (or uplift);
3) erosion of the sediment above sea-level with a slope angle greater than the sediment repose angle (as defined by the user);
4) deposition of the sediment, including that eroded in the previous step;
5) compaction of the sediment according to the weight of the overburden;
6) isostatic subsidence of surfaces in response to the sediment and water loading.

The simulation is carried out for each time step as a sequence of subsidence, sea-level change, erosion, deposition or compaction, etc., as shown schematically in the simplified flow chart of Fig.88.

More details concerning the computer technique and its application are found in Burton, et al.(1987), Scaturo, et al.(1989), Helland-Hansen, et al.(1988).

Clastic progradational wedges developed on passive continental margins were simulated by using the data base derived from the Exmouth plateau, NW Australia (Erskine and Vail, 1988).The initial surface simulates a block-faulted basement. Sea level was assumed to fluctuate by about a maximum of 200 m over a period of 6 millions years.

The eight hypothetical sea-level curves which have been used in the simulation reproduce sea-level stillstand, rapid,slow sea-level falls, and gradual ,rapid sea-level rises. Modelled subsidence is within the range of subsidence rates of the Atlantic margin.Tectonic activity includes landward and seaward tilting, uplift, basculatory tectonics, and rotational subsidence along listric faults.Sedimentation rates are taken as constant, increasing and decreasing.

The graphic simulation generated 71 geometries under defined values of eustasy, subsidence and sediment volume and penetration. Based on different orientation of the PBP and DP,

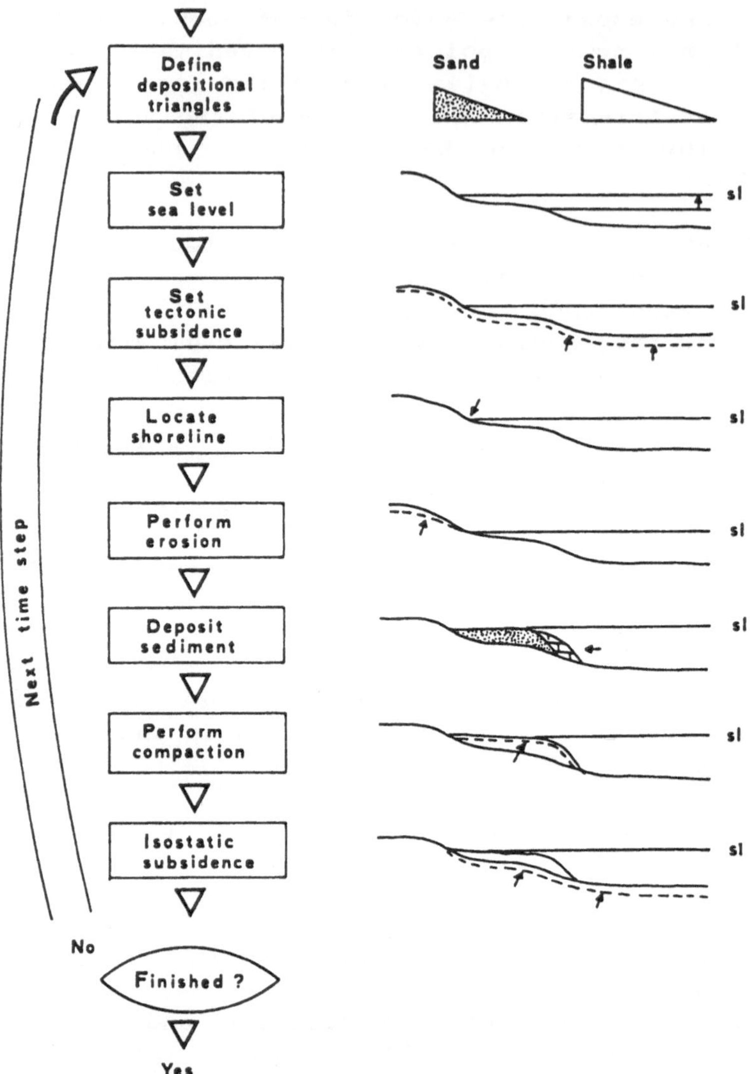

<u>Fig.88</u> - Flow chart of the simulation (from Scaturo,et al., 1989).

as seen in dip-oriented cross sections, the graphical outputs were grouped into three patterns which were formerly defined by Doglioni and Bosellini (1989):

1) parallel;
2) convergent;
3) divergent.

Climbing, descending and parallel directions of progradation produce for each pattern three sub-patterns. The complete

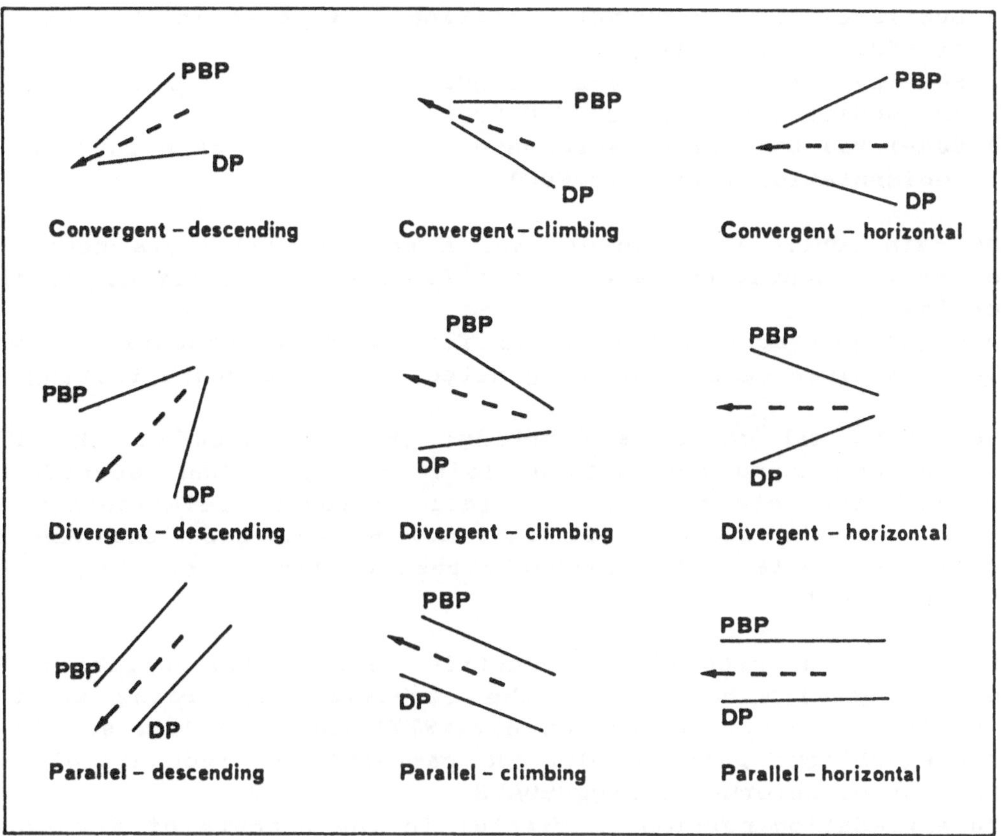

Fig.89 - Types of patterns based on the Platform Break Plane and Downlap Plane, and directions of progradation.

spectrum of patterns (composite patterns were also taken in consideration but are not described here) is shown in Fig.89.

Parallel pattern

In this pattern the clinoforms maintain a constant length because the downlap plane keeps parallel to the platform break plane (Doglioni and Bosellini,1989).

This kind of pattern is controlled by the following three different interrelationships between factors.

1) Sea-level fall - seaward tilting - decrease in the rate of sedimentation (Fig.90A).
2) Sea-level fall - seaward, landward subsidence - decrease in sedimentation rate (Fig.90B).
3) Sea-level stillstand - seaward tilting - decrease in the sedimentation rate (Fig.90C).

The main controlling factors are a seaward tilting,a decrease in the sedimentation rate, and a falling or stillstand of the sea-level.
This pattern does not seem to form when progradation takes place within a depression originated from a landward tilting.

The DP in all cases is climbing; the steepness of the PBP varies and produces offlap relationships when subsidence outpaces the rate of sea-level fall. A toplap relationship is visible in the example of Fig.90A: in this case the PBP is nearly horizontal: this probably results from a stillstand of the sea-level.

The resulting prograding geometries are either climbing or descending with respect to the sea-level. According to the classification by Mitchum, et al.(1977) the clinoform set is a complex-oblique pattern. Oblique patterns are recognisable in the last clinoforms of Figg.90A,B.
The aggradation portion is greater in the outputs of Fig.90C,D and 90A, where sedimentation is controlled by a sea-level fall, and lesser in the examples of Fig.90A which was produced by a sea-level stillstand.
The basinal sediments should record vertically a shallowing-upward sequence.

Examples of parallel patterns from the literature are few as a likely result of the landward tilting undergone by passive margins such as the Atlantic margin, and the NW Australian margin. Parallel patterns may occur in areas which underwent an inversion in the sense of tilting, for example in the Aptian and Oligocene.

A case history displaying a parallel pattern occurs in the Tuscalose Delta and Turbidite (central Louisiana: Berg,1982) (Fig.91) where the deposition appears to have been controlled by a seaward tilting, a decrease in the sedimentation rate and a stillstand of the sea-level; This example is comparable with the graphic simulation of Fig.90A.

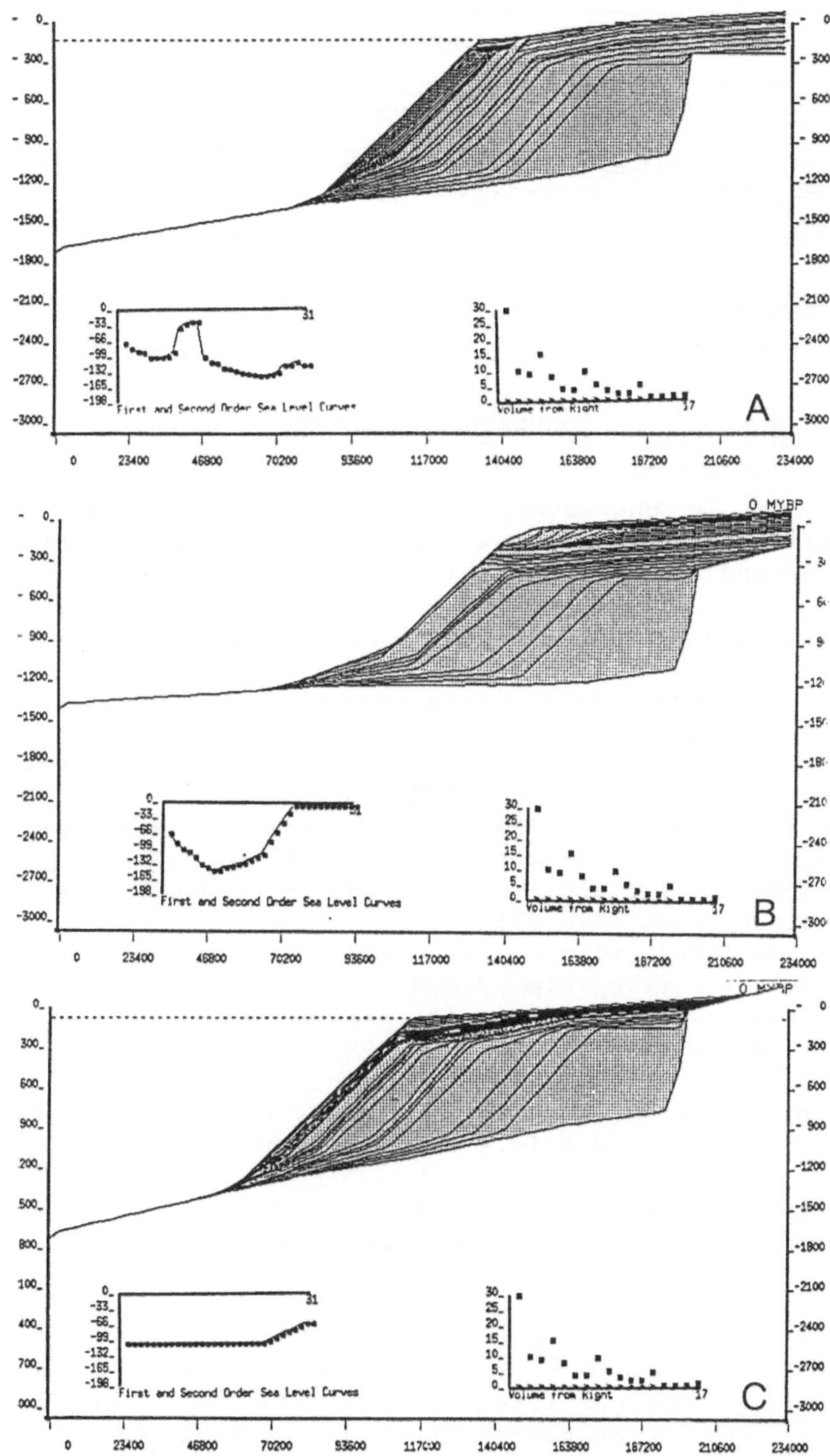

Fig.90 - Examples of computer simulations of parallel patterns.

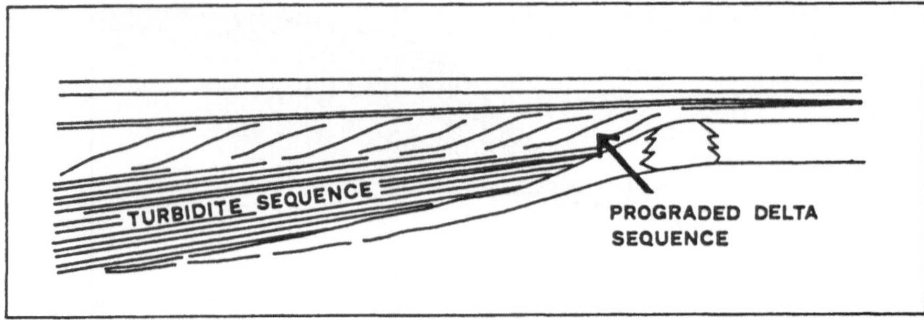

Fig.91 - Louisiana-Tuscalosa deltas and turbidites (Berg,1988).

Convergent pattern

This pattern is characterized by a progressive decrease in length of the clinoforms as a result of the upward convergence of the PBP and the DP (Doglioni and Bosellini,1989).

It was produced by the following combinations.

1) Rapid sea-level fall - landward tilting -constant rate of sedimentation (Fig 92A).
2) Rapid sea-level fall - landward tilting - decreasing sedimentation rate (Fig. 92,B,C,D).
3) Rapid sea-level fall - basculatory tectonics - decreasing sedimentation rate (Fig.92A,E).
4) Sea-level stillstand - no subsidence - decreasing rate of sedimentation (Fig.92F).
5) Slow sea-level rise - landward tilting - decreasing sedimentation rate (Fig.92,G,H).
6) Slow sea-level rise - no subsidence - decreasing sedimentation rate.

The main controlling factors are a decrease in the sedimentation rate and a fall or stillstand of the sea-level.

This pattern is not produced during a rapid sea-level rise, nor by a seaward tilting. Differential subsidence is not always present.

The DP is always climbing.The PBP displays various degrees of steepness. It is steeper when a sea-level rise is accompanied by a landward tilting (Fig.92A), slightly inclined when the sea-level fall is counterbalanced by a landward tilting,or uplift;horizontal to slightly descending under conditions of sea-level fall or stillstand and an absence of subsidence. In most situations the PBP follows the sea level curve (Fig.92A,B).

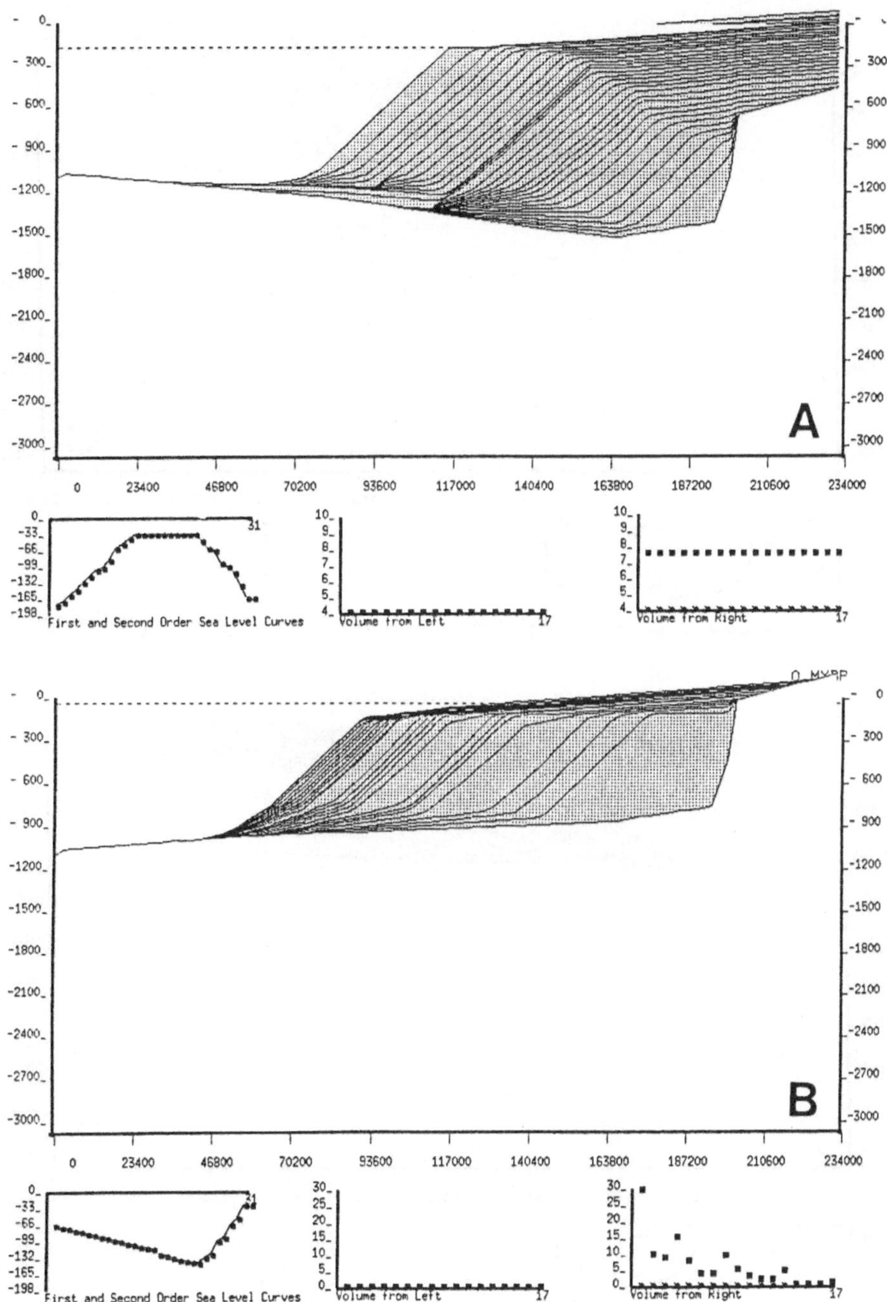

Fig.92 - Examples of computer simulations of convergent patterns.

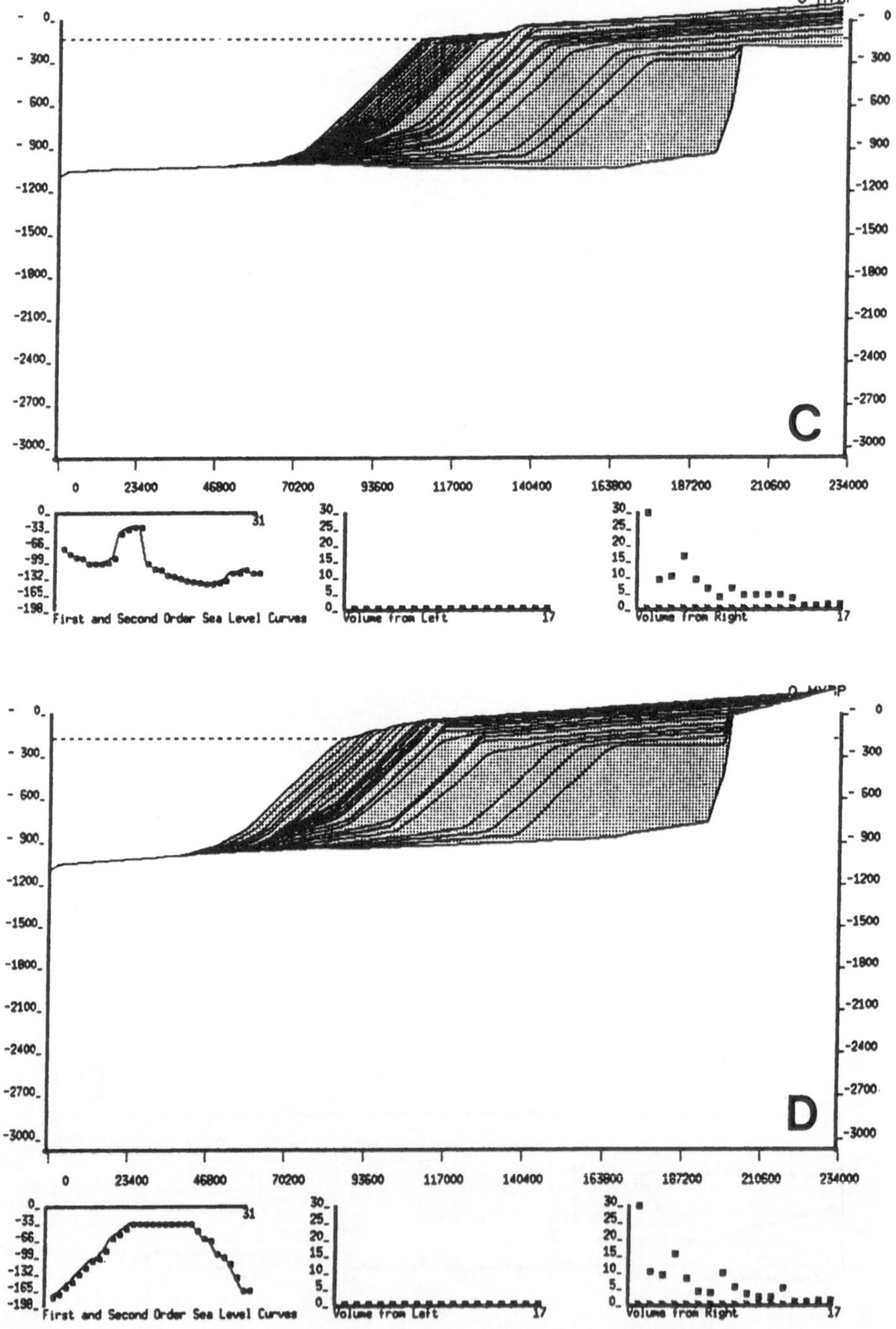

0 MYBP
C
0
23400
46800
70200
93600
117000
140400
163800
187200
210600
234000
First and Second Order Sea Level Curves
Volume from Left
Volume from Right
0 MYBP
D
First and Second Order Sea Level Curves
Volume from Left
Volume from Right

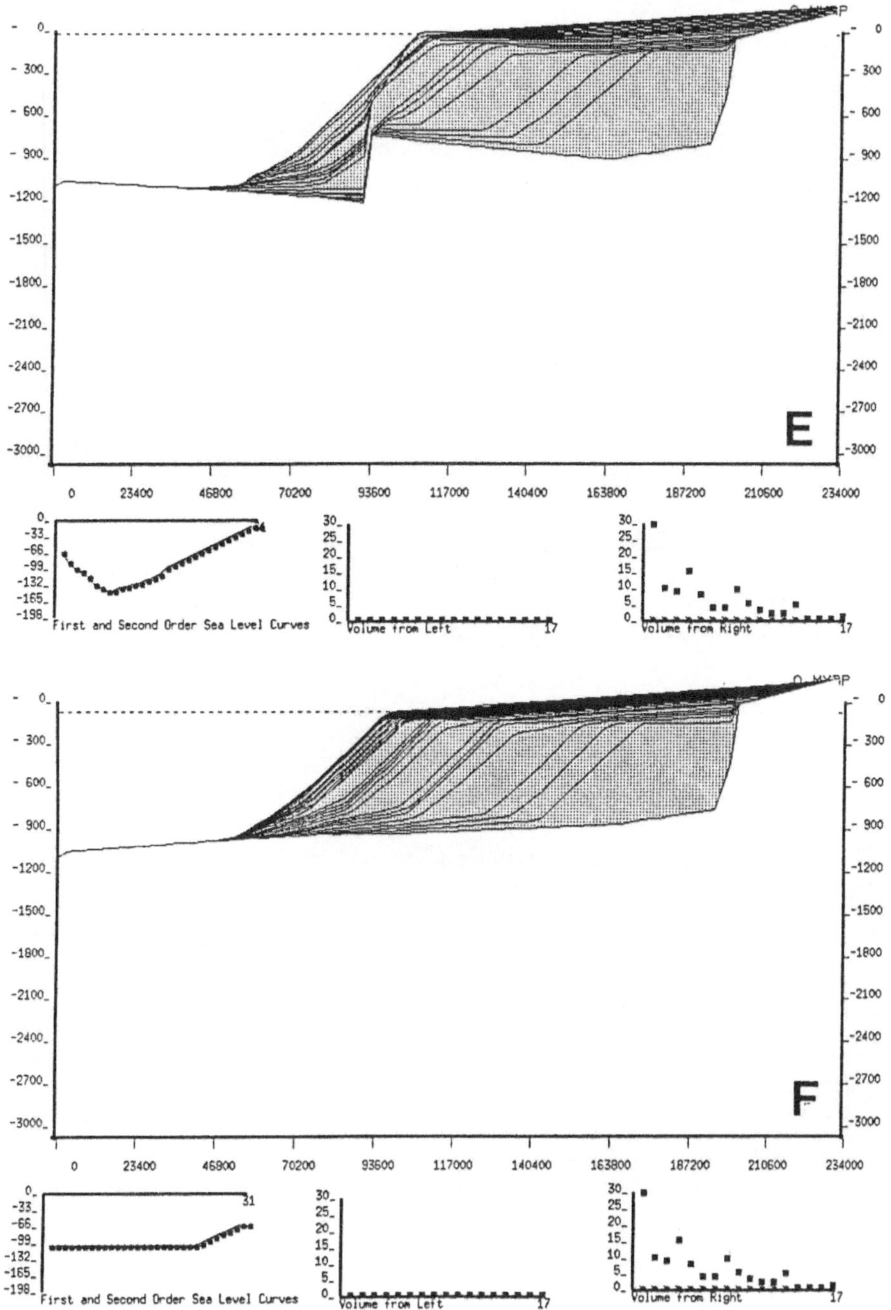

E
F
0
23400
46800
70200
93600
117000
140400
163800
187200
210600
234000
First and Second Order Sea Level Curves
Volume from Left
Volume from Right

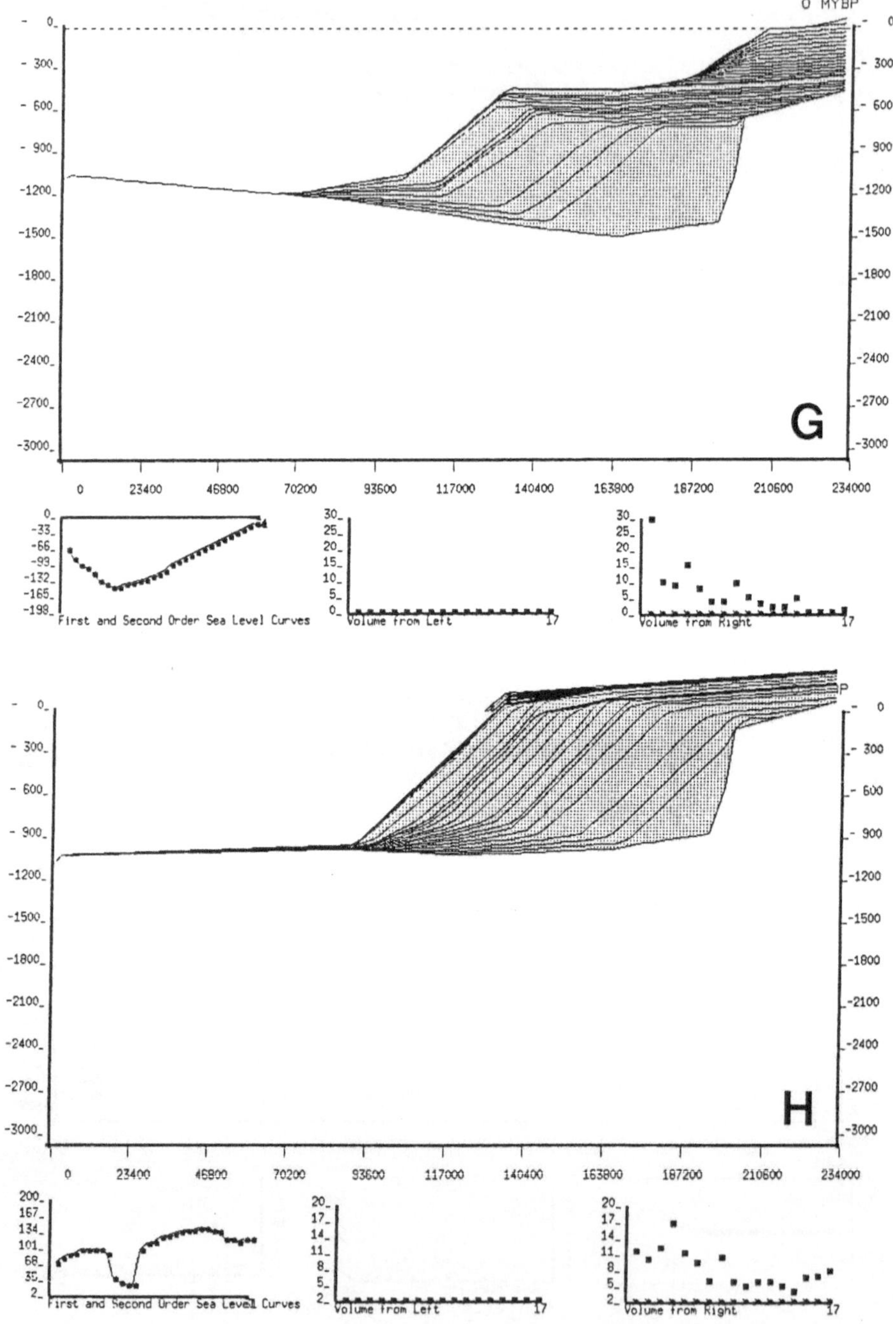

0 MYBP
G
H
First and Second Order Sea Level Curves
Volume from Left
Volume from Right

Commonly, the sedimentary geometry is climbing. A descending pattern occurs only in one graphical simulation (Fig.92C). According to the terminology by Mitchum,et al.(1977) the clinoform geometries are complex-oblique and oblique.

The aggradation portion is significant only in the simulation of Fig.92A,D where the relative sea-level rise resulted from the combined effects of subsidence and a rapid sea-level rise.

Clinoforms thin-out in a basinward direction; they thicken in the case of Fig.92A: this may be explained by a by-passing of sediments consequent to a transition from offlap to toplap upper boundary relationships.

An example of convergent, descending pattern is the Pleistocene occurring in the continental shelf of Israel (Ryan and Cita,1978; Fig.93).The upper toplap boundary is slightly inclined towards the basin and the DP and PBP converge towards the same direction. A comparison with the simulation of Fig.92D indicates that Pliocene strata were laid down during a rapid sea-level fall, an absence of subsidence and a decreasing sedimentation rate.

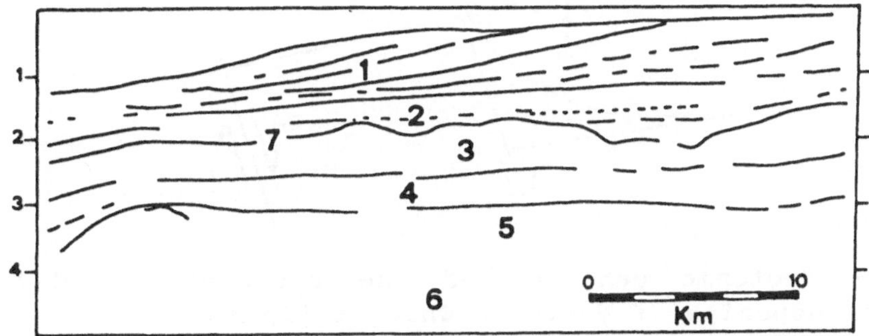

Fig.93 - Descending-convergent pattern in the shelf of Israel (Ryan and Cita,1978).

An absence of tectonic subsidence such as down-to-basin faulting is in accordance with the interpretation given by Ryan and Cita (1978).

An example of convergent, climbing pattern occurs in the Devonian - Mississipian from Ohio (Broadhead,et al.,1982; Fig.94).It is comparable to the graphic simulation of Fig.92E where progradation took place onto two faulted blocks subjected to rotations along listric faults. Each block underwent a rotation with a progressive landward increase in the rate of

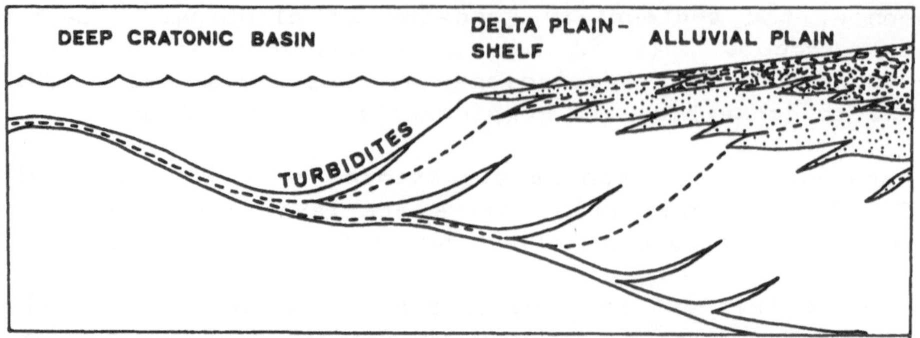

Fig.94 - Basin model of the Devonian-Mississipian turbidites of northern Ohio (Broadhead, et al.,1982).

subsidence and a relative uplift at the basinward edge (Fig.95). The similarity to the simulation of Fig.92E suggests a deposition under a sea-level fall,a decreasing sedimentation rate, and a rotational subsidence of the type sketched in Fig.95. It is possible that the deposition of turbidite into

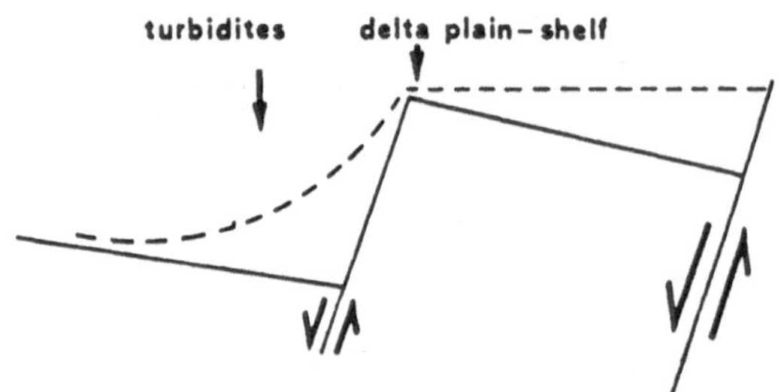

Fig.95 - Tectonic behavior of the basement underlying the turbidite deposits of Northern Ohio of Fig.94.

the deep cratonic basin was triggered by a relative uplift of the delta - plain shelf.

Another example of convergent-climbing pattern occurs in the 1500 m thick, Berriasian-Valanginian clastic wedge located at the Exmouth plateau, NW Australia (Fig.96).

It was studied by Exon and Willcox (1978), Exon,et al.(1982), Stakelberg,et al.(1980) and Erskine and Vail (1988) who interpreted the clastic wedge as a turbidite fan.The Exmouth plateau belongs to a passive margin which underwent rifting

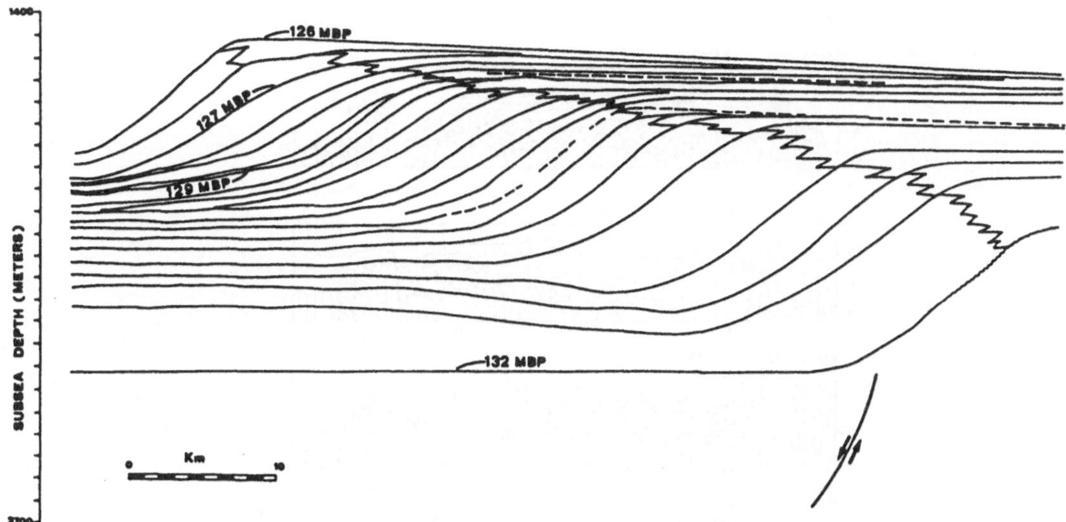

Fig.96 - Clastic wedge of the Exmouth Plateau (from Erskine & Vail,1988).

and block-faulting,following the break-up of Pangaea. According to Exon,et al.(1982) the wedge formed as a result of a gentle, northward, regional tilt from the early Neocomian to the mid-Neocomian, onto a series of faulted blocks. The phase of downfaulting was followed from 125 to 120 m.y. by a phase of thermal uplift, related to movements along transform faults. Fig.97, from Exon and Willcox (1978) shows the evolution of the prograding wedge.

The simulation of Fig 92H is a schematic reproduction of the clastic wedge shown in Fig.96. The wedge formed during a slow sea-level rise and under a decreasing sedimentation rate.

The tectonic control results to have been rather complex: the deposition was affected by basculatory tectonics acting as episodic subsidence and uplift at different locations (Km 3,1 and 163: Fig.92H) and times (subsidence: 132 + 126; 126 + 127 m.y.; uplift from 129 to 128 m.y.). The oblique pattern of clinoforms visible in Fig.92H and 96 results to have been produced by a combination of tectonic uplift (rates: 10 m/200.000 yrs.) and a slow sea-level rise.

Divergent pattern

This pattern displays an upward incraese in length due to the

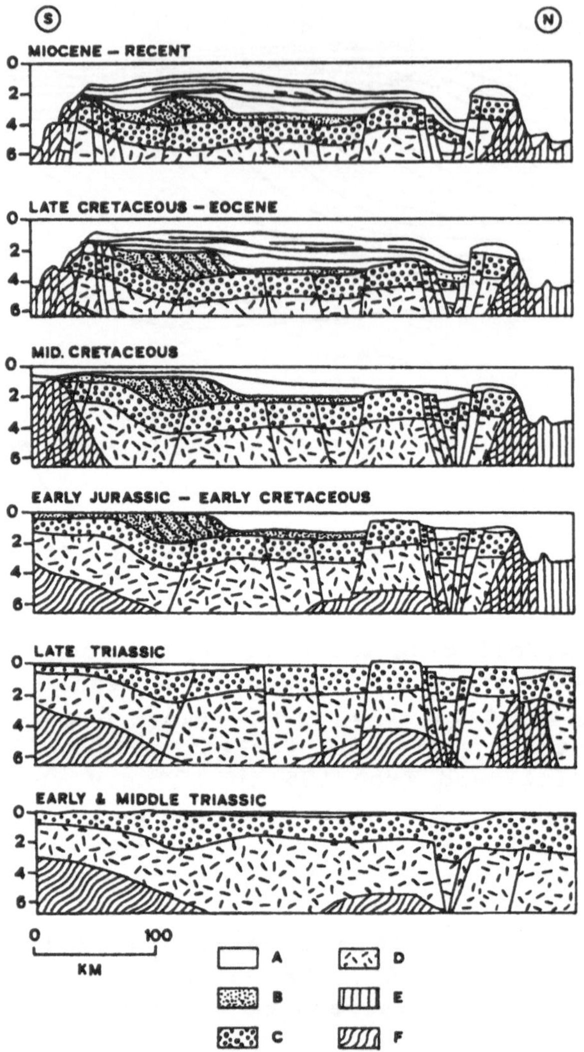

Fig.97 - Schematic cross section showing the evolution of the Exmouth Plateau (after Exon & Wilcox,1978). A: Mid-Cretaceous muds and silt; B:Early Jurassic to Early Cretaceous sands, silts and muds. C: Triassic deltaic fluvial muds and sands. D: Paleozoic sediments. E: oceanic basement; F: continental basement.

basinward divergence of the DP and PBP (Doglioni and Bosellini,1989). Divergent patterns were obtained by the following combinations of variables.

1) Slow sea - level fall - landward tilting - constant sedimentation rate (Fig.98A).
2) Sea-level fall - seaward tilting - constant sedimentation rate (Fig.98B).

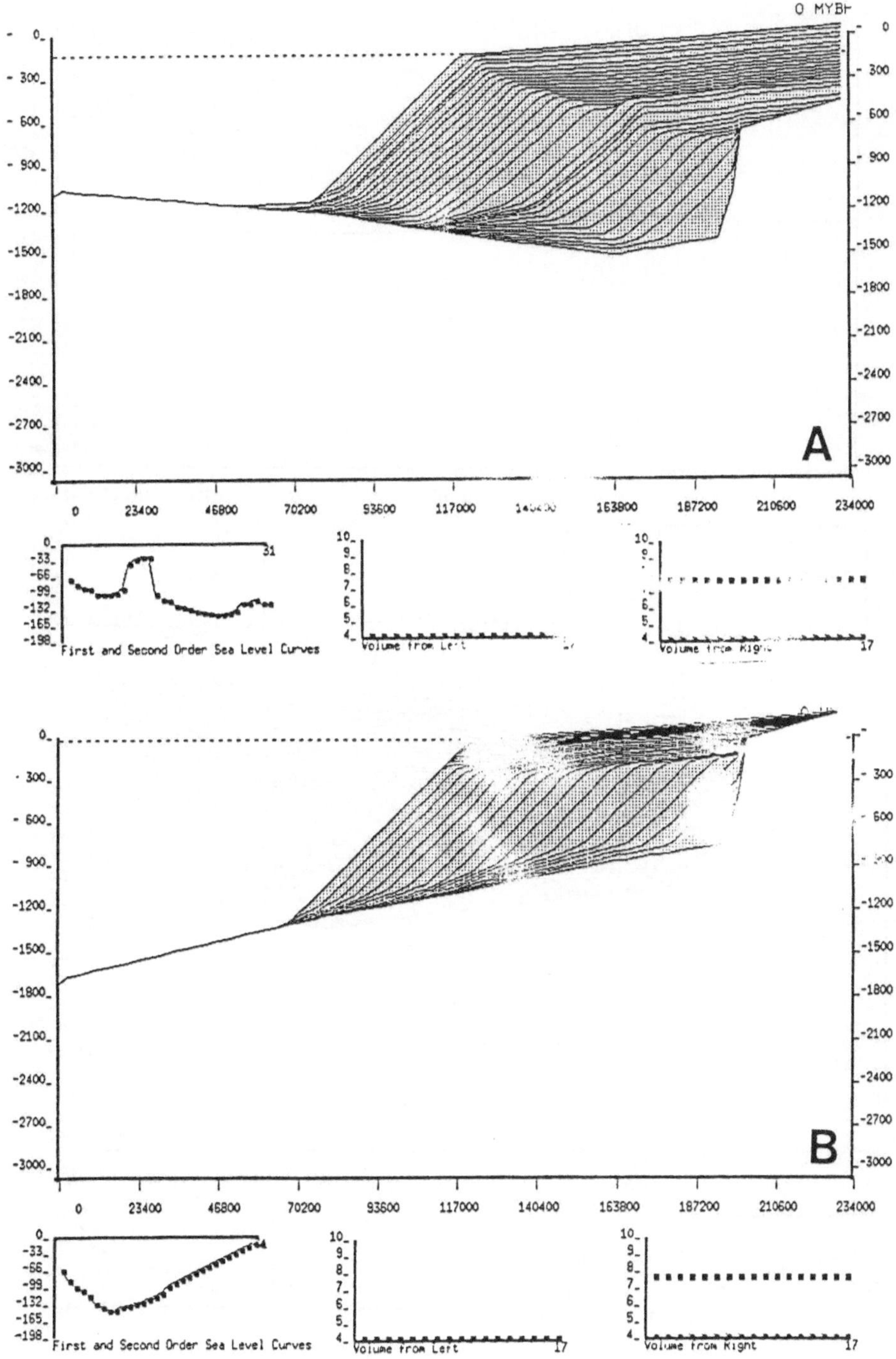

Fig.98 - Examples of computer simulations of divergent patterns.

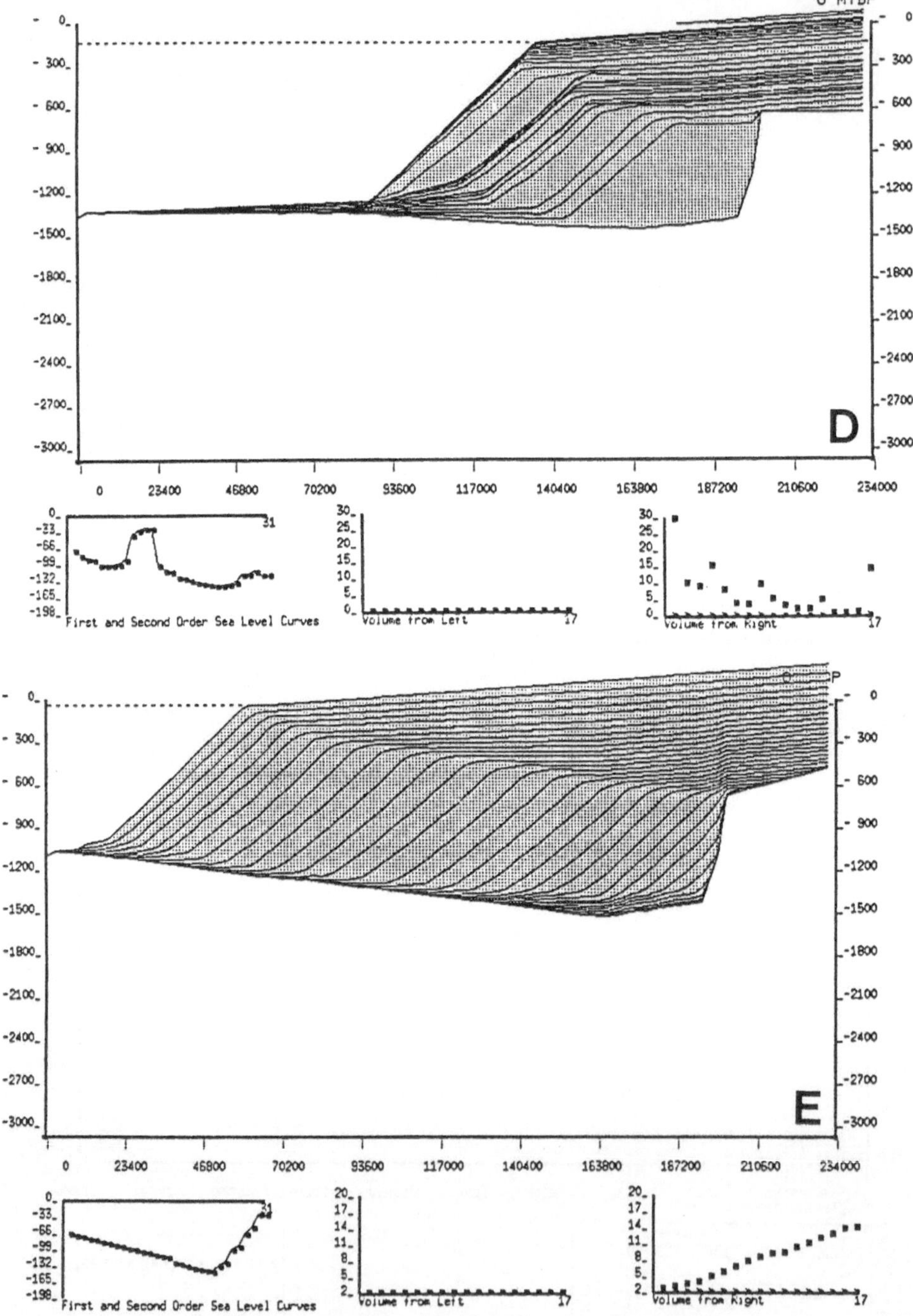
O MYBP
D
0
23400
46800
70200
93600
117000
140400
163800
187200
210600
234000
First and Second Order Sea Level Curves
Volume from Left
Volume from Right
E
First and Second Order Sea Level Curves
Volume from Left
Volume from Right

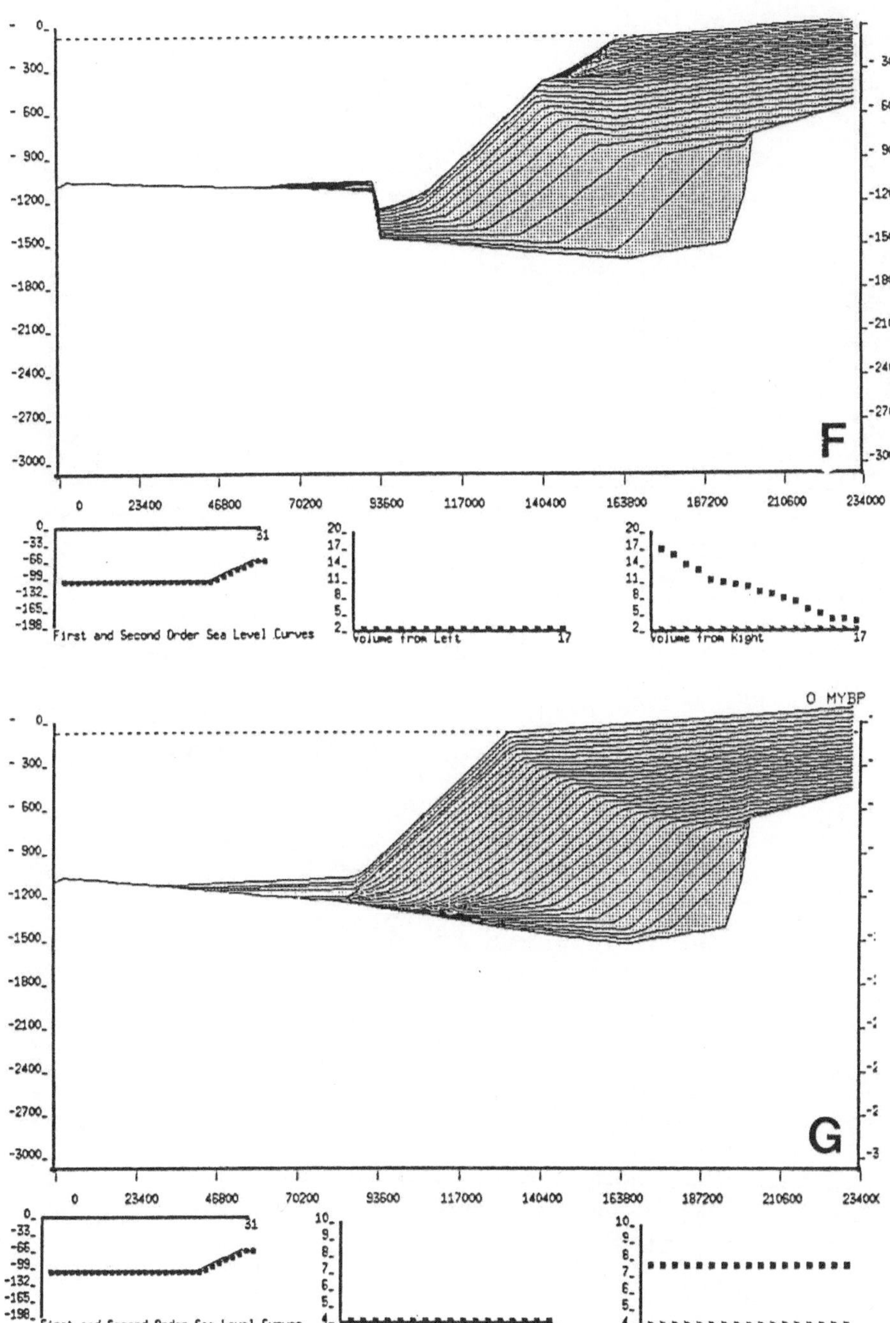
F
G
0 MYBP
First and Second Order Sea Level Curves
Volume from Left
Volume from Right
0
23400
46800
70200
93600
117000
140400
163800
187200
210600
234000

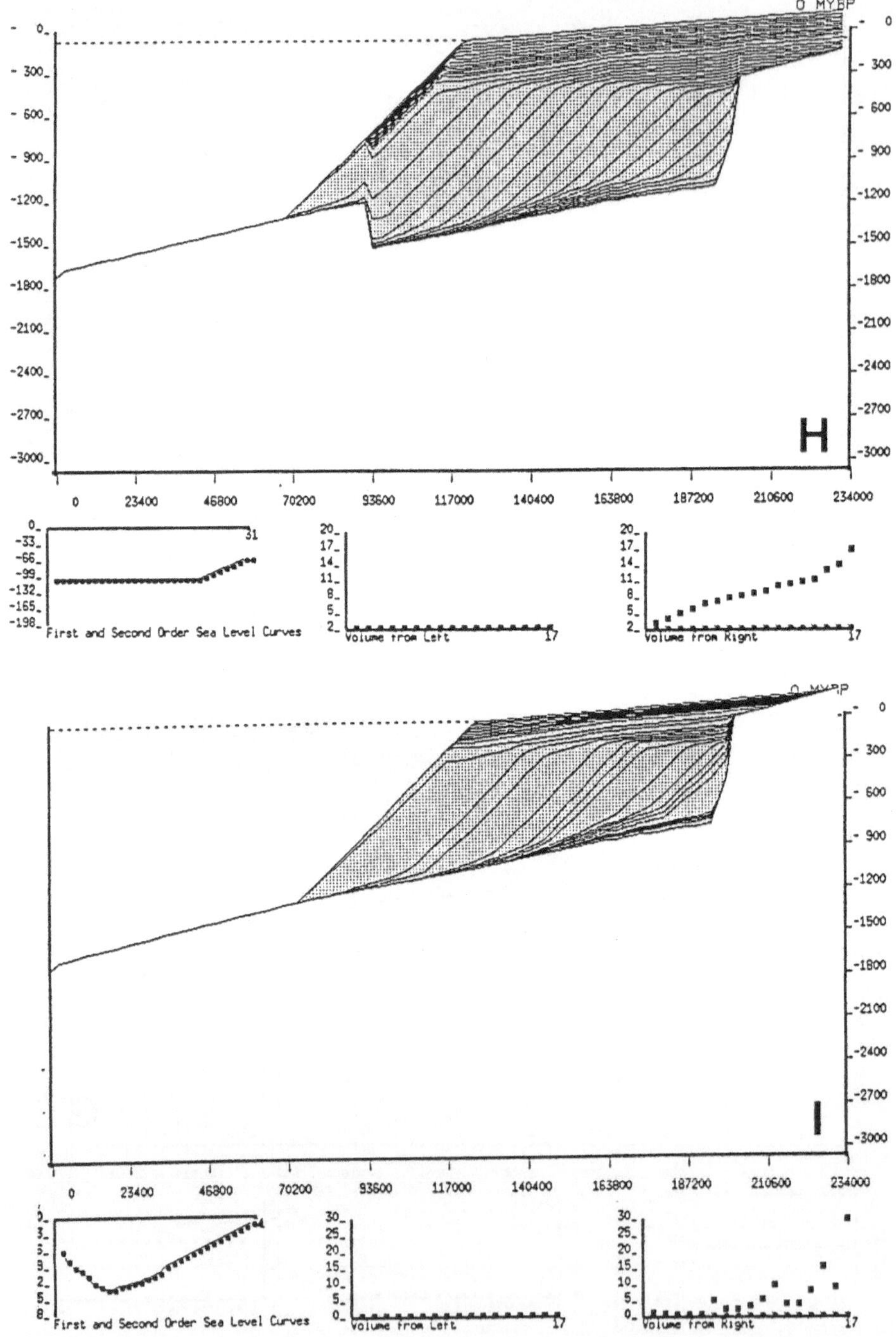
0 MYBP
H
First and Second Order Sea Level Curves
Volume from Left
Volume from Right
0 MYBP
I
First and Second Order Sea Level Curves
Volume from Left
Volume from Right

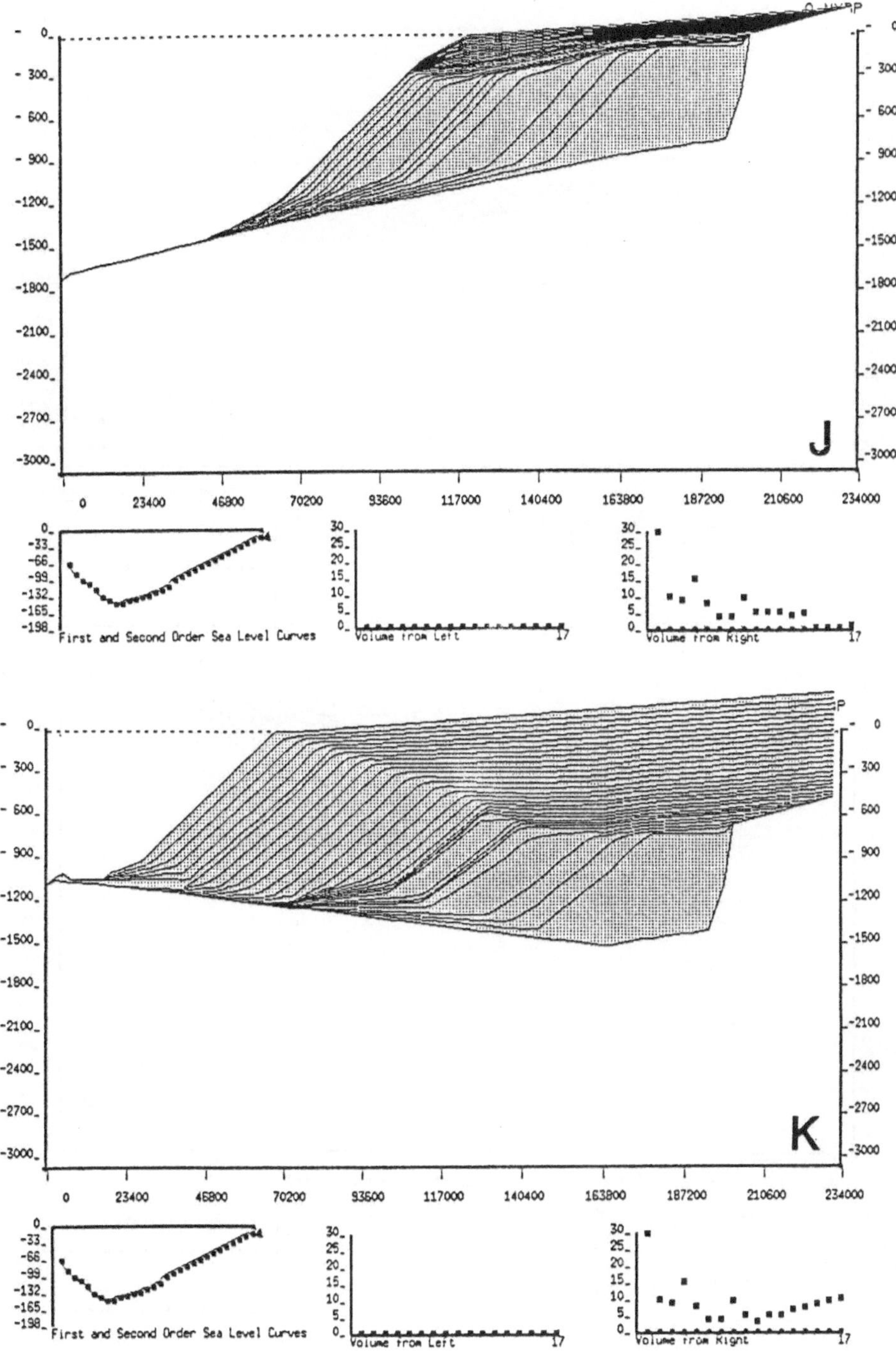
J
First and Second Order Sea Level Curves
Volume from Left
Volume from Right
K
First and Second Order Sea Level Curves
Volume from Left
Volume from Right

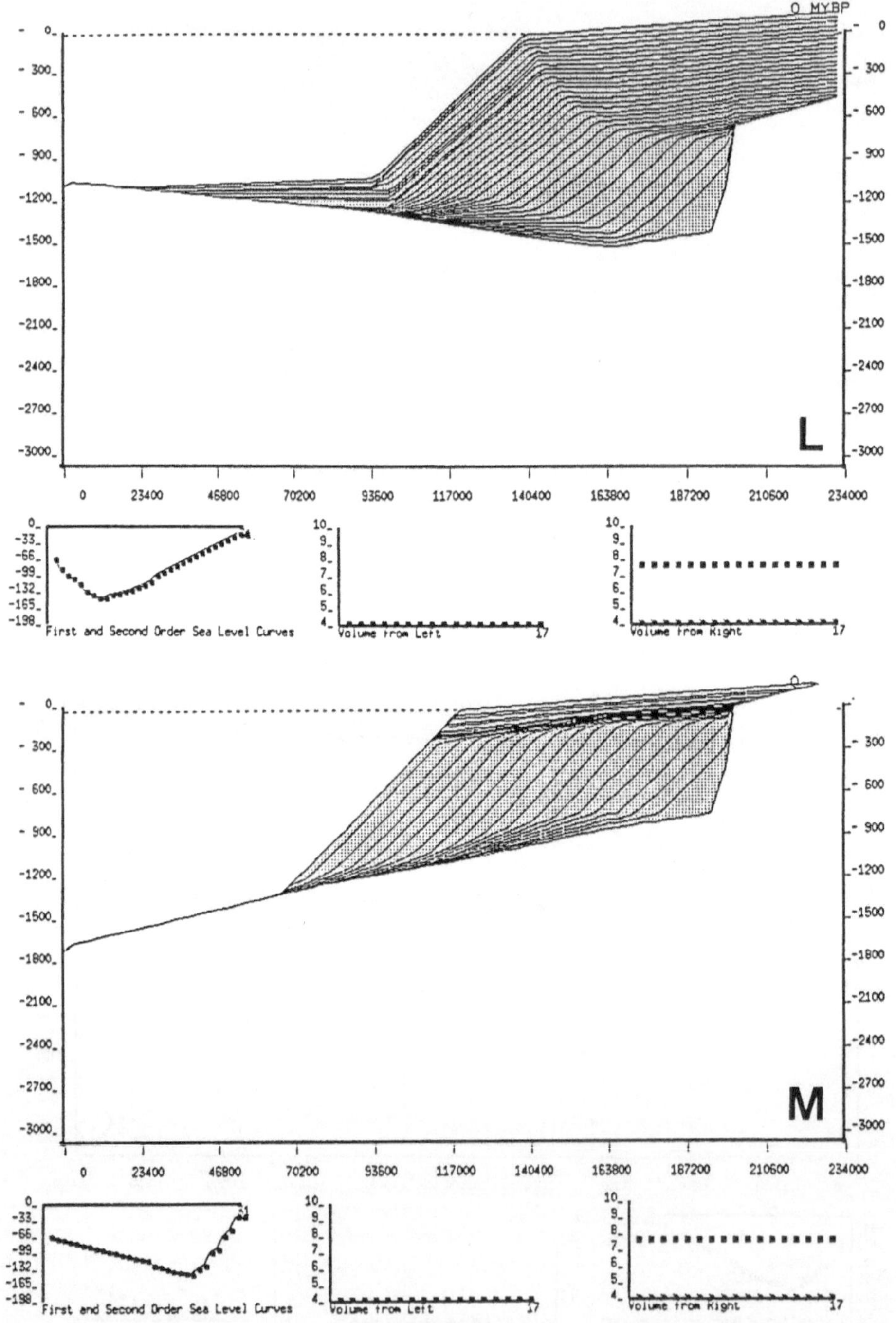

0 MYBP
L
0 23400 46800 70200 93600 117000 140400 163800 187200 210600 234000
0 -300 -600 -900 -1200 -1500 -1800 -2100 -2400 -2700 -3000
First and Second Order Sea Level Curves
Volume from Left
Volume from Right
M
0 23400 46800 70200 93600 117000 140400 163800 187200 210600 234000
First and Second Order Sea Level Curves
Volume from Left
Volume from Right

3) Sea - level fall - landward tilting - decreasing sedimentation rate (Fig.98D).
4) Sea-level fall - landward tilting - increasing rate of sedimentation (Fig.98E).
5) Sea-level stillstand - landward tilting - decreasing rate of sedimentation (Fig.98F).
6) Sea-level stillstand - landward tilting - constant rate of sedimentation (Fig.98G).
7) Sea-level stillstand - landward tilting - increasing rate of sedimentation (Fig.98H).
8) Sea-level rise - seaward tilting - decreasing rate of sedimentation (Fig.98J).
9) Sea-level rise - landward tilting - increasing rate of sedimentation (Fig.98K).
10) Sea-level rise - seaward tilting - increasing rate of sedimentation (Fig.98I).
11) Sea-level rise - landward tilting - constant rate of sedimentation (Fig.98L).
12) Sea-level rise - seaward tilting - constant sedimentation rate (Fig.98M).

This pattern is always formed by a sea-level rise and seaward tilting or a stillstand and a landward tilting.
Differential subsidence appears in all of the above models.
Most of the models were obtained by means of a constant or increasing sedimentation rate; models produced by a decreasing sedimentation rate give way to graphic simulations with very small angles of divergence between the DP and the PBP.
The formation of this pattern appears to be independent of the type of sea-level change.

The DP may be either descending or climbing.

The first case is produced by a constant (less commonly increasing) sedimentation rate and a fall or stillstand of the sea-level.
A climbing DP results mainly from 1) a decreasing sedimentation rate, a landward tilting and a stillstand, rise, or fall of the sea-level, and 2) a decreasing sedimentation rate, a sea-level rise and a seaward tilting.

The PBP may be climbing. The steepness of the PBP is a function of the relative sea-level rise.Offlap upper boundary relationships are produced by a landward tilting; toplap relationships are produced by a seaward subsidence.

The aggraded portion of the clastic wedge is also a function of the steepness of the PBP.

The divergent pattern may be climbing or descending.

The first one is the most common. Climbing progradation produces sigmoidal patterns (Mitchum,et al.,1977).
A climbing progradation results from 1) a landward tilting; 2) a seaward tilting accompanied by a sea-level stillstand and an increase in sedimentation rate,and 3) a sea-level rise and an increasing sedimentation rate.

A descending progradation gives way to oblique patterns (Mitchum,et al.,1977).
A descending progradation results from two combinations of variables: 1) sea-level rise - seaward tilting - decreasing sedimentation rate (Fig.98I,J); and 2) sea-level fall - landward tilting - constant sedimentation rate (Fig.98A).

The great number of examples of divergent patterns found in the literature reflects the wide range of combinations between eustasy, sedimentation rate and subsidence that can produce this pattern.

An example of descending divergent pattern occurs within the 2500 m thick Miocene- Pliocene wedge that laps onto the continental shelf of Sumatra (Fig.99).
It fits the simulation of Fig.98B which was obtained by means of a constant sedimentation rate, a sea-level fall and a seaward tilting.

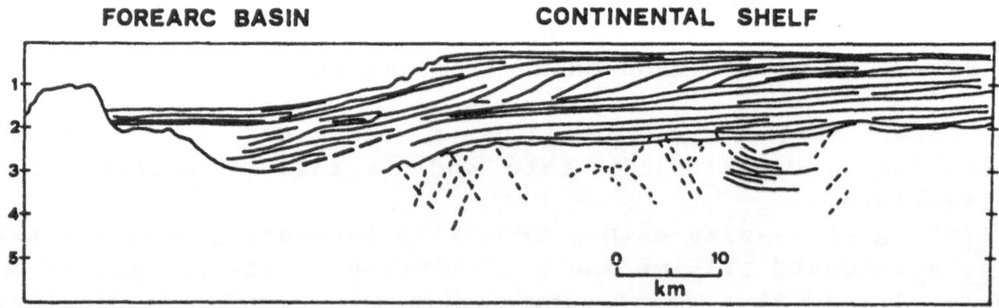

Fig.99 - Descending divergent pattern in the Sumatra continental shelf (from Beaudry and Moore,1985).

An example of climbing ,divergent pattern occurs in the Lower Cretaceous of NE Alaska, studied by Molenaar (1983; Fig.100). It is the product of a sea-level rise, a constant rate of

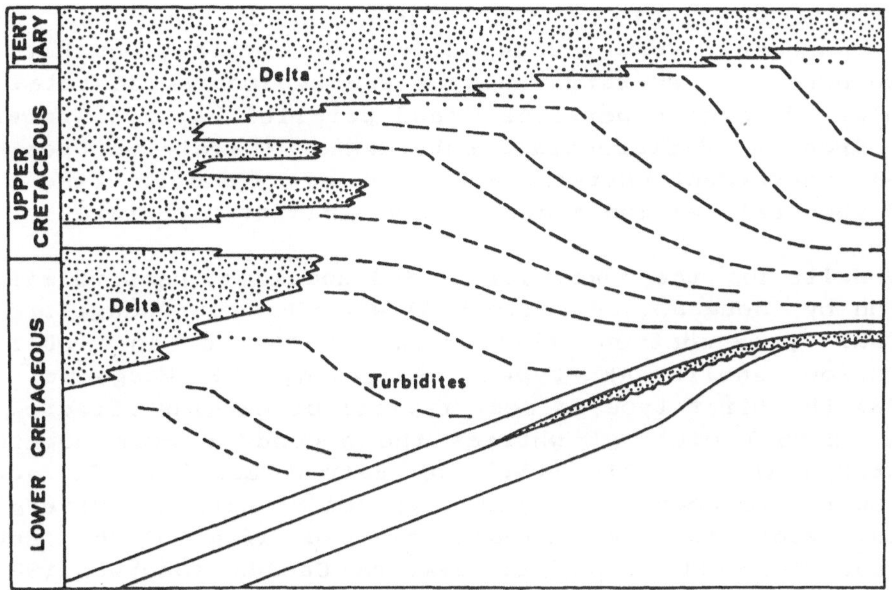

Fig.100 - Cross section showing Cretaceous - Lower Tertiary of NE Alasks (from Molenaar,1983).

sedimentation and a landward tilting (Fig. 98L).

Concluding remarks

The knowledge of the eustatic curve, tectonic history and sedimentation rate for a given area may lead to predictive geometries of clastic wedges.In fact, the forward system of modelling employed here generates a number of two-dimensional representations of sediment geometries that are the product of eustasy,sedimentation rate and type and direction of tectonic movements.Therefore, it becomes possible to separate for each graphical output the eustatic and tectonic component of the relative sea-level curve and to consider the nature of some features such as sequence boundaries, offlap, toplap relationships, thickness of clinoforms, downward shift of the coastal onlap,an so on.I did not describe these featuures here;nonetheless, most of features vary approximately in the same way as the corresponding features of prograding carbonate platforms (cf. Bosellini,1984).Some differences may be attributed to the greater lateral transport occurring in siliciclastic than in carbonate settings.

An interesting aspect of the simulation is that the same results can result from a wide variety of combinations between tectonic behavior, eustatic changes and sedimentation rate. In particular, divergent patterns (and parallel patterns) mostly result from a differential subsidence (mainly a seaward tilting); convergent patterns are produced by a stillstand or a fall of the sealevel and a decreasing sedimentation rate.

These results fit the observations and models of passive margin evolution by Mougenot, et al.(1983) who showed a change from a 'sigmoidal progradation shelfbreak type' to an 'oblique progradation shelfbreak type'. According to Mougenot, et al.(1983) the first type is most typical of margins affected by a rapid thermal cooling, whereas the second is more commonly associated with slightly subsiding mature margins. It seems that Lower Cretaceous margins typically display divergent patterns, such as the Exmouth plateau (Fig.96),the Lower Cretaceous Mannville Group of Western Canada (Jackson,1984), and the Lower Cretaceous of Northeastern Canada (Molenaar,1983; Fig.100).

Slope carbonates (Cretaceous-Paleocene), Gargano massif

Introduction

The stratigraphic interval in the study area represents a transitional zone between Jurassic - Cretaceous shallow-water carbonates,and Cretaceous to Paleocene basinal-hemipelagic deposits (Fig.101).

The carbonate stratigraphic succession of the Gargano peninsula (S.Italy) has been the object of contrasting interpretations. Based on the abundance of shallow-water rudist fragments,it was interpreted as a shallow-water reef by AGIP geologists and the University of Bologna working in the area (Cremononi,et al.,1971; Mattavelli and Pavan,1965; Pavan and Pirini,1965; Martinis and Pavan,1965; Pattera, 1967).

Recently, megabreccias which mark the junction between platform and the basin in the study area were interpreted as generated by a tectonic collapse of the Cretaceous platform along synsedimentary faults (Masse and Borgomano,1987; Sinni and Masse,1986).

This interpretation was contrasted by the results of the field mapping made by Ferioli (1988), which revealed that the traces of synsedimentary faults are in reality submarine slide scars responsible for the formation of megabreccias; Bosellini and Ferioli (1988) and Bosellini (1989) advanced a eustatic interpretation: the formation of the slide scars was correlated with well pronounced lowstands of the sea-level (Haq,et al.,1987),on the assumption that their shedding into the basin could be related to eustatic sealevel falls.

These discordances allowed for the subdivision of the complicated tangle of stratigraphic formations into three depositional sequences (Fig.102) whose description is the object of this section.

This work is based on detailed observations of about 1800 m of spectacular exposures (9 stratigraphic sections) cropping out mainly along road cuts and natural cliff exposures.

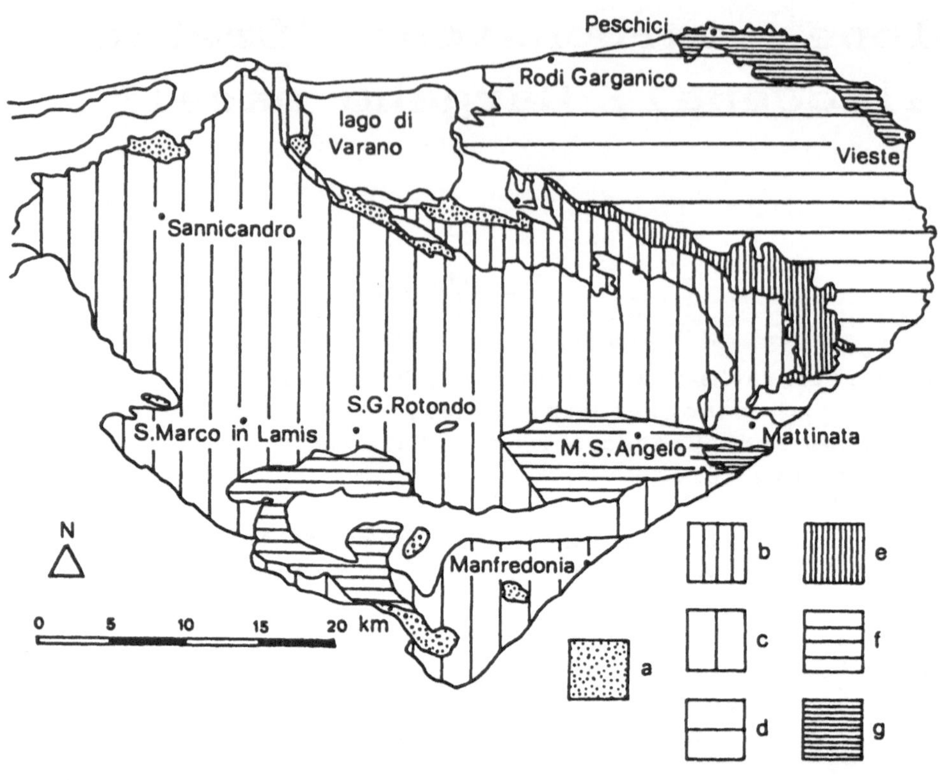

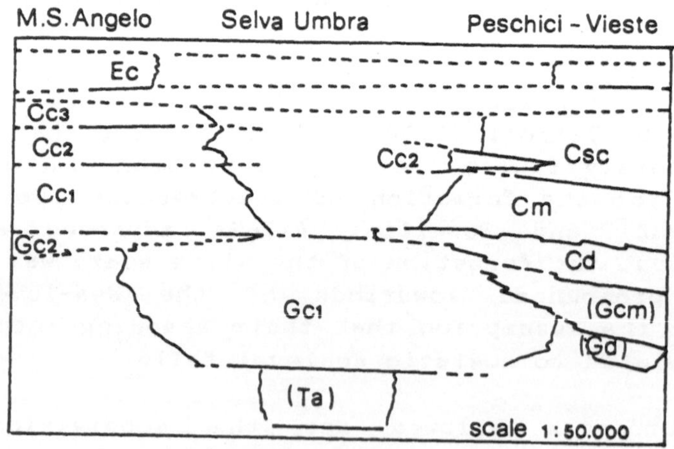

Fig.101 - Location of the study area (Martinis and Pavan,1967) and local stratigraphy (Cremonini, et al.,1971) evidencing for a divergent pattern with a hinge located at M.S.Angelo area (Cc2,Cc1,Cc3).

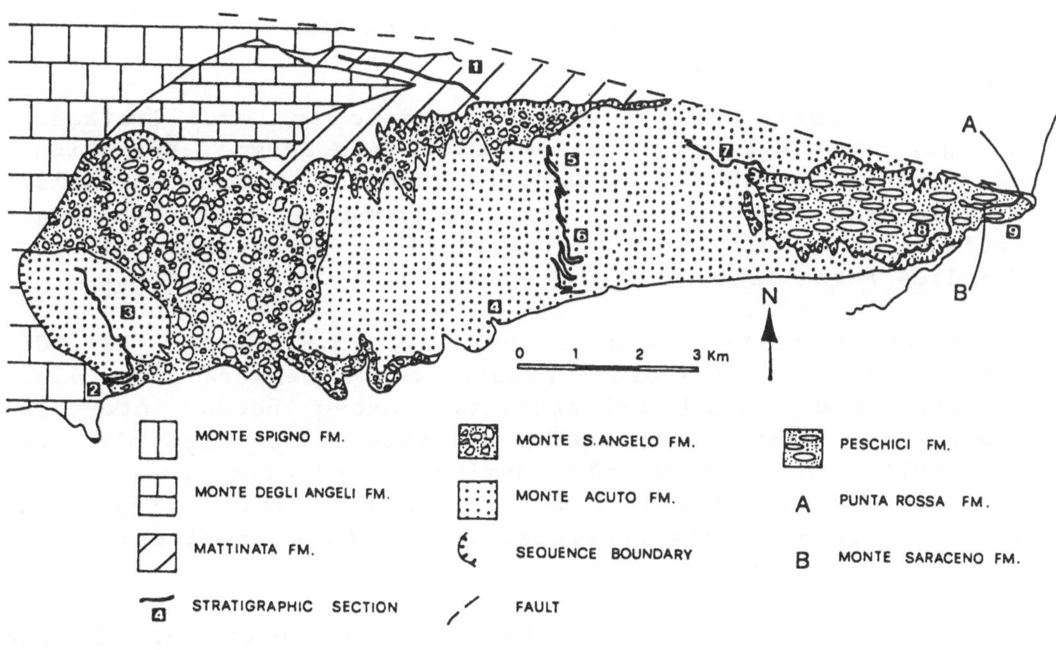

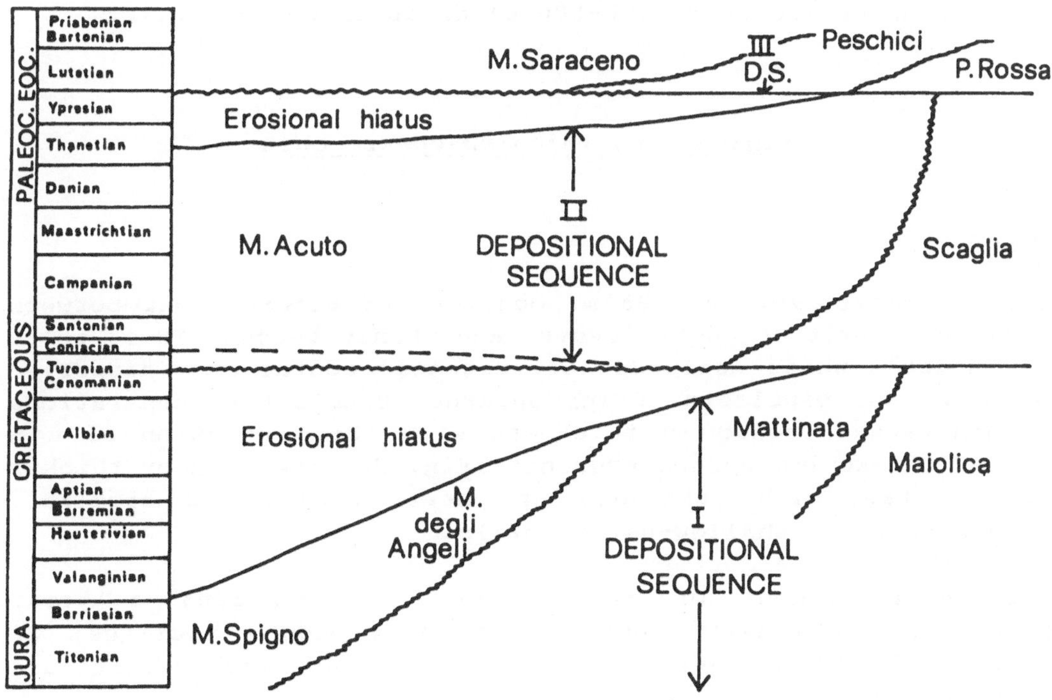

Fig.102 - Geological map ,location of stratigraphic sections and subdivision into depositional sequences (Ferioli,1986-1987;Ferioli and Bosellini,1988).

First depositional sequence

The first depositional sequence, Jurassic to Lower Cretaceous in age, consists of shallow-water ('M.Spigno Formation': Cremonini,et al.,1971; 'Calcari oolitici di Coppa Guardiola': Martinis and Pavan,1967), slope carbonates ('Madonna degli Angeli' and 'Mattinata' Formations) and basinal micrites (Maiolica Formation).

The slope deposits, located north of M.S.Angelo (Fig.101: section #1), contain a great proportion of megabreccias, slump deposits, and graded calcarenites (a-b-c Bouma intervals) intercalated with hemipelagic micrites. The age (Pattera, 1966-1967) ranges from the Aptian (*Bacinella irregularis*; *Cuneolina laurentis*;*Orbitolina sp.*) to the Albian (*Ticinella*; *Globigerinoides*) in the uppermost 170 - 180 m of stratigraphic exposures.

The slope deposits are divisible into a lower fining- and thinning-upward unit (0 - 300 m) and an upper thickening- and coarsening-upward unit (300-500 m),as is shown in Fig.103.

Fining- and thinning-upward unit

Description

The lowermost part (0 - 85 m) consists of alternations between siliceous marls or chert layers, and thinly bedded (10-15 to 50 cm thick) white colored wackestones containing dispersed bivalves and bioclasts. Going upwards, bioclast concentrations within beds increase in thickness with the production of a m scale, thickening-upward sequence (Fig.104) overlain by thicker slumped beds with fragments of chert nodules and bioclast, lithoclasts and shallow-water corals.

The upper segment of this lower unit contains a great proportion of gravity deposits: 1) thin-bedded micrites; 2) graded calcarenite beds; 3) graded - ungraded megabreccias; and 4) slump deposits.

These lithofacies are organized into a general fining- and thinning-upward trend: megabreccias are preponderant at the base; gravity deposits vertically decrease in thickness, grain-size and frequence and eventually pass to thin-bedded

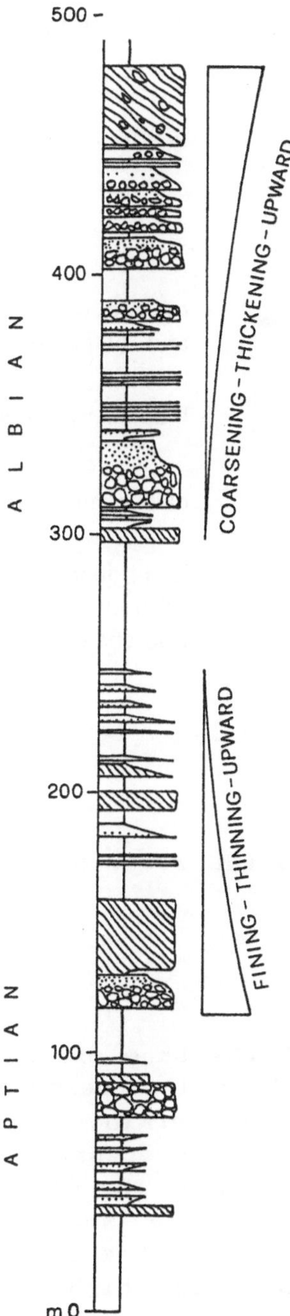

Fig.103 - First depositional sequence.Discordances and slumped breccias are common in the lower unit. Conversely, the upper unit contains a great proportion of 1 to 25 m thick graded megabreccias and pebbly mudstones whose vertical increases results in a coarsening and thickening- upward trend.

pelagic micrites organized into packages separated by intraformational truncation surfaces.

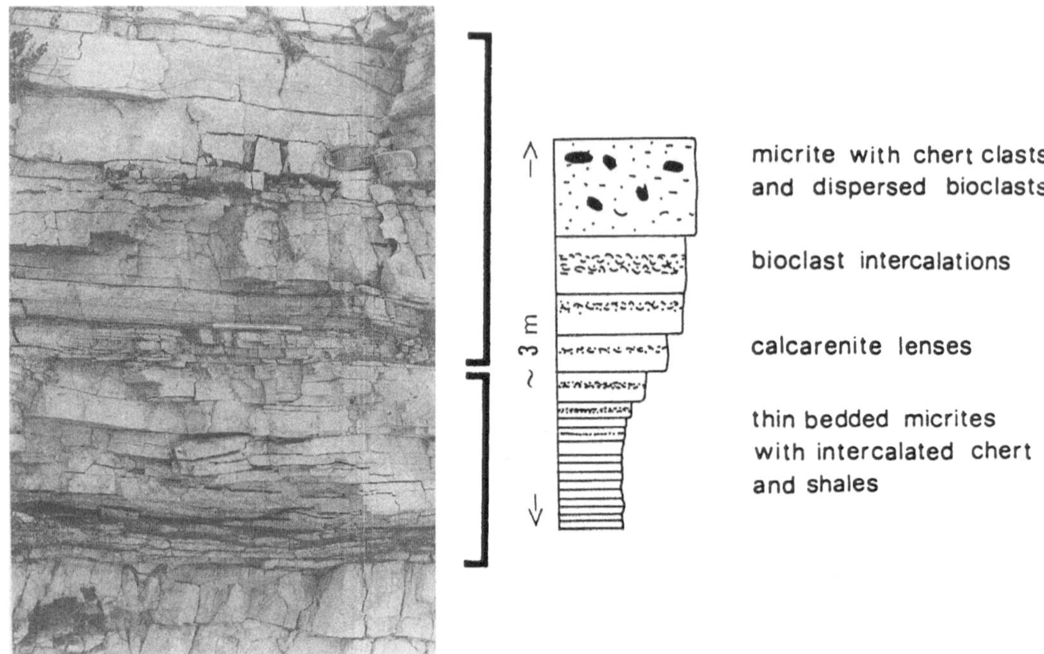

Fig.104 - Thickening-upward sequences in the lower part of the first depositional sequence.

Interpretation

Thickening-upward sequences occurring at the base of the lower unit, similar to the thickening-upward cycles described by Aigner (1985), are interpreted as the basal terminations of an apron or bioclast wedge of sediment prograding downslope, on a gently sloping surface.The coarse grain-size of the thickening-upward sequences suggest a proximity to the shallow-water platform. Slumped, thick micrite beds on top of the thickening-upward sequence may have originated from high-frequency sea-level fluctuations.

The fining-upward and thinning-upward trends displayed by the lower unit suggest a mechanism of vertical aggradation, rather than a progradation of the slope. The composition of the slope suggests a change in the slope character through geological time, from a ramp-type setting, to a depositional margin (Fig.105).

The abundance of slumped breccias at the top of the unit reflects a relatively unstable sedimentary interface which contrasts with undisturbed bedding styles of the basinal micrites occurring at the base of the section. Coeval basinal limestones (Maiolica Formation) in the Gargano peninsula,450 m thick, consist of white, well bedded micrite layers, 10 - 40 cm thick, with slumped horizons sandwiched between undisturbed beds. Neptunian dykes are also frequent.The Maiolica Formation was deposited on a steep slope subjected to synsedimentary tectonics and frequent seismic activity.
Therefore,it is possible that the change in declivity of the slope , rather than being a response to a eustatic change (the sea-level curve shows a stillstand in the Aptian: Haq,et al.,1987) was a consequence of synsedimentary downfaulting which produced a depression, infilled by a mechanism of vertical sediment aggradation (cf. Eberli,1987).

Coarsening- and thickening-upward unit

Description

The upper unit is composed of the following lithofacies: 1) hemipelagic micrites; 2) graded calcarenites and calcirudites; and 3) megabreccias.

The base of this unit consists of m-scale thickening-upward bioclast calcarenite cycles analogous to those occurring at the base of the underlying fining-upward and thinning-upward unit.These cycles and the hemipelagic micrites occur as interbeds within ten-m thick,graded megabreccias which become progressively thicker higher-up the section. The uppermost part is composed of pebbly mudstones including rounded clasts containing lithofacies indicative for an inner lagoonal environment.

Interpretation

The predominance of thick megabreccias towards the top suggests some corresponding change in slope character through the geological time. The upper unit marks a progradational stage of the platform which led to the development of a by-pass gullied slope margin (Fig.105).It was the response to a eustatic sea-level rise which took place in the Albian (Haq,et al.,1987).

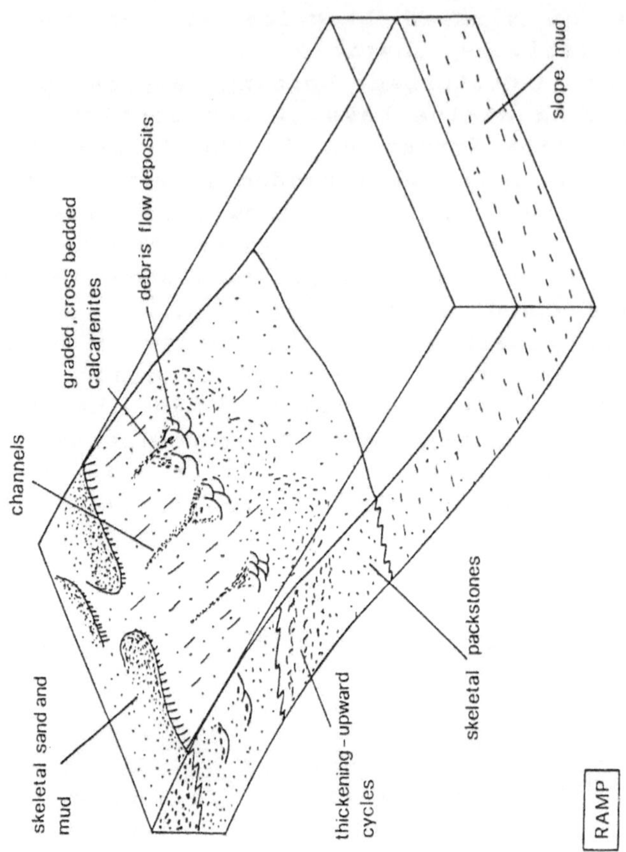
skeletal sand and mud
channels
graded, cross bedded calcarenites
debris flow deposits
slope mud
thickening - upward cycles
skeletal packstones
RAMP

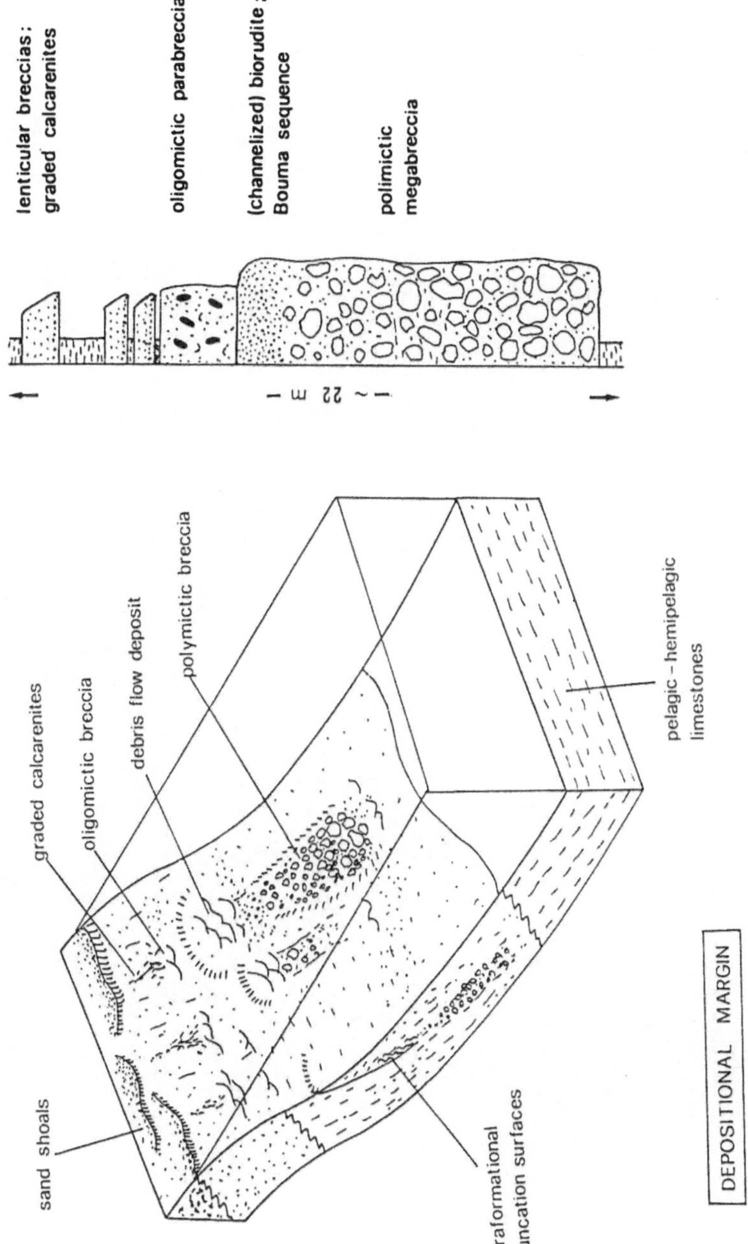

lenticular breccias;
graded calcarenites
oligomictic parabreccia
(channelized) biorudite;
Bouma sequence
polimictic
megabreccia
~ 22 m
sand shoals
graded calcarenites
oligomictic breccia
debris flow deposit
polymictic breccia
pelagic - hemipelagic
limestones
intraformational
truncation surfaces
DEPOSITIONAL MARGIN

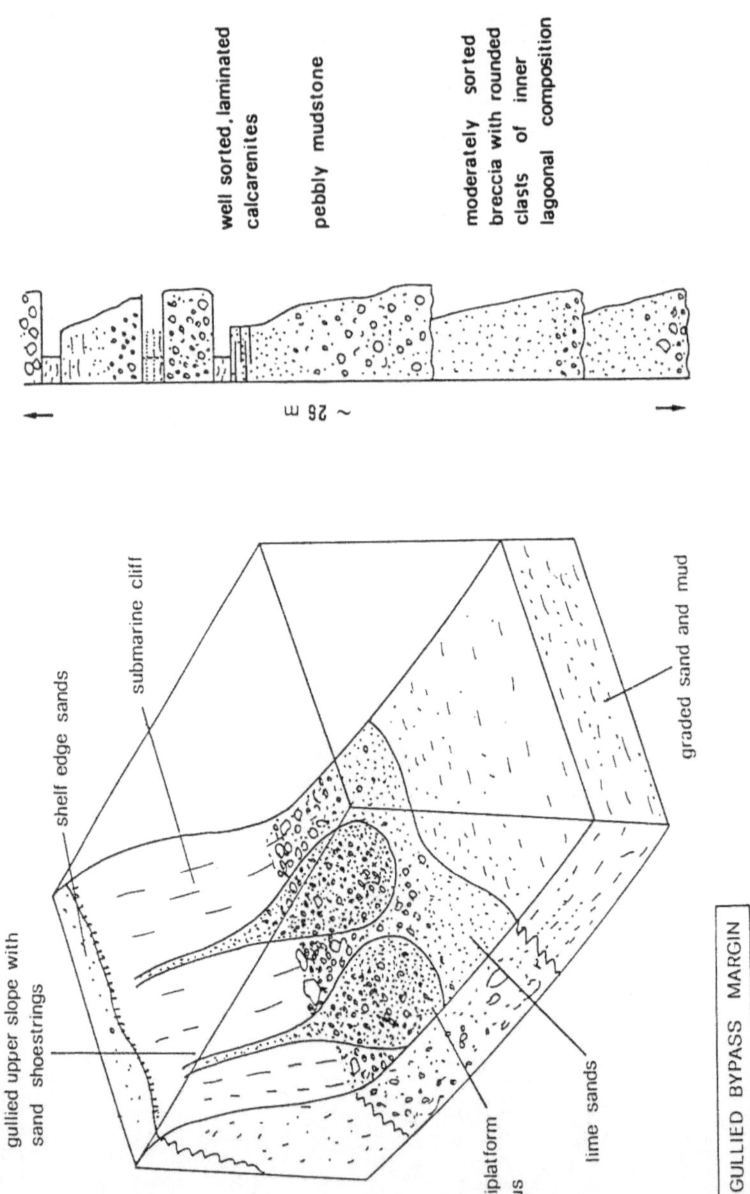

Fig.105 - Phases of evolution of the first depositional sequence.

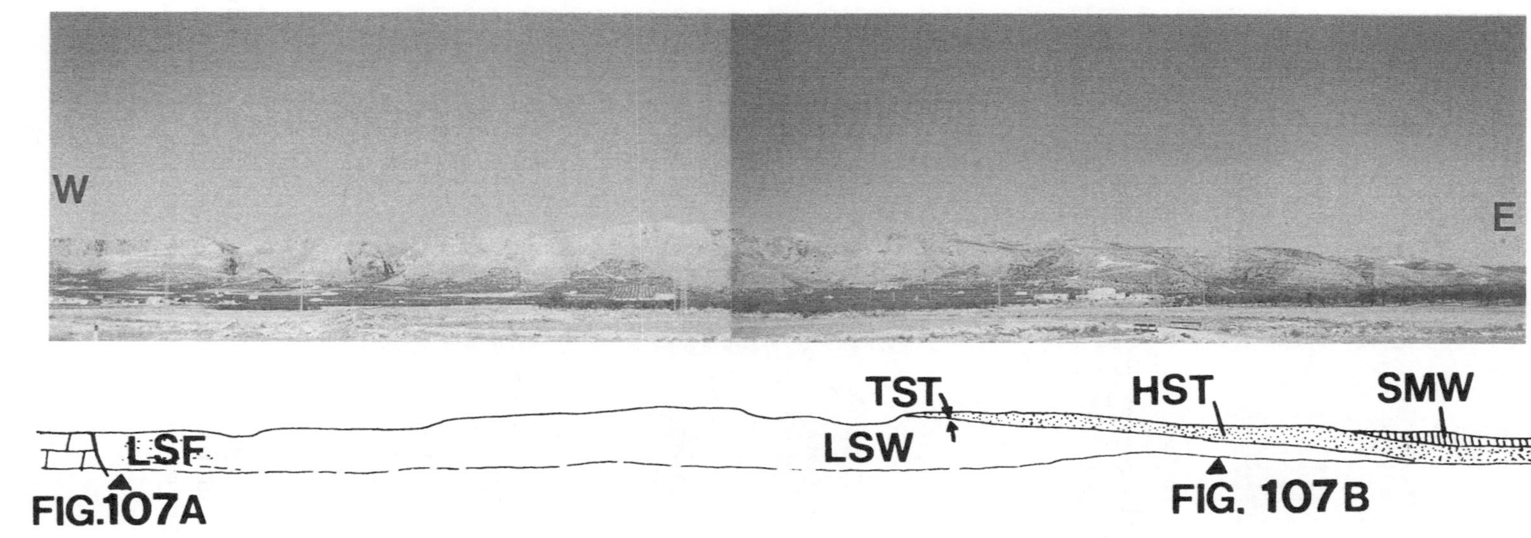

Fig.106 - Panoramic view of the second depositional sequence.

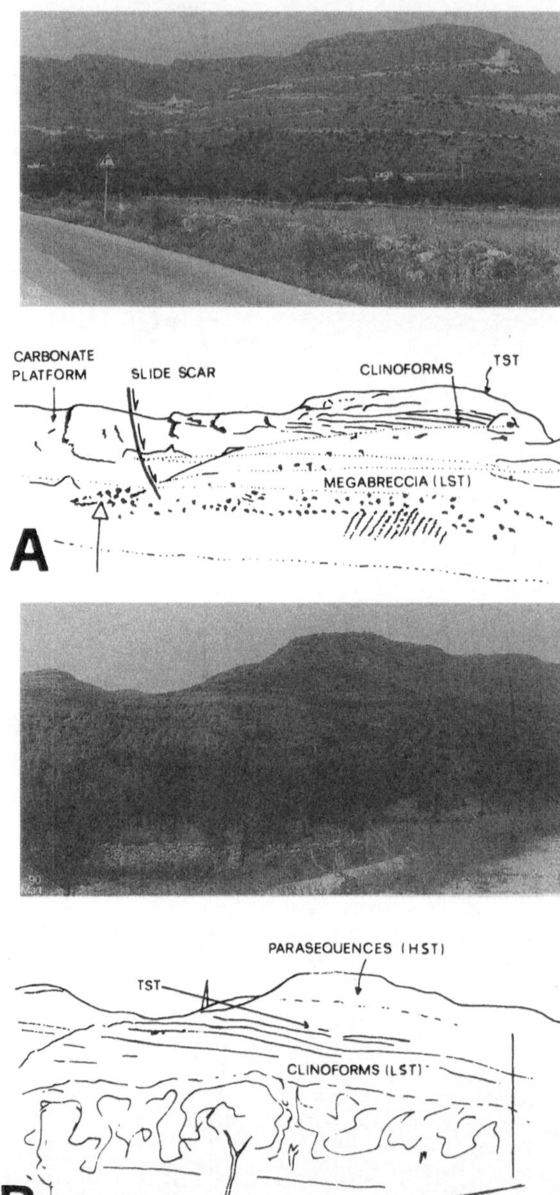

Fig.107 - A Panoramic view of locality section #2 showing the trace of slide scar, and megabreccia (lowstand system tract). B Flattening-out clinoforms composed of rudist grainstones and rudstones.

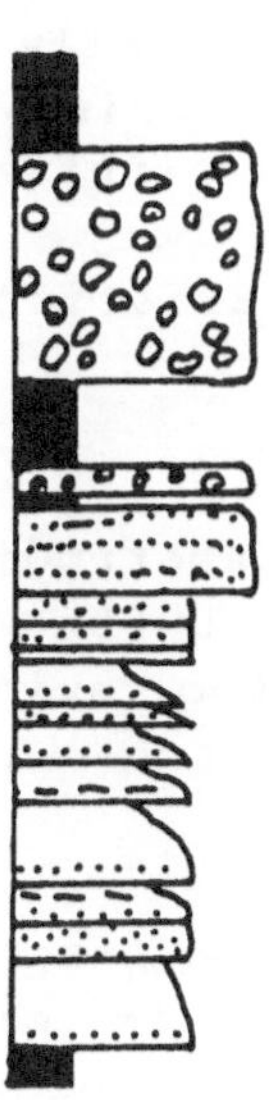

Fig.108 - Transgressive system tract overlain by a parasequence, and (below) example of a parasequence of the highstand system tract of the second depositional sequence.

Second depositional sequence

The second depositional sequence,Albian to Senonian in age (Pattera,1966-1967),is composed of the following facies tracts:

1) lowstand facies tract;
2) transgressive facies tract;
3) highstand facies tract; and
4) shelf margin facies tract.

Some panoramic views and details of these facies tracts are shown in Fig.106 - 108.

Lowstand facies tract

Description

The lowstand facies tract is a gravity deposit, approximately 220 m thick, which extends from the amphiteater-like slide scar, basinward for about some hundred of meters. The stratigraphic section (section #2: Fig.102; Fig.109) consists of about 300 m thick lithoclast megabreccia laterally and vertically transitional to clinoforms composed mainly of bioclast (rudist) detritus. Along the vertical,at a first sight the megabreccia appears to be a uniform ,unlayered body. In a basinward direction however, it records repeated variations in size, packing and composition of the clasts.

The basal 60 m of the succession are composed of well packed, dm - cm size clasts of mainly oobiosparites and biosparrudites. The remaining part consists of repetitions of three units which form a modal sequence .Its recognition is not too evident: these three units are recognized by subtle increases in the percentage of different types of clasts.

1) The base consists of densely packed dm to sm-size unrounded clasts of oolite lithofacies.The megabreccia here is clast-supported, poorly sorted, has a chaotic fabric, angular to sub-angular clast boundaries,absence of interstitial mud and grading.
2) The intermediate part is a less densely packed parabreccia with a biosparrudite matrix. The upper limit of clast size is on the order of 1/2 m. Clasts of rudist rudstones are particularly frequent in this part.It is characterized by a

poor sorting and a matrix support.There are also basal zones of inverse grading.

3) the upper unit is a breccia with clasts of oncolites, supratidal laminites and of skeletal wakestones of lagoonal provenance containing structures of subaerial exposure.

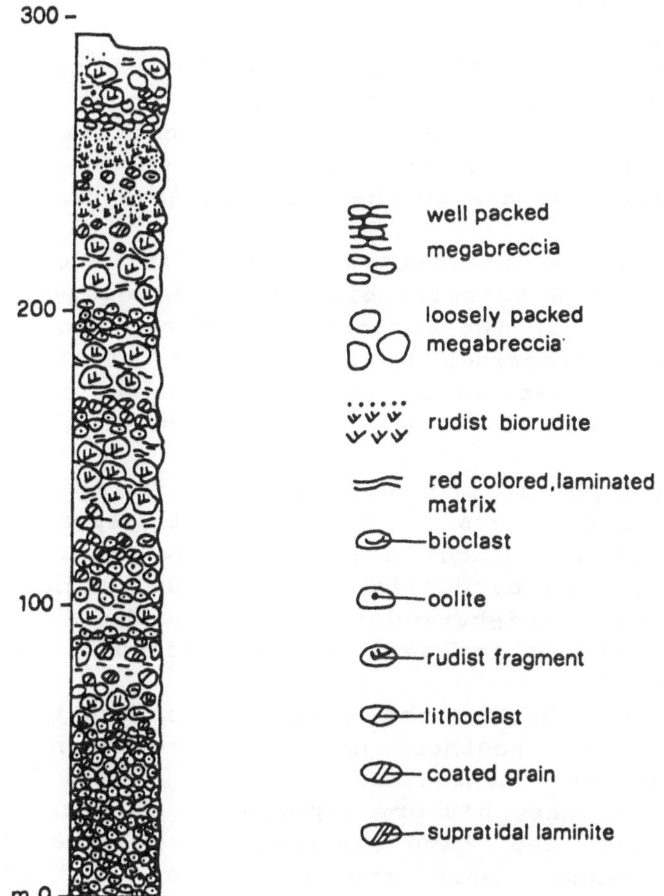

Fig.109 - Lowstand facies tract (section #2).The lowermost 60 m represent a lowstand fan,produced by a catastrophic slope failure. The remaining part of the section is interpreted as a lowstand wedge facies tract.

Going basinward, sequences record a progressive decrease in the proportion of oolite lithofacies and an increase in the lithofacies of the upper unit.Clasts of rudist rudstones also become more common upwards, especially in the uppermost 50 m. Parabreccia fabrics are more frequent in the upper part of the section.

Interpretation

The megabreccia is a gravity deposit laid down at the base and against the platform. Bauxite clasts dispersed in the megabreccia, laminations ponded in depressions, rounding of clasts and the yellow-colored matrix between clasts suggest that some areas of the platform were emergent when the slope failure took place.

According to Sarg (1988),during the formation of a type-1 sequence boundary there is a significant amount of slope front erosion with downslope input of large quantities of material. In fact, the formation of the megabreccia was correlated with a coeval subaerial hiathus in the southern Appennines.

The megabreccia was interpreted by Bosellini and Ferioli(1988) as the result of a catastrophic collapse induced by a sudden eustatic lowstand in the Turonian (90 m.y.: Haq,et al.,1987). The slide scar described by Mullins, et al.(1986) from the Florida plateau was also taken by these authors as an actualistic example of the situation encountered here.

This interpretation is problematic. In fact, the occurrence of a modal sequence points to a multievent deposition for the megabreccia which excludes a single catastrophic collapse.The slide scar reported by Mullins,et al.(1986) originated at the peak of a sealevel highstand,in the early Middle Miocene, not during a lowstand of the sealevel as claimed by Bosellini and Ferioli (1988).
It is not known whether the type-1 unconformity recorded in other areas in the southern Appennines represents a regional uplift of the Appennines, or a eustatic sea-level fall.By analogy with the case history reported by Mullins,et al.(1986) the megabreccia may have originated at the peak of a progradation phase, when the relief of the platform was highest: the megabreccia would represent the formation of an erosional margin from a by-pass gullied slope which is represented by the top of the first depositional sequence. Exposure of the platform, documented by the occurrence of bauxite clasts within the megabreccia, may have been produced during a regional phase of uplift which is recorded in several places in the southern Appennines.

The clast-support of the first unit of the modal sequence reflects a rock-fall type of gravity transport, in contrast with the matrix-support of the intermediate unit which originated by debris flows .Rockfall deposition requires a steep slope typical of by-pass or erosional margins, whereas

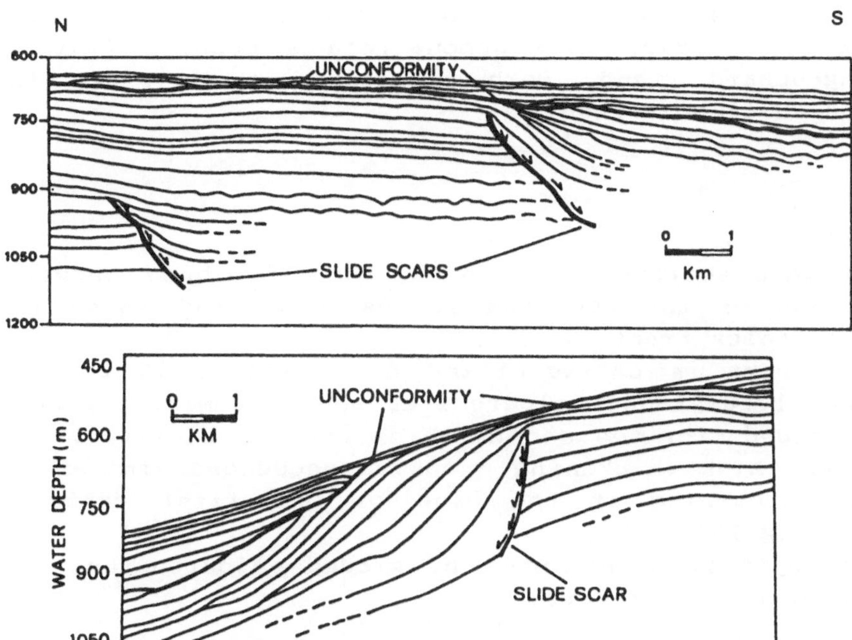

Fig.110 - Slide scars in the west Florida carbonate platform margin reported by Mullins, et al.(1986).

debris flows take place along less inclined slopes.

The modal sequence reflects different episodes of a process of landward retreat of the carbonate escarpment, typical of high-relief carbonate slopes,which is compatible with a period of slow sea-level rise.The catastrophic collapse is represented only by the lowermost 60 m thick, clast-supported megabreccia.

The geometry of the lowstand facies tract of the second depositional sequence, which can be perceived from an examination of natural exposures, is similar to that of the depositional sequence described by Mullins, et al.(1986), where sequences are organized into a flattening-out clinoform pattern and a divergent-climbing pattern (Fig.110).

Transgressive facies tract

The transgressive facies tract overlies and onlaps the megabreccias and rudist biorudites of the lowstand facies tract. It is a finer grained deposit that stands out clearly on outcrop faces.It is exemplified by two stratigraphic sections located shelfward and basinward (sections #3 and 4: Fig.111) from the slide scar. They demonstrate a general fining- and deepening-upward trend. Both sections are divisible into a lower and an upper segment.

Description

The basinward section, 9 m thick, is sandwiched between the megabreccias of the lowstand facies tract and those of the highstand facies tract.
The lower part is characterized by ungraded, 10 cm to 1 m thick, mostly ungraded calcarenites and predominating couplets of thin bedded micrites and cm-thick shales. Micrite beds form 30 ÷ 50 cm thick thickening - upward sequences similar to the sequences occurring at the base of the first depositional sequence (Fig.104).
The upper part is constituted by graded calcarenites, micrite beds and thin chert layers.

The shelfward section,75 m thick,is also divisible into two segments (Fig.111).
The lower is 20 m thick and consists of ungraded megabreccias, calcarenites, and thin micrite layers (Fig.112).
Megabreccia layers (Fig.112,113) are 4 m thick and about 70 m wide. The bases exhibit concave-up outlines, load casts and gutter casts (as much as 60 cm wide), infilled with rudist bioclasts. The clasts composing the megabreccia are mainly of rudist rudstones and to a lesser extent of biomicrites, especially at the base of the section. Larger clasts are 4 x 4 m wide in cross section, but the most common dimensions are of dm-size. Ovoidal shapes are predominating. Clast imbrications are especially frequent where breccias are poor with matrix. The megabreccias are not graded. The upper surface is channalized (Fig.112,113). The only internal structures within megabreccias are amalgamations and channel-fills. Gutter casts indicate a 20°÷35° N current direction.
Calcarenites and micrites are cm- to 30 cm thick: they are interlayered with megabreccias. Most of the calcarenites are graded; they are also typically hummocky cross-bedded (Fig.113). A thickening- and coarsening-upward trend is recognisable at the bottom of the section, where blocks and clasts are dispersed and floating within the calcarenites (Fig.113).
The upper half of the section consists of predominating, graded hummocky cross-bedded rudist grainstones (Fig.113), organized into 50 cm to 2 m thick sequences (Fig.113) with a tripartite division:

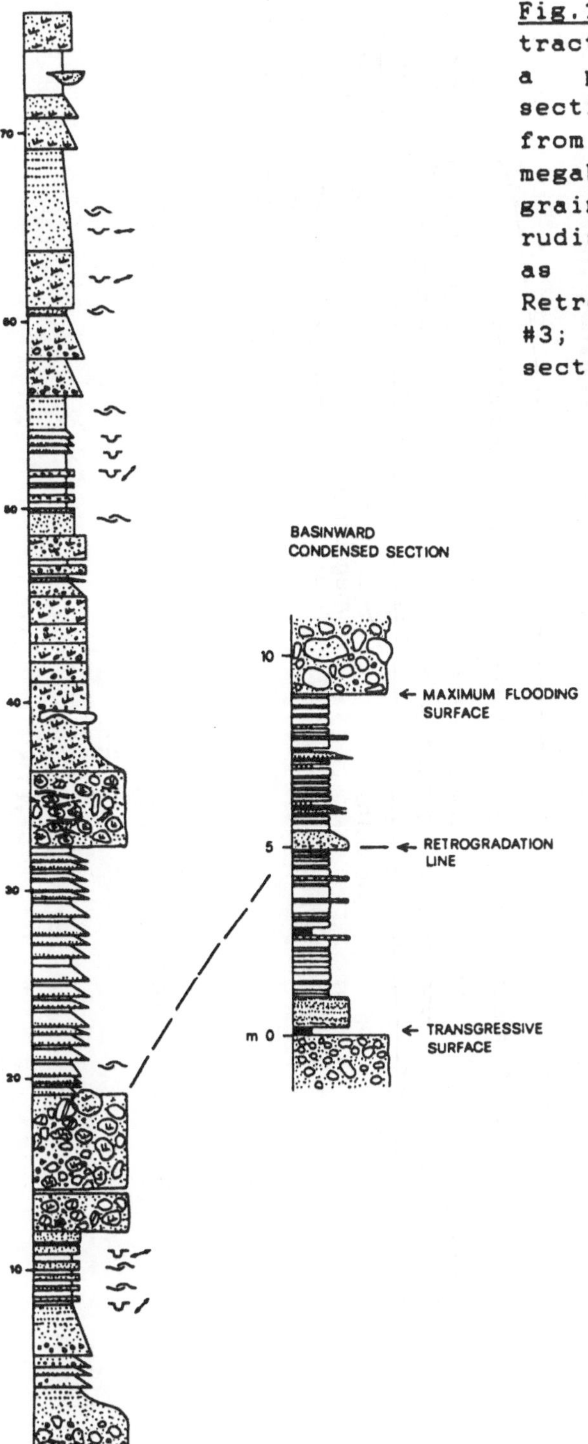

Fig.111 - Transgressive system tract. The panoramic view shows a part of the shelfward section. The sharp transition from partly channelized megabreccias to the finer grained hummocky cross-bedded rudist beds is interpreted as a retrogradation line. Retrogradational units: section #3; basinward condensed section: section #4.

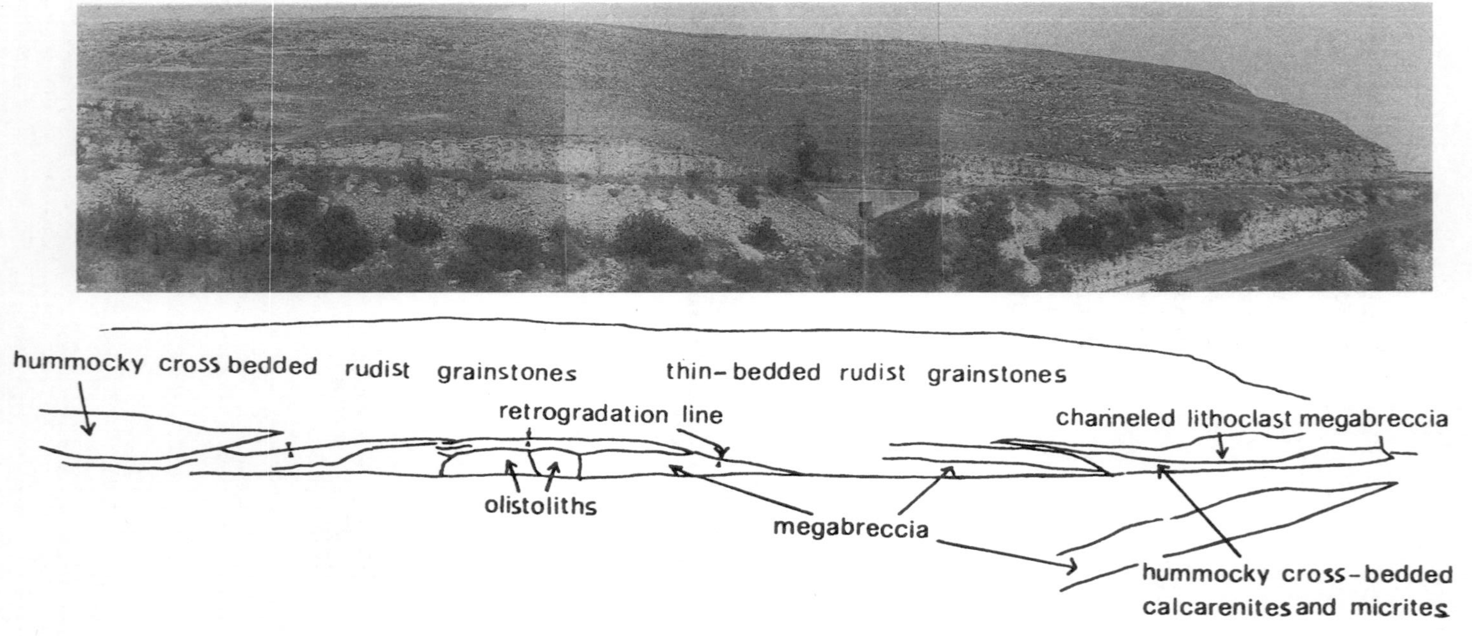

<u>Fig.112</u> - Panoramic view showing important features of the transgressive system tract.

1) a lower lithoclast base;
2) an intermediate unit consisting of well sorted sets of laminated 1÷2 mm rudist bioclasts; and
3) an upper micrite unit.

Grain sizes of these sequences decrease towards the top of the section. Wavelengths of hummocky-cross bedding are of some meter. Gutter casts and other erosional features at the base of these sequences indicate a 60°-85° N current direction.

Interpretation

Channeled megabreccias are interpreted as ephemeral, anastomozing gullies which incised the upper slope.
The hummocky-stratified, thin-bedded calcarenites and micrites into which channels were cut indicate a deposition in the upper shelf located above storm wave base. Thickening-upward sequences occurring in basinward and shelfward sites indicate that sedimentation rate could keep pace with an initially slowly rising sea-level.
The sedimentation took place mainly by aggradation.

The depositional facies of the upper part of the shelfward section is interpreted as a retrogradational transgressive facies tract developed on a sloping surface. It reworked rudist bioclast material delivered to the slope. Its deposition was concomitant with a sharp vertical decrease in grain size which may be related to a process of progressive sorting and loss of the coarser fraction, with extraction of fine sediments from storm flows in the upper slope.Sediments were reworked into sandwaves and wave- generated structures.
The absence of coarsening- and thickening-upward trends rules out a progradational trend.

The sharp transition recorded in the shelfward section from megabreccias to fine-grained, better sorted biorudites and calcirudites with hummocky cross-bedding reflects a rapid deepening consequent to a relative sea-level rise.

It is correlated with the change in bathymetry and increasing upward deepening tendency recorded in the upper half of the basinward section,where chert layers replace shaley micrites.
The surface delimiting the two segments of the transgressive facies tract is termed <u>retrogradational line</u> (Fig.112)

It is traceable from the slide scar to the basin and marks an abrupt increase in the deepening-upward trend within the

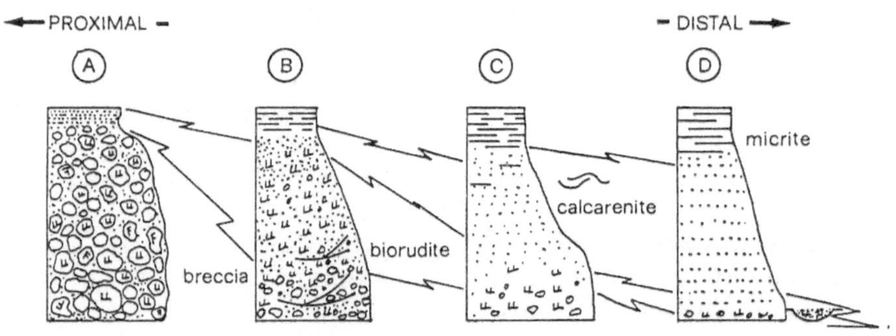

Fig.113 - Transgressive system tract. (section #3: Fig.102).A: Thickening-upward trend at the base of the measured section with boulders floating within the calcarenites.B: m-scale, graded and thinning-upward sequence with a thick basal biorudite bed overlain by thin beds of calcarenites. C:large-scale hummocky cross-bedding within rudist biorudites and calcarenites.

transgressive facies tract. The retrogradational line separates a lower aggradational to slowly retrogradational facies tract, from an upper tract characterized by a more pronounced retrogradational character.This surface corresponds to an inflexion in the profile of the transgressive facies tract which could be seen in seismic profiles.

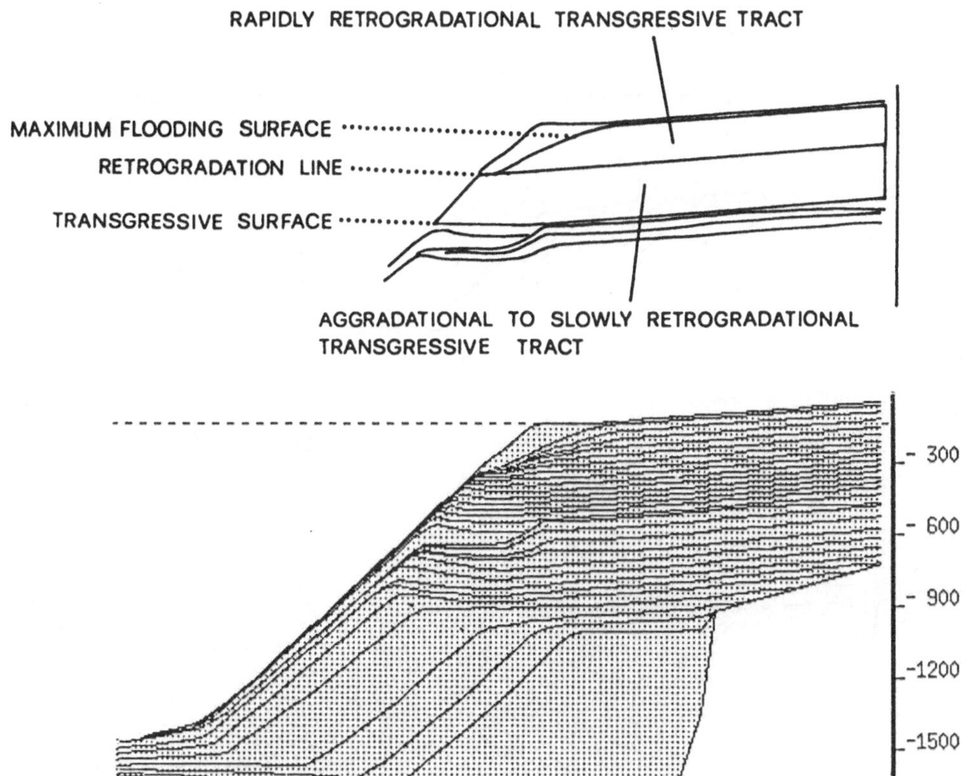

Fig.114 - Computer modelling of a transgressive facies tract.A retrogradational line (surface) results from a slight increase in subsidence or sealevel rise when the relative sea level rise is a compound effect of subsidence and eustasy.

The computer modelling (Fig.114) demonstrates that retrogradational lines (surfaces) are produced by a slight increase either in the rate of subsidence or eustatic sea-level rise, when the relative sea-level rise is the summation of subsidence and eustasy.

Stronger increases in the rate of either the two components of the relative sea-level curve lead to a backstepping (Fig.115) which would have been documented in the field by hardgrounds above the transgressive surface.

Correlations between the two sections indicate that the

transgressive facies tract is a wedge that thickens shelfward and thins basinward into a condensed section, as a consequence of offshore sediment starvation.The lithosome geometry of the TST, simulated by computer modelling (Fig.115),originated from tectonic tilt (seaward and landward subsidence), a rise in the eustatic sea-level and a decrease in the sedimentation rate.

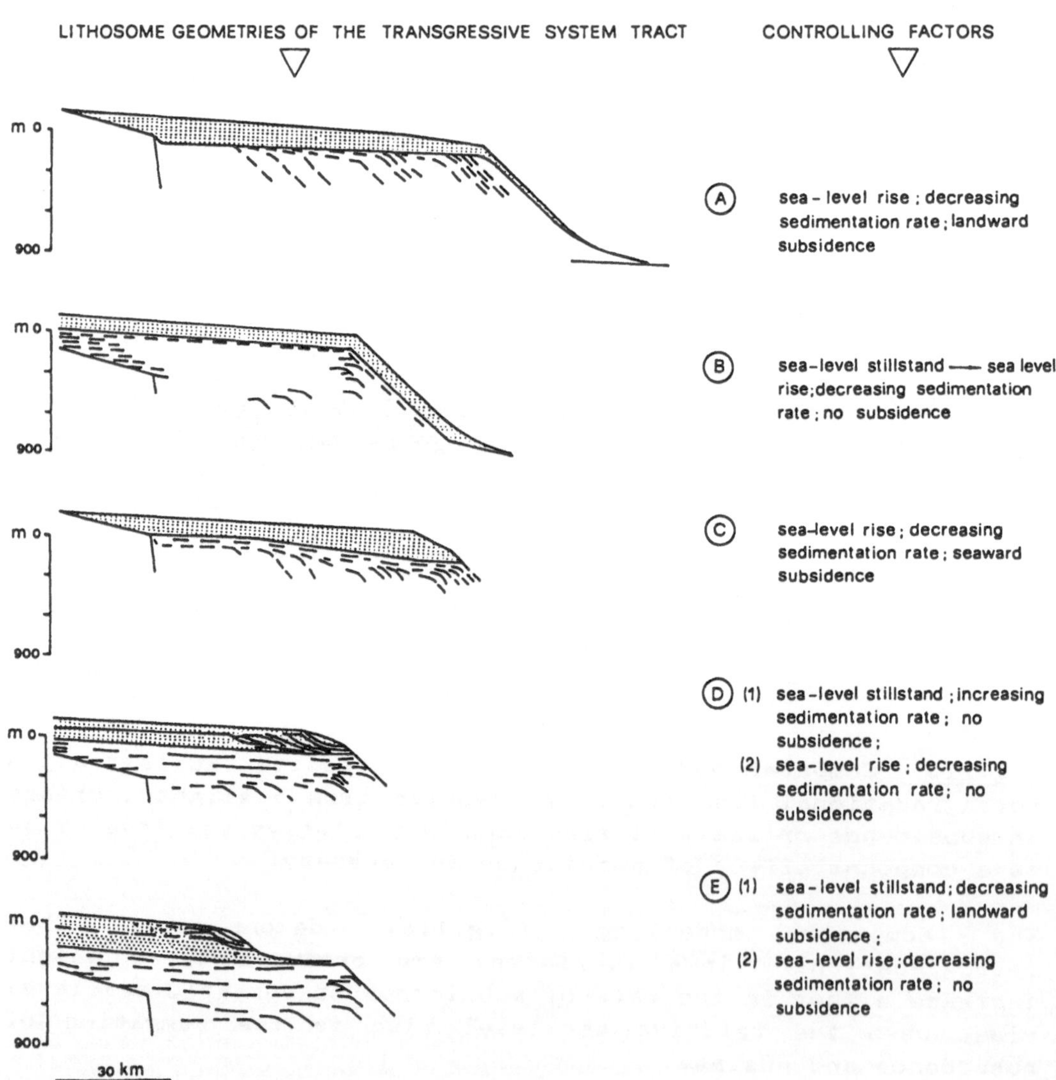

Fig.115 - Computer modelling of different geometries of the transgressive facies tract obtained by various combinations of subsidence, eustatic sea-level rise and sedimentation rate.

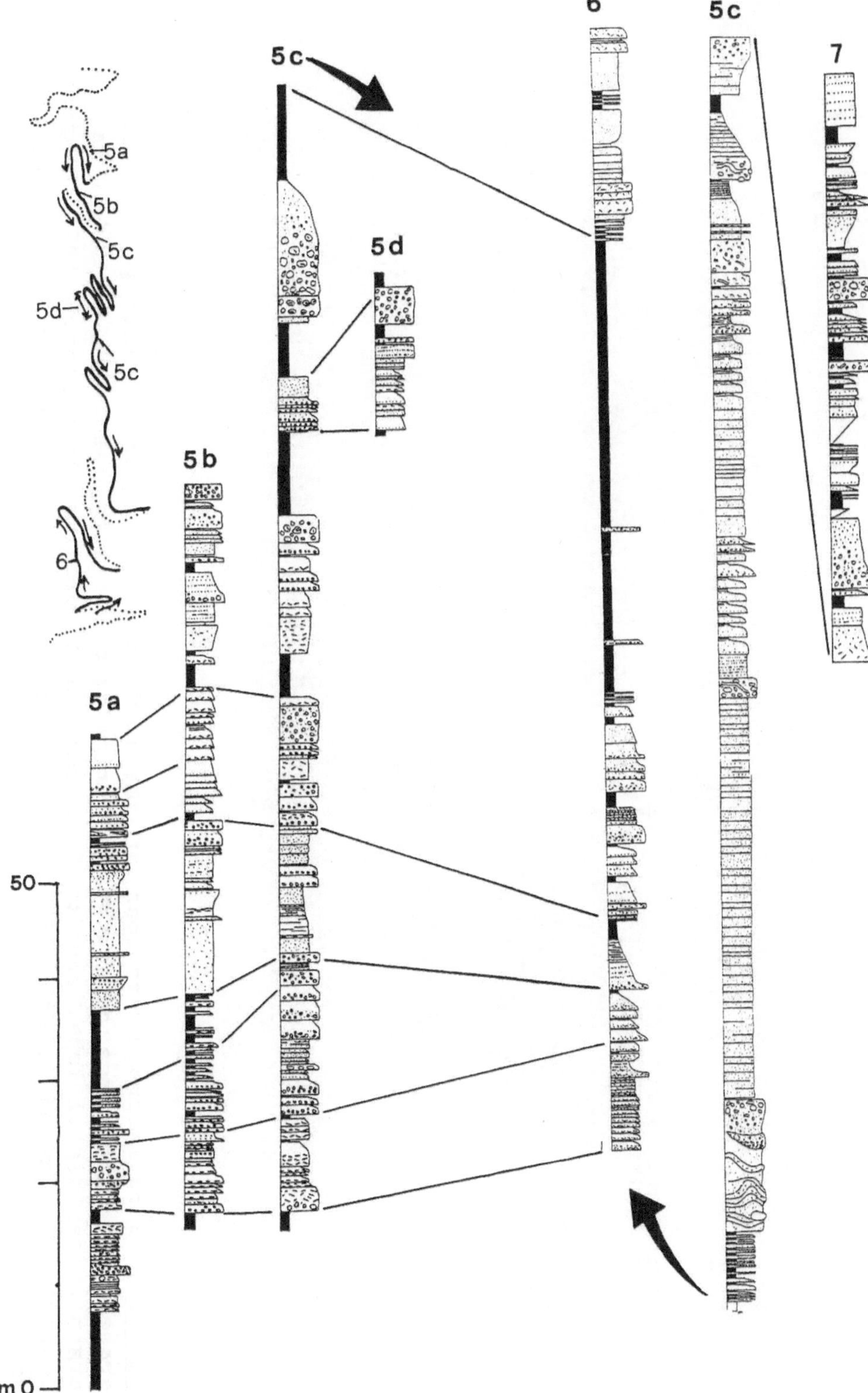
5a
5b
5c
5d
5c
6
5c
5d
5b
5a
6
5c
7
50
m 0

<u>Fig.116</u> - Stratigraphic sections of the highstand system tract showing sigmoidal-shaped parasequence lithosomes separated by hemipelagic micrites (sections #5 and 6) and shelf margin wedge (distal lobe: section #5c: upper part; proximal lobe: section #7).The panoramic views show two parasequences with interposed slumped micrite horizons and the distal lobe far on the right.

Highstand facies tract

Description

The highstand facies tract is bounded below by the transgressive facies tract and above by the shelf margin facies tract.It is 140 m thick and consists of several 10-14 m thick parasequences separated by hemipelagic micrites or/and intraformational discordances (Fig.116).

These parasequences have a dominant oblique sigmoidal progradational geometry. They are characterized by abrupt bases,amalgamations, basinward decreases in grain-size and distal offlapping patterns.
Compositionally, they are compound sequences formed by three lithofacies:

1) bioclast beds;
2) lithoclast conglomerates; and
3) breccias.

The last lithofacies is not always present.These three lithofacies may be stacked onto another but more commonly are separated by thin,often slumped hemipelagic micrite horizons.

Bioclast beds consist of coarse-tail graded, 1 to 0.5 m thick calcirudites and calcarenites.It is possible to recognize proximal (northern) and distal (southern) end-members.

In the north,proximal beds are 0.40 as much as 2.60 m thick and coarse grained. They consist of a lower, channelized breccia or trough cross-bedded biorudite, an intermediate unit composed of dm-thick, graded calcarenites and an upper unit given by laminated micrites.
In the south individual beds are between 1 and 1.30 m thick. The lower part is up to 0.4 m thick and is a calcirudite lag including rudist bioclasts and micrite lithoclasts. Some bioclasts are imbricated.Gutter casts and channels ,visible at places, are infilled with angular lithoclasts or rudist fragments.In most cases however the basal lag unit is only few cm thick.The middle part of the cycle is composed of finer-grained, laminated calcarenites with undetermined shell debris which in some instances is organized into sets of laminae of progressively decreasing thickness and grain-size. The uppermost unit consists of plane parallel to small-scale undulated lamination.Laminae both thin and fine upwards.Along the strike, these beds are 20-15 m thick and have a

wide,concave-up lower surface or a lower planar and upper concave-down upper boundary. Laterally,they pinch-out into finer-grained lithofacies and mudstones.

These bioclast beds are organized into a sequence constituted by a stacking of 5-7 beds to a minimum of 3 layers with an asequential, irregular, or thinning-upward trend.

Conglomerates occur as lenticular 1 to 0.5 m thick beds with sharp erosional bases and sharp tops. They display a crude grading in the upper part, but most of the bodies however are asequential. They contain dm-cm, as much as 30 cm size, packed, rounded clasts of micrite and some fragments of rudist lithofacies. Some clast imbrications were observed. The matrix is either calcarenite-calcirudite, or micrite.

Slumped breccias occur at the top of the parasequences, (above lithoclast conglomerates) and are more commonly found in the upper part of the highstand facies tract.They sharply truncate underlying hemipelagic sediments. Clasts in these deposits are of shallow-water and hemipelagic provenance. Shallow-water clasts are composed of rudist rudstones and floatstones. Clasts are mainly ovoidal and tabular in shape, as large as 80 cm and with rounded clast boundaries. They have a chaotic fabric and are vertically organized into coarsening-upward trends. Their topmost part is channelized and infilled with rudist fragments and clasts. The matrix is a mixture of clasts and bioclasts. There is no evidence of stratification, except where clasts are imbricated; m-size olistoliths are floating within the matrix.

Interpretation

There are two basic types of vertical bed organizations within parasequences:

1) Thinning-upward, asequential;
2) thickening-upward, coarsening-upward.

The first, found in bioclast-rich beds and the second characterizing lithoclast lithofacies, reflect respectively aggradational and progradational trends (infilling of depressions followed by frontal accretion).

The parasequences reflect episodes of accretion driven by relative high-frequency sea-level oscillations (tectonics and eustasy).A possible mechanism for the formation of parasequences is shown in Fig.117.

The whole highstand facies tract is a coarsening- and thickening-upward megasequence, as a result of an increase in proportion and thickness of lithoclast lithofacies and breccias. The occurrence of a megasequence trend suggests some control by differential subsidence.

Shelf margin facies tract

Description

The shelf margin facies tract is constituted by a 70 m thick compound depositional lobe (Fig.116).Layers are vertically stacked,without hemipelagic micrite interbeds.
It is composed of a distal lobe at the base, and a proximal lobe at the top.

Distal lobe

The lower contact with the underlying hemipelagic micrites is gradational. The first 8-10 m of the succession consist of dm-thin layers (a-b Bouma intervals: Fig.118A). It is followed by a 20 m thick chaotic deposit (slump breccia) consisting of a disordered structure with variable amounts of inclusions of rudist clasts, distorted intraformational calcarenite beds floating within a calcirudite - calcarenite matrix. The remaining part of the lobe is an aggradational deposit. Apart from few slump horizons, this part of the distal lobe is a stacking of m-scale thick, thinning- and fining-upward cycles (Fig.118B). Individual cycles pass from biocalcirudites with fragments of rudists to fine-grained with plane parallel lamination. The upper part shows evidence for bioturbation (*Planolites*). Most of the calcarenite beds are ungraded and structureless, with sharp bases and tops.

Proximal lobe

The upper part of the lobe consists of 15 - 20 m thick deposits bounded or developed above slumped horizons or intraformational discordances (traces of translational slides).The deposits are thickening and coarsening-upward sequences (Fig.118C) characterized by an upward transition from thinly bedded calcarenite - micrite alternations to coarse- grained, thick

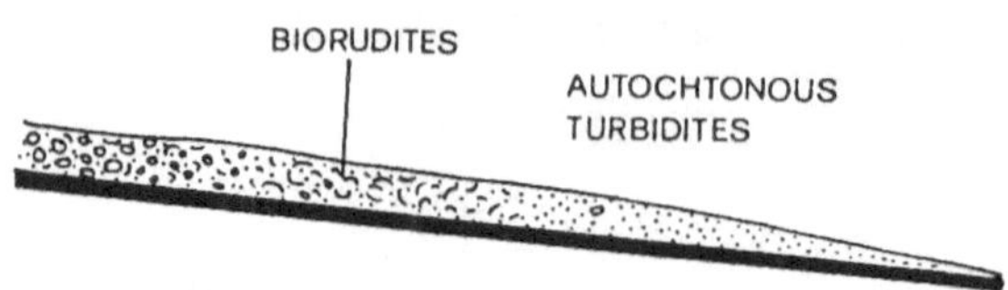

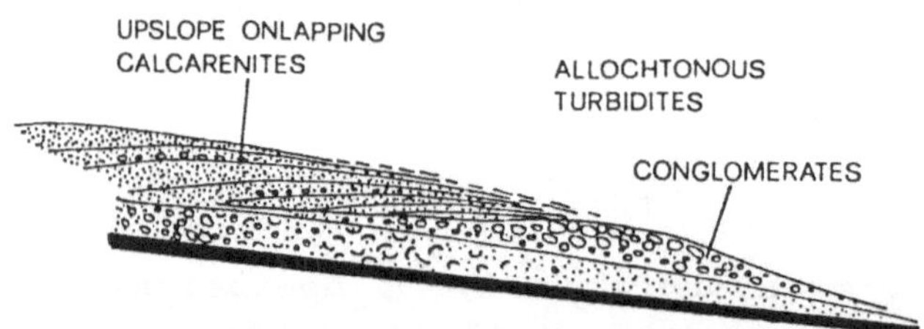

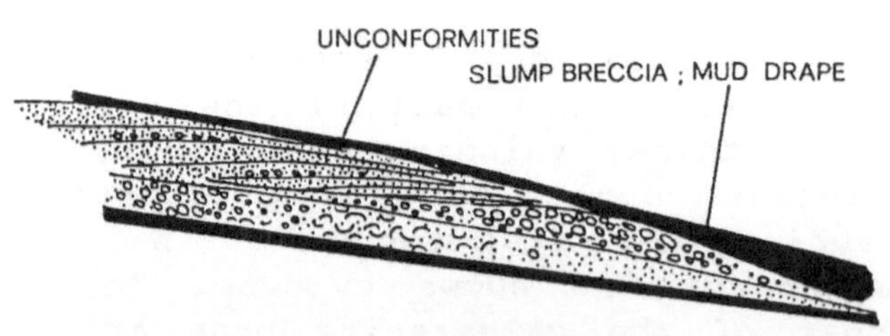

Fig.117 - Mechanism of formation of parasequences by an interplay of high-frequency sealevel fluctuations and varying subsidence rates. Allochtonous,lithoclast-rich turbidites and upslope onlap of calcarenite bodies formed during highstand periods which followed a stage of emergence of the platform which favoured the formation of lithoclast beds.

thick skeletal calcarenites, to 7 - 8 m thick sigmoidal beds (Fig.118D) and eventually to megabreccias containing lithoclasts and fragments of rudist-bearing rudstones.

<u>Fig.118</u> - Shelf margin system tract. **A)** dm-thick layers at the base of this system tract (distal lobe: section #5c). **B)** Meter-scale thinning and fining-upward beds (distal lobe:section #5c). **C)** Thickening-upward trend developed above a slide scar (proximal lobe: section #7). **D)** Sigmoidal beds (proximal lobe: section #5c).

Fig.119 - Cross section of a portion of the lobe showing lower thin beds draping an irregular topography (distal lobe) and upper, thicker convex-upward beds (proximal lobe).

Interpretation

The lobe is a sigmoidal body characterized by a major episode of aggradation (distal lobe) capped by progradational deposits (proximal lobe).
The absence of hemipelagic interbeds in the lobe reflects an increase in the average sedimentation rate.

The rate of aggradation was controlled by the accomodation potential (subsidence rate). Initially,the increase in the sedimentation rate on the shelf due to a slow sea-level rise led to the fill-up of the depression on the slope with the formation of thin, concave-up irregular beds (Fig.119); successively,the decrease in the accomodation together with the increase in the rate of relative sealevel rise determined a progradation of the lobe with an increasing upward deepening tendency.

Third depositional sequence

This sequence is lower to medium Eocene in age. The lower sequence boundary has a tight amphiteaterlike outline (Fig.102). It truncates the underlying deposits of the second depositional sequence. The upper sequence boundary is not visible.

Fig.120 - Panoramic view of M.Saraceno showing clinostratification (Punta Rossa Formation) overlying basinal sediments (Peschici Formation).

The third depositional sequence represents the progradation of a *Nummulite* buildup (M.Saraceno Formation) on slope sediments (Punta Rossa Formation) above basinal sediments (Peschici Formation).

The progradational character of the sedimentation is well evident from an examination of the measured stratigraphic sections and panoramic views which show clinostratification developed above basinal sediments and in turn overlain by a shallow-water *Nummulite* buildup. (Fig.120).

This depositional sequence is characterized by the stacking of the following facies tracts (Fig.121):

1) lowstand facies tract (thickness: 50 m);
2) transgressive facies tract (thickness: 20 cm);
3) highstand facies tract (thickness: 30 m).

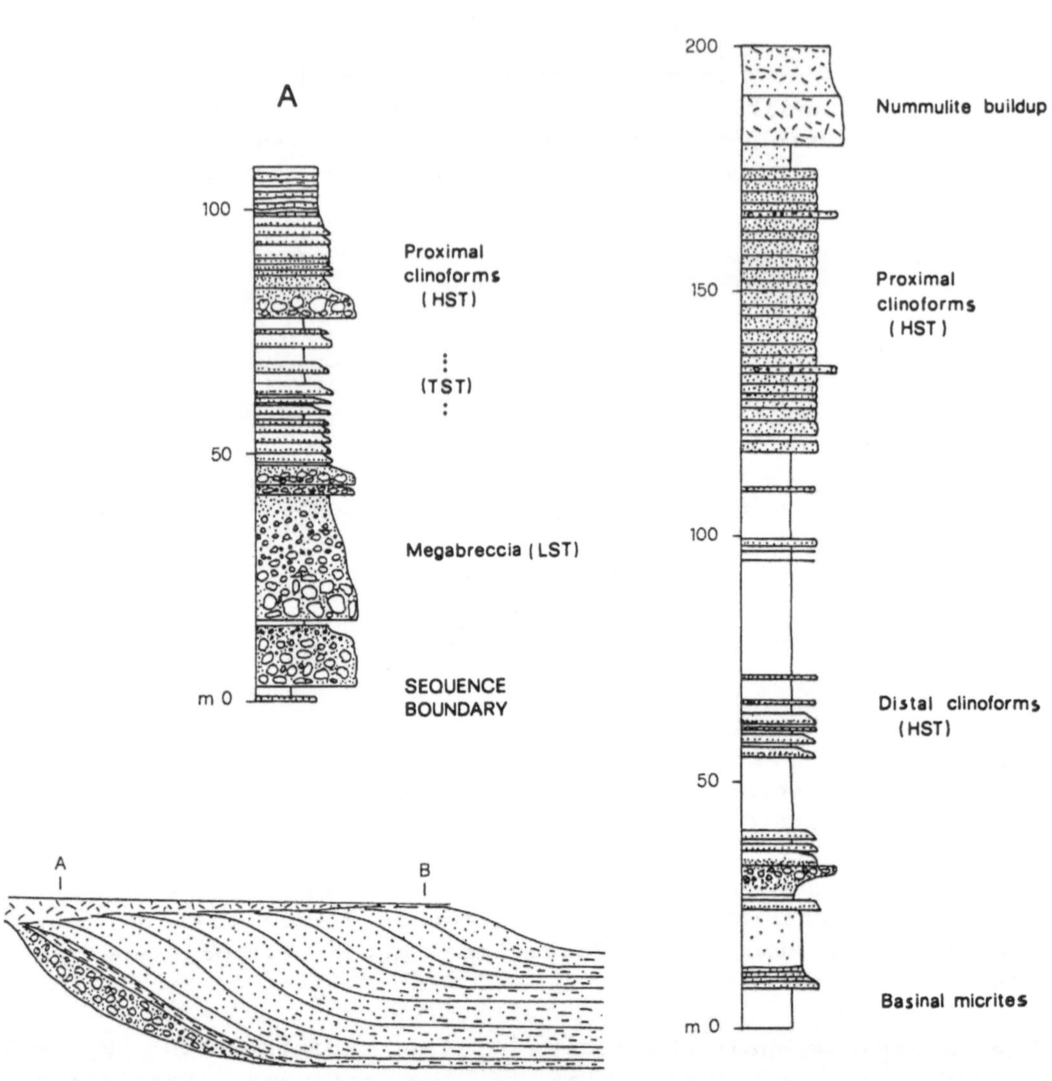

Fig.121 - Third depositional sequence.

Lowstand facies tract

Description

It is constituted by a 50 m thick megabreccia. The lower contact above thin-bedded, hemipelagic limestones of the second depositional sequence is erosional. The megabreccia is a multi-event deposit consisting of several channalized bodies separated by amalgamation surfaces. These bodies are a parabreccia containing m-size blocks and clasts of biocalcirudites, mainly of rudist material and open lagoon sediments, embedded within a micrite matrix. Smaller micrite clasts are of dm-cm size. Clasts are poorly sorted and angular. Upwards, the megabreccia displays a normal grading and clasts become more enriched in fossils,mainly in *Nummulites*.

Interpretation

The megabreccia represents a series of episodes resulting from an erosion of a carbonate slope. It is analogous in texture and fabric to the rudist parabreccia occurring at the base of the second depositional sequence,which have been interpreted as a lowstand fan. The modality of downslope movement took palce by debris flow, as is shown by the clastic, chaotic texture supported by a fine matrix.

Transgressive facies tract

Description

This is constituted by a thin stratigraphic thickness consisting of thin micrite horizons intercalated with a few, thinning-upward graded beds composed of *Nummulite* bioclasts,a thin hardground horizon and large-scale symmetrical undulations (Fig.122).

Interpretation

The thin micrite beds interbedded with wackestone graded beds represent a condensed section formed during a sealevel rise which drowned the lowstand wedge.The thinning-upward trend displayed by graded beds resulted from sediment starvation. Symmetrical undulations may represent traces of slide scars or the result of smoothing of the seafloor by currents or waves,as is also suggested by the occurrence of hardgrounds on top of

Fig.122 - Symmetrical undulation and hardground in the transgressive facies tract.

the undulations.
Slide scars result from gravitational instability produced by sediment overloading more commonly during periods of increased sedimentation rate, for example during a highstand facies tract, than during periods of sediment starvation.

Highstand facies tract

Description

The highstand facies tract consists of a series of clinoforms: proximal and distal clinoforms have been observed in the two measured stratigraphic sections.

Proximal clinoforms

Proximal clinoforms are thickening- and coarsening-upward sequences formed by m-scale, lens-shaped, sigmoidal megabreccias interbedded with hemipelagic sediments (Fig.123A). Thicker megabreccias have a tripartite division:the base consists of micrite lithoclasts (5-10 cm in size), gastropods, *Nummulite* shells. The intermediate unit is a biorudite containing gastropods and *Nummulites*.The upper unit contains *Nummulite* shells oriented parallel to the bedding plane (Fig.123B). The inclination of *Nummulite* shells reflects that of clinoforms: going upwards, stratigraphically, the angles of inclination decrease from 35°- 20° to 10° or less.

Distal clinoforms

In the measured area distal clinoforms display a downlap

Fig.123 - Highstand facies tract.A: proximal clinoforms showing a megabreccia (right) passing to sigmoidal clinoforms (left).B:detail of proximal clinoform showing a basal lag composed of gastropods and disoriented *Nummulite* shells overlain by a thicker unit composed of *Nummulites*, mainly inclined at a low angle with respect to the bedding plane.C: dm-thick distal clinoforms composed of *Nummulite* shells.

termination against the lower hemipelagic basinal sediments (Fig.124).Clinoforms are 1÷2 m thick,contain abundant *Nummulite* bioclasts with a parallel to random orientation to bedding planes, and lesser amounts of other shell debris.In the distal positions *Nummulite* beds occur as dm-thick intercalations within hemipelagic micrites (Fig.123C).

Interpretation

The two sections represent shelfward and basinward parts of sigmoidal clinoforms of an oblique progradational pattern (Fig.125).

Fig.124 - Distal clinoforms downlapping hemipelagic micrites (section #9).

The sigmoidal shape is suggested by: 1) upslope flattening and downlap terminations of clinoforms, and 2) upward changes in the shell inclinations.

Rather than homogeneous, these clinoforms have a multistored geometry resulting from various types of gravity mechanisms (debris flows,slump, avalanching of *Nummulite* shells and grain flows in the distal areas).The progradational and frontal accretional character of the clinoform is indicated by abrupt bases at the proximal areas, amalgamations, and gradational terminations near distal ends.

Concluding remarks

The geometry of the second depositional sequence corresponds to the sigmoid progradation configuration described by Mitchum,et al.(1977): "A sigmoid progradational configuration is a prograding clinoform pattern formed by superposed sigmoid (S-shaped) reflections interpreted as strata with thin, gently

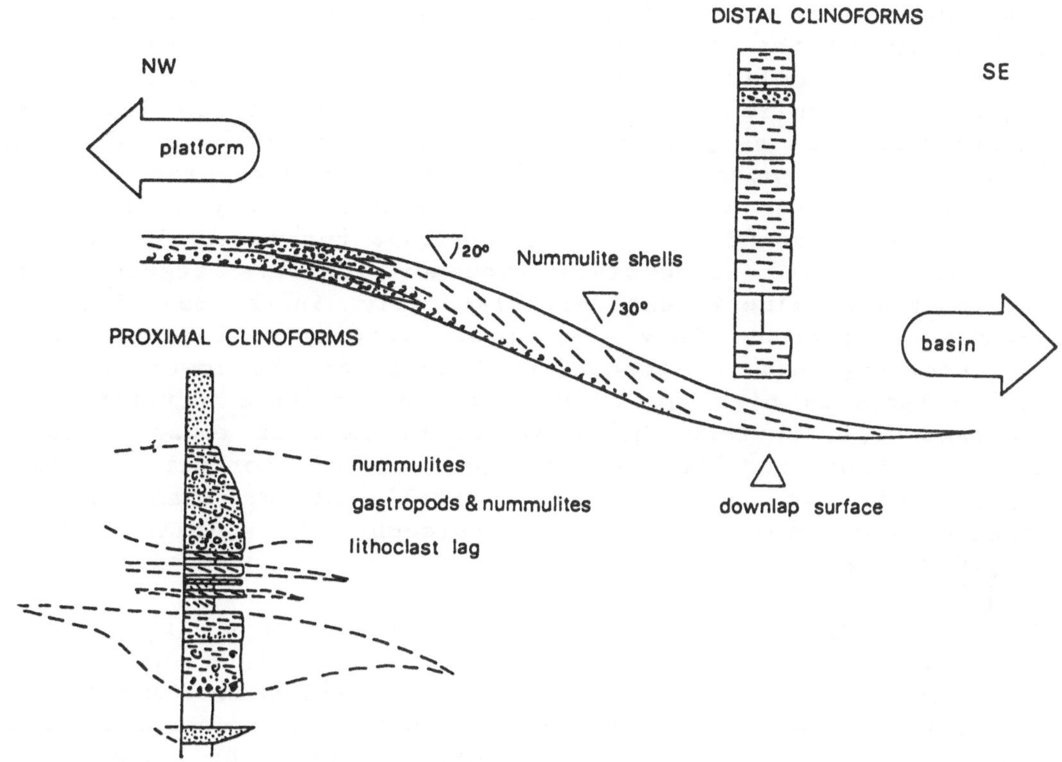

Fig.125 - Highstand facies tract. Inferred geometry of clinoforms.

dipping upper and lower segments, and thicker, more steeply dipping middle segments. The upper (topset) segments of the strata have horizontal or very low angles of dip and are concordant with the upper surface of the facies unit. The thicker middle (foresets) segments form lenses superposed to allow successively younger lenses to be displaced laterally in a depositionally downdip direction, forming overall outbuilding or prograding patterns... The most distinctive feature of the sigmoidal configuration is the interpreted parallelism and concordance of the upper stratal (topset) segments, suggesting a degree of continued upbuilding (aggradation) of the upper segments coincident with prograding of the middle segments. This configuration implies relatively low sediment supply, relatively rapid basin subsidence, and/or rapid rise in sea level to allow deposition and preservation of the topset units".

The types of clinoforms, the upper toplap boundary relations with the shallow water *Nummulite* buildup and the lower downlap termination of the third depositional sequence are geometrical features of an oblique progradational pattern (Mitchum, et al., 1977): "An oblique progradational configuration ... is interpreted as a prograding clinoform pattern consisting ideally of a number of relatively steep-dipping strata terminating updip by toplap at or near a nearly flat upper surface, and a downdip by downlap against the lower surface of the facies unit. Successively younger foresets segments of strata build almost entirely laterally in a depositional downdip direction. They may pass laterally into thinner bottomset segments, or terminate abruptly at the lower surface at a relatively high angle. They build out from a relatively constant upper surface characterized by lack of topset strata and by pronounced toplap terminations of foreset strata. Depositional dips are characteristically higher than in the sigmoid configuration, and may approach 10° " (Mitchum, et al.,1977: page 125).

The third depositional sequence formed during a first-order sea-level fall.The progradation during the highstand facies tract was the response to a sealevel fall which forced the sediment to expand laterally, as a result of a decrease in the vertical space available for sedimentation. The relatively minor thicknesses of the facies tracts,compared to those of the second depositional sequence, imply a corresponding minor rate of accomodation potential.Tectonic subsidence was negligible to absent during the deposition of the third depositional sequence.

The geometries of the first and second depositional sequence (sigmoidal or climbing-divergent pattern) and of the third depositional sequence (oblique or climbing convergent pattern) are the product of different combinations of allocyclic factors.The second depositional sequences formed under conditions of differential subsidence, an increasing sedimentation rate and a sea-level curve characterized by a rapid relative sea-level fall --> a rapid sealevel rise---> a stillstand to slow sea-level rise.
The third depositional sequence formed under conditions of negligible subsidence, a decreasing sedimentation rate and a stillstand or fall of the sealevel.

Geometries similar to that of the second depositional sequence (sigmoidal or divergent patterns) simulated by computer modelling (see above), occur in the Exmouth plateau clastic wedge (Erskine and Vail,1988; Fig.96,92H) and in the post-middle Miocene sediments filling the slide scar described

by Mullins et al. (1986; Fig.110).This last example shows a divergent pattern composed of a set of depositional sequences thinning away from the slide scar; the eustatic sea-level curve (Haq, et al.,1987) indicates that these sediments were deposited during a sealevel highstand.
The results of the computer simulation indicate that a divergent pattern which is generated during a sealevel fall forms under a landward or seaward subsidence control. The computer simulation of the Exmouth plateau also suggests a subsidence control acting by means of a basculatory tectonics and a seaward tilting.

The recognition of an interplay of differential tectonics during the deposition of the second depsitional sequence casts a shadow of doubt on the eustatic interpretation furnished by Bosellini & Ferioli (1988).This also by considering that the Turonian (the time of formation of the lowstand facies tract of the second depositional sequence) is a non-glacial time interval; also the sea-floor spreading does not show to have undergone significant changes during this time.
On the other hand, the correlations of the two lowstand facies tracts with well pronounced sea-level lowstand of the curve by Haq, et al.(1987) is well evident.

This apparent contradiction leads to reconsidering the significane of these sharp lowstand of the sea-level that are visible from the eustatic curve.

In the Gargano peninsula the Eocene succession at some localities (Tremiti Islands; Vieste area: Cremonini, et al.1971) records the direct deposition of shallow-water sediments (Formazione di San Domino) above the basinal succession (Formazione di Caprara). The amount by which sealevel had fallen to produce such a sedimentary record is unrealistic; a tectonic uplift in these cases is more likely.That tectonics was active during the sedimentation of the Cretaceous and Eocene slopes is apparent when one examines the nearby coeval bedding disturbances of pelagic deposits consisting of recurrent, spectacular slumped horizons sandwiched between undisturbed layers.

The relative sea-level fall on the platform and the differential subsidence in the basin is in accordance with a model of tectonic tilting and flexure, and tectonic inversion characterizing the initial stages of the evolution of a marginal geosyncline (e.g. Wezel,1988).This mechanism, quite common, has been recognized for example along the continental platform bordering the Tyrrenian basin, in the Mediterranean Sea.

The interpreted reconstruction of the first and second depositional sequence combines the different thicknesses of the two sequences, the location of the third depositional sequence at a lower topographic level than the second depositional sequence, and the types of patterns of the two sequences (Fig.126). It is similar to the geometry described by Mullins,et al.(1986).It differs from the interpretative cross section of Bosellini (1989) which assumes that the third depositional sequence formed on top of the second depositional sequence.In his reconstruction, Bosellini (1989) did not take into account the general paleogeographic data, which indicate that the most part of the Gargano paeninsula was already emerged at the time of deposition of the third depositional sequence (Azzaroli Cita,1967).Also,the third depositional sequence is located topographically at a lower level than the second depositional sequence.

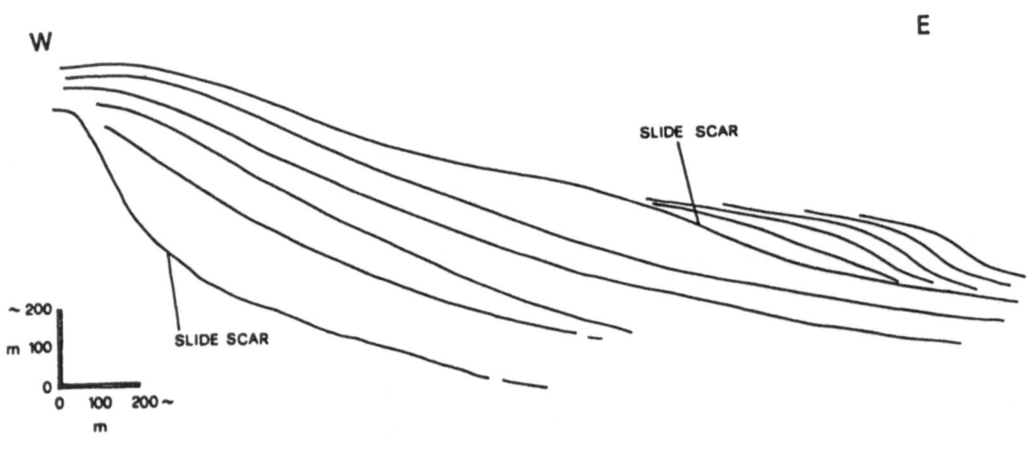

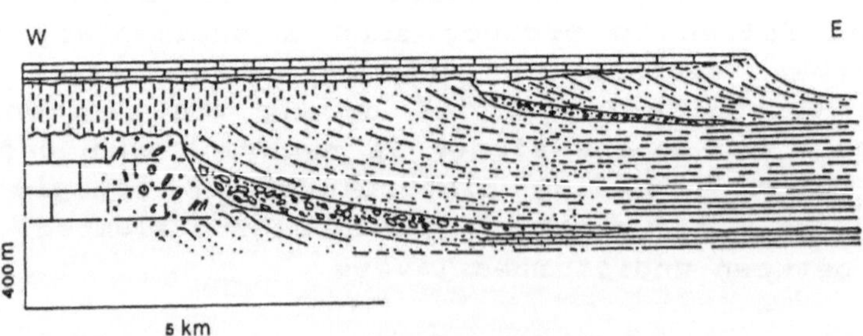

Fig.126 - Geometries of the second and third depositional sequences (above) and comparison with the cross section of Bosellini (1989) (below).

Middle Triassic carbonate buildups, Dolomites

Introduction

The Middle Triassic of the Dolomites records a period of extensional tectonics attributed to intracontinental rifting (Brandner,1984) or diapiric uprise (De Jong,1967).In the Anisian the carbonate platform underwent a complex deformation leading to its break-up and the formation of a complex buildup mosaic (buildups separated by pelagic basins).

The differential tectonic subsidence and uplift led to the formation of a ridge, the "dorsale badioto-gardenese" (Pisa,et al.1980,) which is basically a ramp structure elevating the basin in the west, as results from an examination of the cross section compiled by Bosellini (1965).

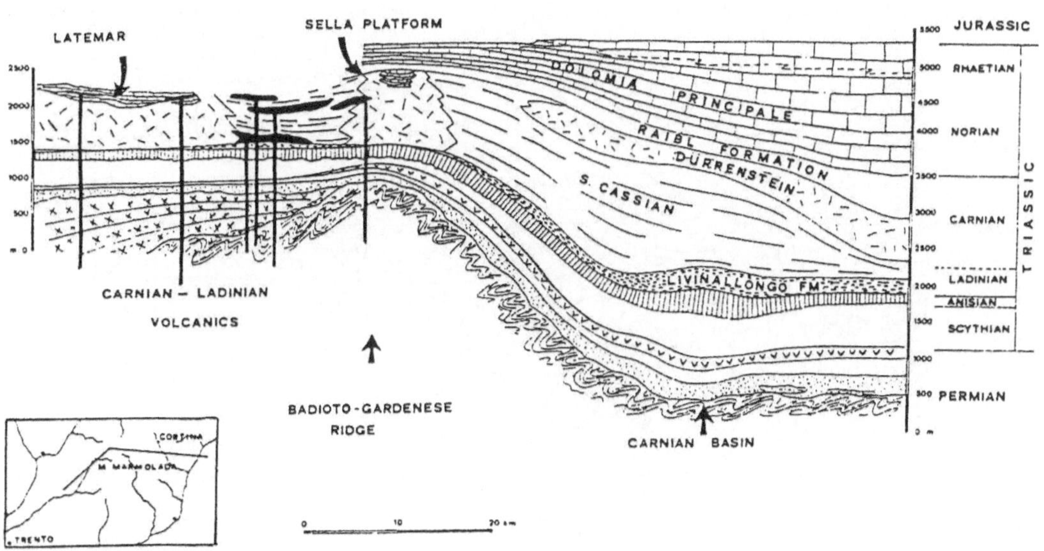

Fig.127 - Schematic reconstruction of the geometrical relations between the Permian and Triassic Formations of the Dolomites (simplified from Bosellini,1965).

Due to the late Anisian uplift, Ladinian buildups grew from slightly elevated areas in the west.Thicknesses of buildups are of about 800-1000m. The estimated subsidence rate was on the order of 400 m/Ma (Bosellini,1984). These buildups were surrounded by a starved, anoxic basin. Basinal sediments of the Livinallongo Formation (0-200 m thick) which record a vertical deepening-upward trend (Cros,1974).Thicknesses of the Formation are 800 m in the East, 600 to 500 m in the West.

The development of shallow-water carbonate buildups display a strong antecedent topography control, as they developed on uplifted areas by keeping pace with a strong subsidence. They are: the Schlern, Sciliar,Rosetta, Latemar, Cernera, Marmolada, Catinaccio buildups (Fig.128). In the west individual carbonate buildups, separated by pelagic deposits, grew vertically and

<u>Fig.128</u> - Distribution of Ladinian - Carnian buildups of the Dolomites (simplified from Leonardi,1967).

laterally. The progradation is rather consistent in the west (9-10 km: see Catinaccio: Bosellini, 1984), much less in the East (1-2 km). Sedimentation rates were also variable.
The Upper Ladinian was a time of intense tectonic and volcanic activity (Viel,1979) that led to the emersion of carbonate buildups (Blendinger,1982; Wendt & Fursich,1980). Carbonate sedimentation was resumed in the Carnian. The formation of a continental flexure was surmised by Blendinger (1982) and Sarntein (1967).
The buildups in the West in the Dolomites were recently studied by Bosellini (1984); those in the East by Blendinger (1984;1985,1986).Whereas the first (1984) produced a model of progradation, the second working in the East proposed a model of stationary to retreating platform. This led to some discussion centered on the neeed to delineate the spatial limits of validity of the two respective models.

The difference in the evolution between the buildups located in the western and eastern sectors may be integrated by extending the intrashelf ramp model to the Dolomite region, which results therefore subdivided into the following sectors (Fig.129):

1) a 'shallow ramp', exemplified by the Latemar-Catinaccio buildups;
2) an 'intermediate ramp', constituted by the Marmolada Limestone; and
3) a 'deep ramp',exemplified by the Cernera massif.

The sedimentary trends in the three sectors were respectively characterized by progradation (Latemar-Catinaccio), aggradation (Marmolada Limestone) and retrogradation (Cernera buildup).

The basin progressively deepened: whereas carbonate buildups became widespread in the West and merged one into another,deeper basinal conditions became dominant in the east (Gaetani,et al.,1981).

Shallow ramp

The Latemar buildup,800 m thick, Late Anisian to Ladinian in age, has been studied in detail by Goldhammer,et al.(1990) who recognized the following facies in ascending order: 1) open subtidal platform (250 m thick); 2) restricted peritidal platform (90 m thick); 3) tepee horizon (120 m thick); and 4) restricted peritidal platform with thinner cycles (210 m thick).

The Latemar represents a third-order depositional sequence.The vertical changes in lithology reflect a progressive decrease in the accomodation. Progressive increase in subaerial features on top of individual m-scale cycles records a shift from submergent conditions to some more punctuated by subaerial exposure. The overall trend of the megasequence is shallowing-upward.

Goldhammer,et al. (1990) distinguished: 1) a transgressive system tract (aggradational, amalgamated megacycles); and 2) an early highstand system tract constituted by thinner cycles which record third order sea level falls of the sea level and a decline in the third order accomodation.

The Latemar platform underwent a progradational and aggradational centrifugal growth. The model of growth is a climbing divergent pattern ('Ladinian model' by Bosellini, 1984).

Apart from differences in size of the individual buildups,the situation in this sector is broadly comparable to that occurring in the shallow ramp of Florida Bay if individual buildups are compared to banks of Florida Bay which form an anastomosing network of island and banks,and are characterized by aggradation followed by progradation (mainly directed towards the basin or deep ramp),varying sedimentation rates and merging of banks.

Intermediate ramp

The Marmolada Limestone, 100 m thick, does not record a cyclicity. Conversely, Blendinger,et al (1982) observed at several places 100 m of superimposed sandwaves (e.g. Sasso Piatto) and the stacking of homogeneous facies. Multidirectional sandwaves occur in nearby areas, for example at Viezzena.
According to Blendinger (1986) the Ladinian platforms at this area record a continuous subsidence; progradation was prevented by synsedimentary faults; subsidence rate here was stronger than in the western area (2:1 ratio). The resulting platform was similar to a tectonic horst. Clinoforms here display a descending pattern (Bosellini, 1984).

The Marmolada Limestone represents an area of gentle connection of the western side of the 'shallow ramp' to the basin in the east.It developed under prevailing subtidal conditions (Gaetani, et al.,1981).

The facies organization is analogous to that of Lili Bank (Hine,1987) consisting of : 1) sand bar facies; 2) sand cay; 3) sand flat; and 4) reef facies. There is also an analogous environmental differentiation into a leeward, low-energy sand flat and a windward margin.

Deep ramp

The deep ramp consists of basinal sediments, thickening upwards, and isolated buildups. Both display a deepening-upward trend.The Cernera buildup,700 m thick (Blendinger, et al.,1982) is characterized by upward converging margins, as a result of a strong subsidence.The platform eventually was drowned and covered by pelagic ammonoid facies.
The general trend is similar to deepening-upward sequences developed in the deep ramp.

In the Upper Ladinian the Dolomites were subjected to deformation. Platforms in the west were uplifted and underwent a subaerial exposure, as a result of rotational subsidence. In the Carnian carbonate deposition was resumed: the Dürrenstein Formation records a progradational trend developed as an offlap ramp (Galli,1989; chapter #).

The Ladinian - Carnian stratigraphic interval therefore represents an onlap - offlap geometry.

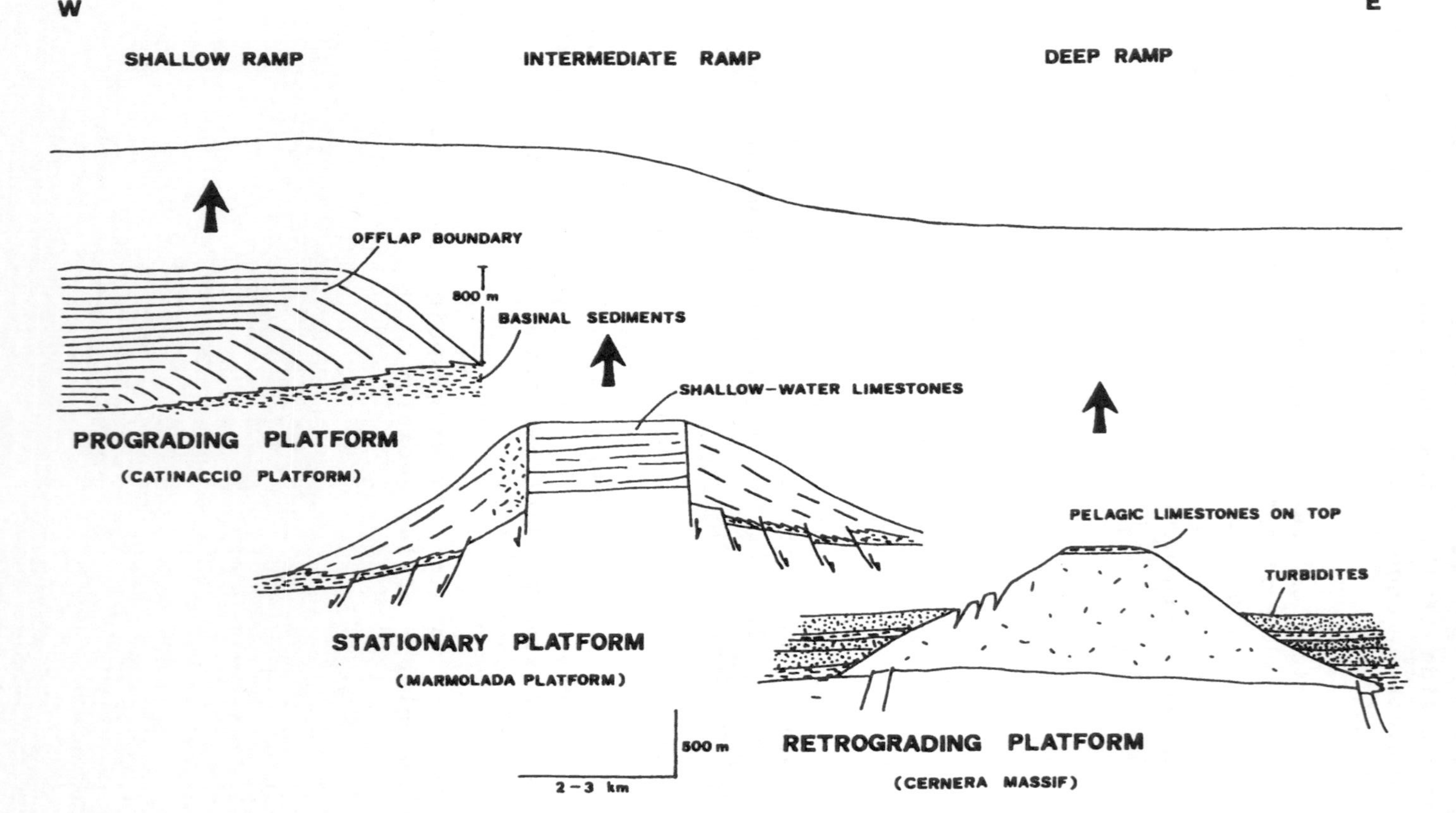

Fig.129 - 'Onlap ramp' interpretation of the carbonate buildups of the Dolomites (sketches from Bosellini,1984, Blendinger,1986;Blendinger, et al.,1982).

Part III

A sequence common to shallow-water and slope sequences examined in the two preceeding parts is a thinning-upward and deepening-upward (transgressive) hemicycle overlain by a thickening-upward and shallowing-upward (regressive) hemicycle. This type of sequence is the 'leit-motiv' or the dominant theme of the sequences described in this work.

Modal sequence

Onlap ramps (and divergent patterns) are the product of tectonic tilting and are associated with relief inversions and deepening-upward sequences (hinge sequences). Rotational tilting appears to be linked by Mörner (1986) to geoidal eustasy. The association of an oligotypic fauna in the deep ramp and the frequent intercalations of black shales or poorly fossiliferous mudstones in the ramp reflect the poorly oxygenated conditions existing in the basin ("AOE" events). The onlap geometry is thought to be found mainly during krikogenetic rejuvenation periods.It is a praecursor episode of the drowning of a platform.

Offlap ramps (and convergent patterns) are progradational units; they have opposite characters to those of the onlap ramp geometry; they are linked to krikogenetic quiescence periods.

The two main geometries of intrashelf ramps described in the previous chapters are not restricted to shallow-water carbonate settings. Along the vertical,offlap geometries normally are developed above the onlap geometries. The resulting geometry (onlap ---> offlap) is recorded in several carbonate palimpsests.

One example is represented by the Mesozoic carbonate platforms in the Eastern N.America Atlantic margin where a Lower Jurassic carbonate aggradational stage formed during an episode of

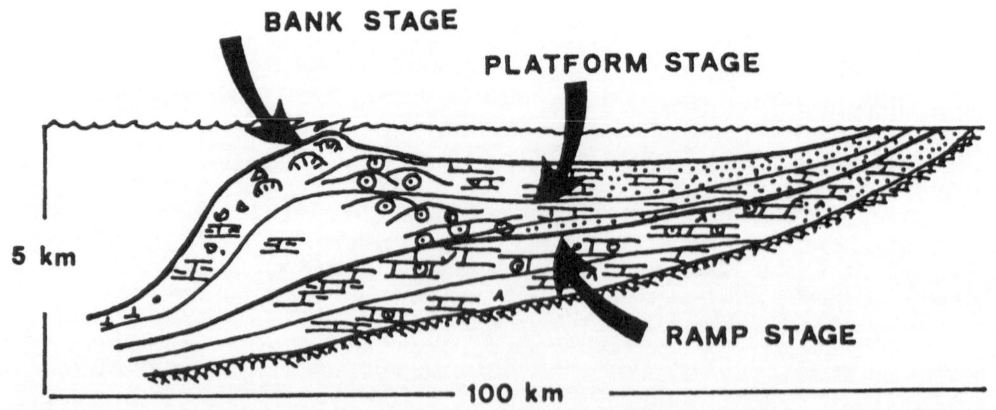

Fig.130 - Jurassic platform of the north America continental margin (after Jansa,1981).

rifting-drifting as a result of a rapid subsidence. In the Upper Jurassic the resumed progradation led to the development of a platform and later to a bank (Jansa,1981;Fig.130).

Another similar transition occurs in the Devonian carbonate shelf, in the Canning Basin, where a Frasnian aggradational margin is overlain by a Famennian progradational platform (Playford,1980).

The Great Bahama Bank studied by Eberli and Ginsburg (1987) shows an aggradational platform formed in the Middle Cretaceous to Miocene underlain by a prograding platform in the Pliocene.

All the above aggradational stages fall into krikogenetic rejuvenation periods. Such an architectural organization is also found in siliciclastics.

The onlap-offlap geometry results from the following succession of episodes:

1) tectonic tilting (leading to the emersion of the platform and deepening of the basin and shedding of megabreccias) ;
2) formation of the onlap ramp by vertical sediment growth.
3) formation of the offlap ramp by progradation.

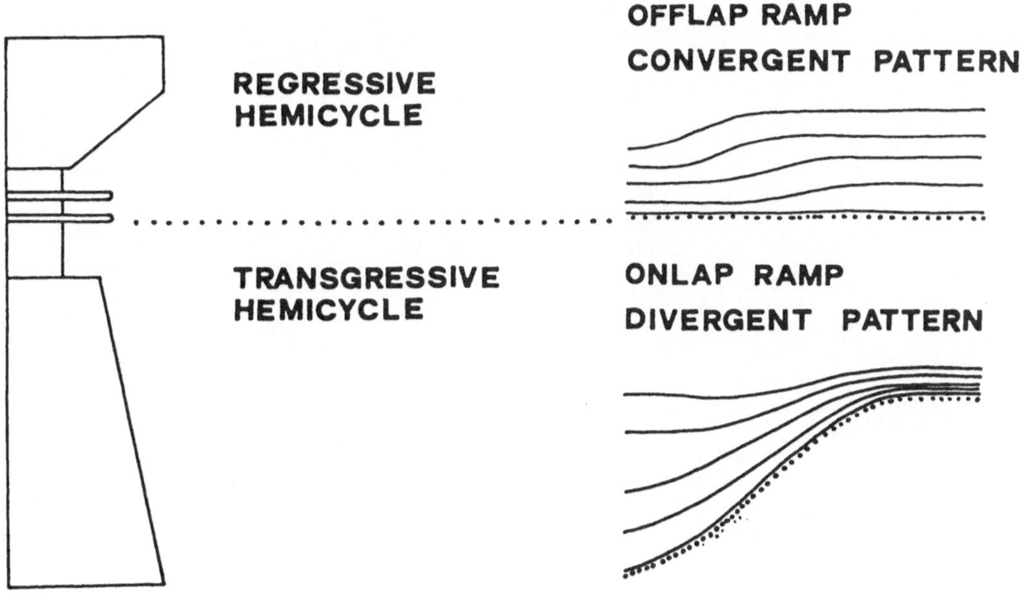

Fig.131 - Modal sequence.

The resulting modal sequence (Fig.131) is composed of a thinning-upward - and deepening-upward hemicycle overlain by a thickening-upward shallowing-upward hemicycle.It ranges in thickness from tens to hundred of meters as it occurs at a small, parasequence scale and at a bigger, megasequence scale, (Figs. 31,32,36,49,56,86?,103,108,116,121) therefore suggesting a hierarchy of the controlling factors at different levels.

Part IV

Most of the divergent patterns and onlap ramps described here fall into periods of short-term sea-level fall.Short-term sea level falls represent short-term phases of geoidal deformation.Onlap ramps,divergent patterns, megabreccias, drownings of carbonate platforms,relief inversions, and seismoturbidites are linked to these short-term periods of sea-level fall.

The plotting of these significant time intervals in the 'eustatic' chart by Haq, et al.(1987) shows that lag-times between short-term sea-level fall periods are not spaced at random, each lag-time being the summation of the following two lag-times.

Short-term sea-level falls: an indicator of geoidal pulses?

The case history of the Gargano massif (chapter #4) has arisen a problem regarding the interpretation of the nature of the short-term Turonian and Ypresian sea-level falls responsible for the shedding of megabreccias into the basin.

The chart of global sea-level (Haq et al.(1987) shows abrupt falls of the sea-level associated with type-1 unconformities and lowstand system tracts (lowstand fan;lowstand wedge), as shown by the scheme of Fig.132.

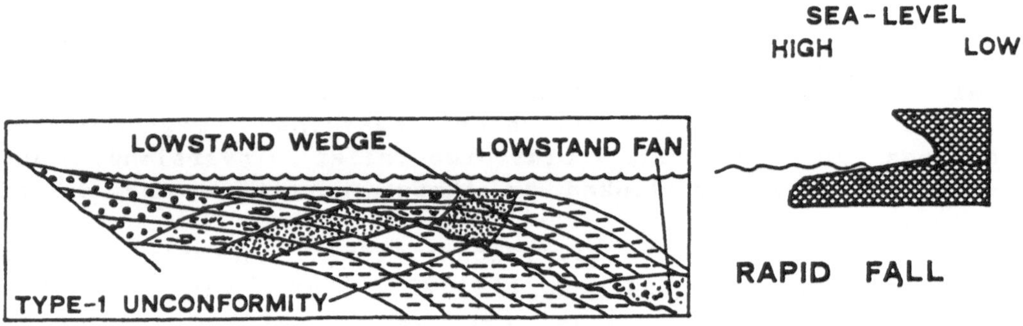

Fig.132 - Elements of sequence stratigraphy associated with a rapid sea-level fall (after Helland-Hansen,et al.,1986).

The derivation of the eustatic chart by Haq, et al.(1987) from the analysis of passive margin sequences is taken by some authors as an evidence for its source from global subsidence behavior, rather than eustasy (Miall,1984).In fact, the chart of global subsidence built up by Guidish, et al.(1984;Fig.133) mirrors the global eustatic curve. According to Sloss (1984) unconformities bounding sequences (type-1 unconformities) result from globally synchronous episodes of cratonic uplift. Stille(1924) considered unconformity surfaces to be related to peaks of maximum regression corresponding to tectonic phases.

A possibility discussed in this chapter is that short-term sea-level falls represent geoidal pulses.

Mörner (1986) believes that the short-term eustatic falls, of brief geological duration that punctuate the sea-level chart by

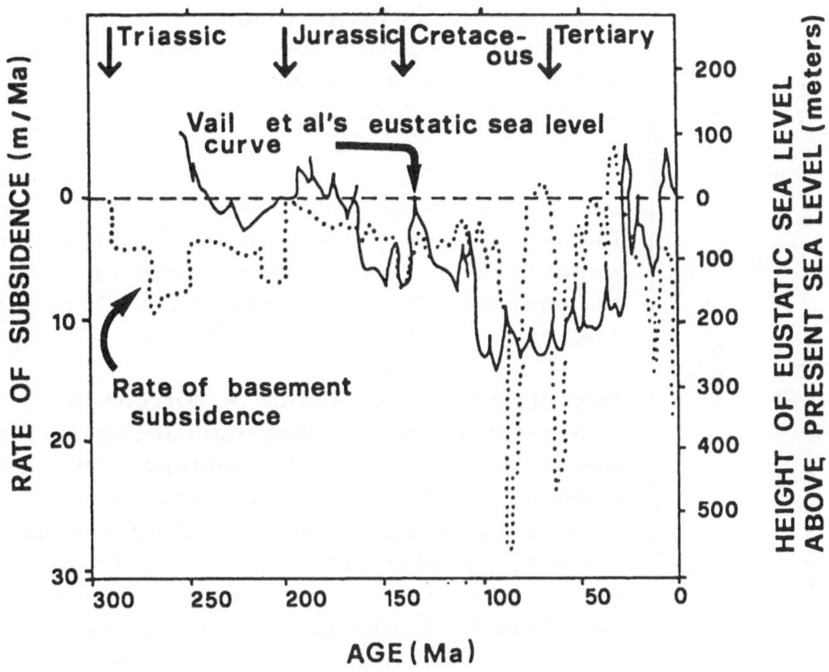

Fig.133 - Relationships between the eustatic sea-level changes and the changes of basement subsidence rate (Guidish, et al.,1984).

Haq, et al.(1987) are produced by changes in the geoidal configuration.According to Mörner (1986) geoidal eustasy consists of sea-level changes caused by deformation of the geoidal surface, controlled by mantle convection mechanisms.

The following aspects of carbonate sedimentology can be viewed as key elements to the recognition of short-term (<500.000 yrs) phases of geoidal pulses:

1. onlap ramps;divergent patterns;
2. megabreccias;
3. drowning of platforms;
4. seismoturbidites;
5. relief inversions.

The Cretaceous system offers a good opportunity to study geoidal eustatic changes: in fact, it is a non-glacial systems, hence, eustasy was the product of the interplay of tectonics and geoidal eustasy. Mörner (1981) recognized Cretaceous latitudinal, north - south migrations of the geoidal eustasy. The Upper Cretaceous was a period of intense geoidal deformation. The most spectacular example of geoidal change

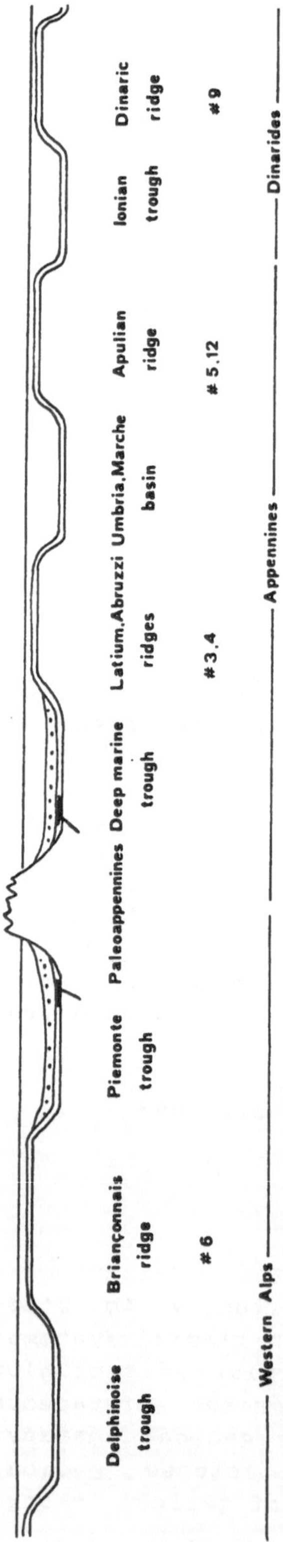

Fig.134 - Paleogeographic scheme of the central Mediterranean area (simplified after J.Auboin and Brousse, 1977) showing a series of ridges separated by basins.

is represented by the foundering of the Darwin Rise in the Pacific Ocean, concomitant with the rising of the Melanesian Rise and the East Pacific Rise (Menard,1964; Wezel,1988).

During the Cenomanian - Turonian interval the rate of sea-floor spreading decreased; eustasy was mainly caused by geoidal changes. This interval records at several places complex facies characterized by slumps, facies changes, megabreccias and unconformities. The Mediterranean region consisted of elongated basins separated by intrageosynclinal ridges (Fig.134) which became the sites of carbonate platform development.Slumps, megabreccias and unconformities were produced by uplift and deformation of the intrageosynclinal ridges. Deformation phases were concomitant with the establishment of anoxic facies in the basins (Jenkins,1991 ;Wezel,1985). During the Cenomanian - Turonian there may have been tectonic phases superimposed on geoidal eustasy; nevertheless, the inference of a global episode of deformation of the geoidal surface is based on the worldwide extent of the 'events'.Geoidal deformation was not synchronous everywhere: emersion was not a generalized phenomenon: some areas remained submerged by a thin veneer of water, others were drowned and transformed into gujots and seamounts; pelagic,basinal conditions persisted in the geosynclinal troughs. Anoxic conditions ('AOE' events) were established in the basins. It is possible that uplift caused by the geoidal deformation 'event' contributed to the development of anoxic conditions: as a matter of fact, the whole proto-Atlantic ocean appears to be a semiclosed basin surrounded by shallow-water carbonate platforms in the Turonian -Cenomanian, as shown in Fig. 135.

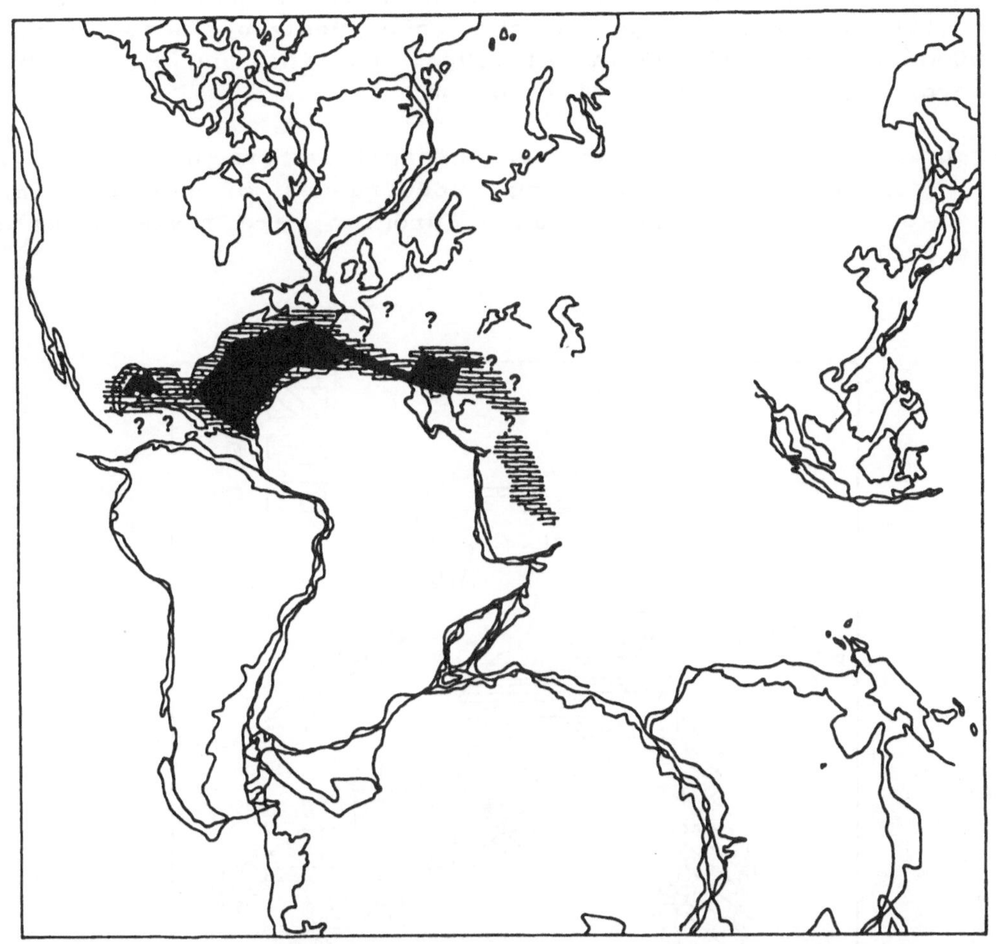

Fig.135 - Jurassic-Cretaceous situation showing the distribution of oceanic crust (black area) and carbonate platforms (from Jenkyns,1991).

Megabreccias

Megabreccia debris sheets are common features in ancient reworked carbonate sequences. They are interpreted in general as the product of large-scale collapse of carbonate platform margins.Carbonate megabreccias occur widely throughout the Phanerozoic, especially but not exclusively in the Mediterranean area.

The triggering mechanism of their formation is open to

discussion. Possible causes are: gravitational instability produced by overloading (Crevello and Schlager, 1980); faulting and seismic shocks (Mutti and Ricci Lucchi,1984; Mullins and Hine,1989); undercutting of the escarpment by seeps and/or bottom currents (Paull,et al.,1984; Mullins and Hine,1989); bioerosion; and chemical dissolution.

Megabreccias and associated unconformities are ubiquitous in the Mediterranean region. Some occurrences of megabreccias, dated mainly to the Cenomanian - Turonian, are reviewed below (Fig.136).

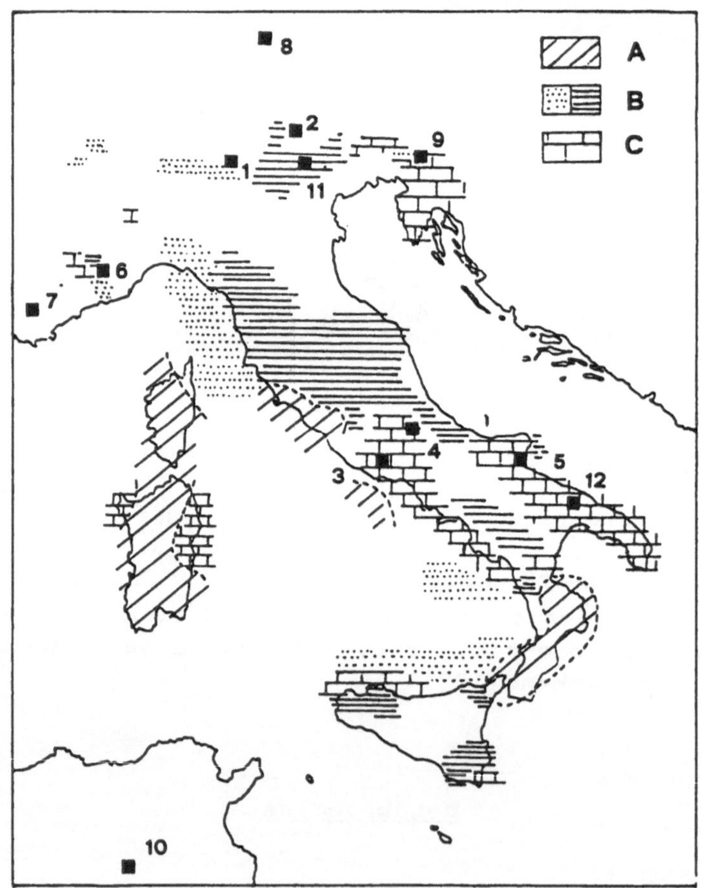

Fig.136 - Location of Cenomanian-Turonian megabreccias and related unconformities described in the text.A: emergent areas;B:basinal siliciclastics;C:carbonate platforms.

Lombardy basin (Fig.136:#1)

The Cretaceous sequence is characterized by basinal sediments passing from pelagic carbonates to a siliciclastic flysch. The Middle/Late Cenomanian sequence interval is represented by a succession of two chaotic layers (0 to 40 m thick) interbedded with thin bedded turbidites. Debris flow deposits occur also in the overlying Turonian interval (pers. obs.). They are analogous to debris flows interpreted as slope deposits described in nearby areas (Bichsel and Haring,1981). The two chaotic layers delimit a wedge onlapping onto the northern margin of the basin (Bersezio and Fornaciari,1987). Paleocurrents indicate a northern provenance.

The formation of these conglomerates was related to an eoalpine compressive phase in the southern Alps. These conglomerate layers would be a stratigraphic evidence for the beginning of Cretaceous activity during the Cenomanian-Turonian, as is indicated by Doglioni and Bosellini (1987). Castellarin (1982) related the formation of megabreccias to the action of incipient transcurrent faults (extensions of oceanic transforms) located north of the area, between the African and the European margins. According to Bersezio and Fornaciari (1987), the unconformity records a first closure of the Lombardy basin towards the north.

Insubric flysch (Fig.136: #2;Fig.137)

A 600 m thick flysch succession crops out in the area comprised between Malè and Rumo (Castellarin, et al., 1976; Castellarin, 1982),along a 12 Km long belt. The 'Upper conglomerates' (Insubric flysch) studied in detail by Castellarin, et al.(1976) was delivered from the north by high-density currents. The textural features of these conglomerates indicate that phases of meteoric degradation took place prior to transport. These conglomerates would represent a narrow turbidite fan located on the direct prosecution of the northern slope. It is generally accepted that a northern continental shelf gave origin to these clastic deposits by uplifting and/or rejuvenation. Tectonic movements producing these uplifts took place in the Albian-Turonian time interval, in the northern Calcareous Alps.

The beginning of the flysch deposition in the Turonian was interpreted as the remote effect of the Pregosau tectonic phase. The source of these clastics is located within the Austroalpine domain (Castellarin,1982).

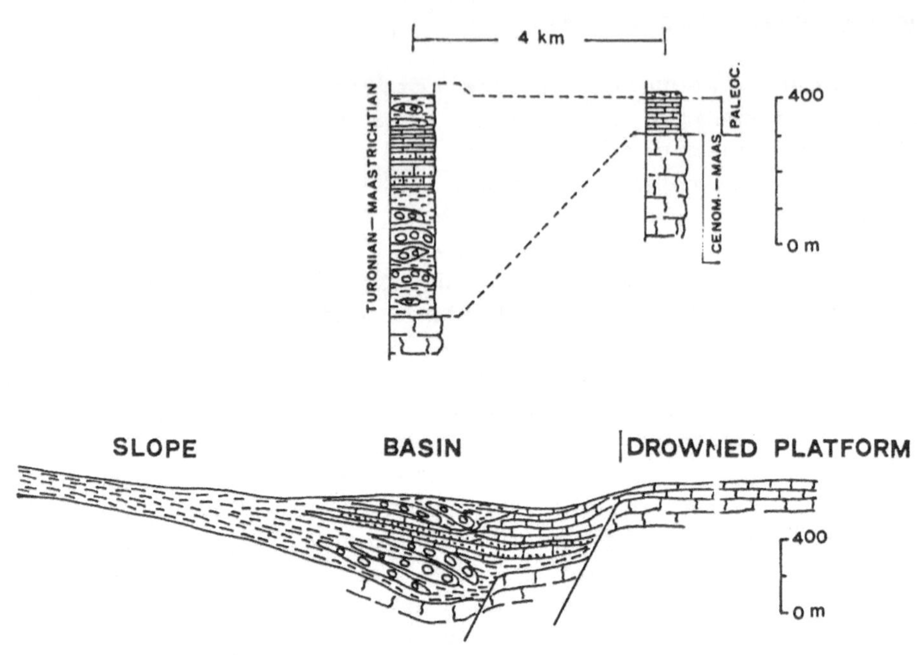

Fig.137 - Hypothetical relations between the Venetian platform and the Flysch basin during Upper Cretaceous (from Castellarin, et al.,1976).

Northern Calcareous Alps (Fig.136:#8).

Gaupp,et al. (1981) described a Cenomanian-Turonian sequence developed above Triassic-Jurassic carbonate rocks, passing from shallow-water calcarenites to a turbidite flysch, deposited during two major tectonic phases of the early Alpine orogeny (Austrian - Mediterranean phases). A clast-supported polymictic megabreccia ('Blockbreccia: Gaupp,et al.1981) developed during the Cenomanian-Turonian transition, along the northern margin of the northern Calcareous Alps, for more than 100 Km. This megabreccia is thought to have originated from blockfalls at submarine fault scarps, due to intrabasinal uplift.

Gaupp,et al. (1981) suggested a combination of nappe imbrication, followed by emergence, blockfaulting and isostatic subsidence of the thrust sheets.

Lombardy basin - Venetian platform transition (Fig.136: #11; Fig.138,139).

One of the best documented examples of tectonic synsedimentary evolution of a carbonate margin is that described by Castellarin (1972). The investigated area separates the Lombardy basin to the east (deep water,cherty limestones), Jurassic to Cretaceous in age, from the Venetian platform (Fig.138).Since the beginning of the Jurassic, synsedimentary block-faulting was active in this transitional area.A fault was activated at the beginning of the Liassic, contemporaneously with the deposition of the Trento platform, in the Pliensbachian, Pliensbachian-Toarcian transition, and in the Aalenian.

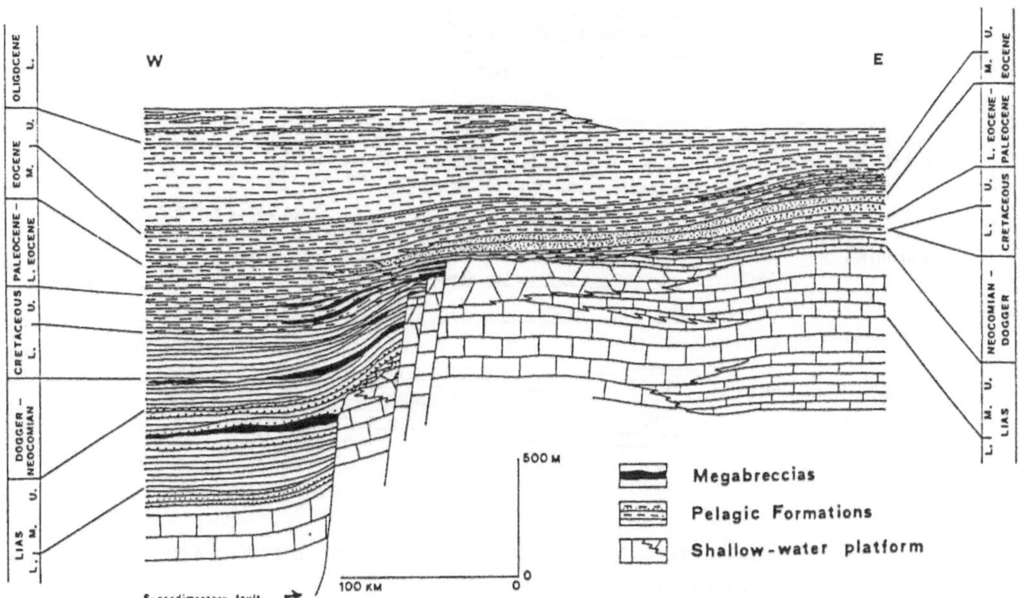

<u>Fig.138</u> - Cross section of the Lombardy Basin - Venetian platform transition (from Castellarin,1972).

The eastern side of the Trento platform,bordering the Belluno Trough, has a zig-zag outline due to the control of synsedimentary faults. The margin was cut by canyons which delivered megabreccias (Pelf Breccia:Bosellini, et al.,1981) into anoxic,poorly oxygenated basin. During the Liassic the margin underwent a retrogressive migration by fault activity.

The reactivation of faults in the western border of the Trento platform took place in the Valanginian - Hauterivian interval, as is shown by the accumulation of poligenic megabreccias. In the following time intervals (Turonian-Maastrichtian) a retrogressive, platformward migration of fault zones took place (cannibalistic development of poligenic breccias). In the Turonian the 'M.Gurlo' paleofault was individuated, contemporaneously with the uplift of the platform, as is demonstrated by an inversion of the inclination of blocks (Fig.139).

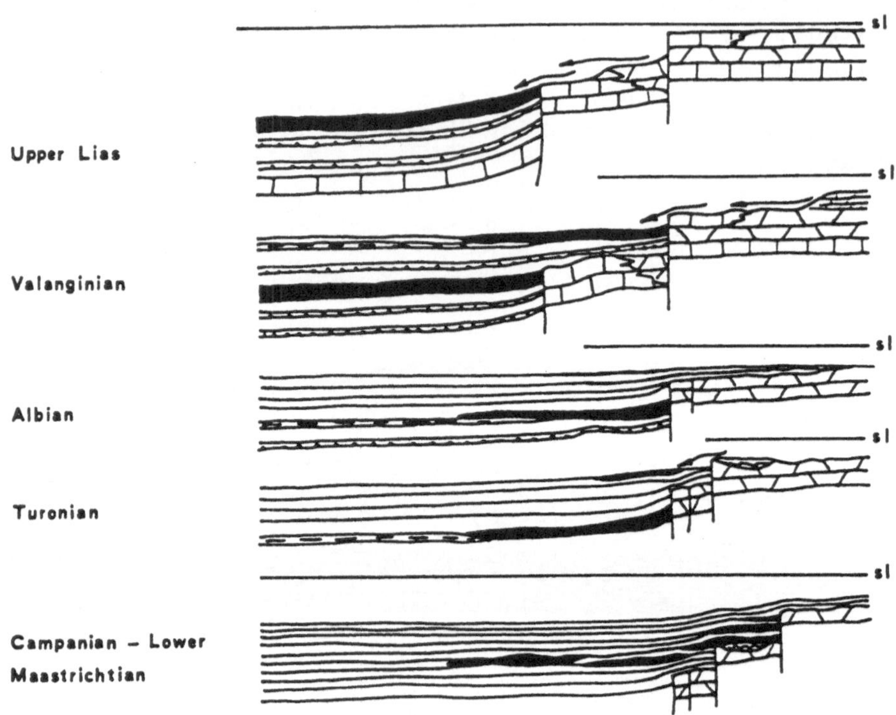

Fig.139 - Evolution of the margin in the Jurassic-Cretaceous (from Castellarin,1972).

The 'Bellino-Garda' paleofaults were reactivated in the Upper Cretaceous, with the formation of impressive, huge detachment scarps and megabreccia accumulations,as a result of a general uplift of the Venetian platform (Castellarin,1972).

According to Castellarin (1982), the reactivation of these fault scarps was a consequence of transcurrent movements during the Gosau orogenic phase.

Friuli platform (Fig.136:#9;Fig.140)

The sedimentary and tectonic evolution of the eastern part of the Friuli platform was studied by Tunis and Venturini (1986) who proposed a model consisting of a complex series of stepwise retrogradation episodes of the platform margin resulting from transcurrent tectonics. Figure 140 shows a wedge with unconformities dated to the Toarcian, Valanginian-Hauterivian, Cenomanian-Turonian and Campanian.

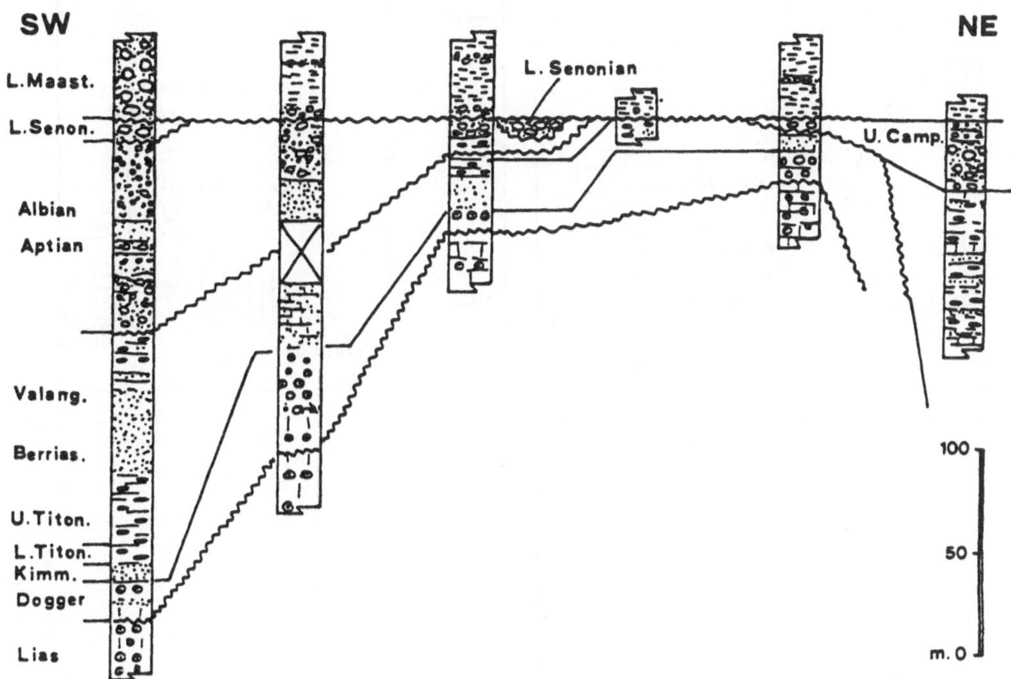

Fig.140 - Correlations of stratigraphic sections of the eastern part of the Friuli platform.

In the Turonian-Cenomanian the platform underwent a period of non deposition and emersion followed by subsidence.

Cretaceous buildups, south of France (Fig.136:#7;Fig.141).

The data by Masse and Philip (1981) show a platform in the south of France separated by the Alpine Basin by a ridge area ("Bombement durancien") trending east-west which formed as early as the Albian (Fig.141). An overall deepening-upward tendency, initiated in the Cenomanian, gradually established pelagic conditions during the Upper Turonian and determined the reduction of the extent of the shallow-water carbonate deposition in the platform.This deepening-upward tendency in some areas was counterbalanced by deformation of the "Bombement durançien" which led to the development of shallow-water carbonates in the Turonian. Modest uplifts are recorded in the Valanginian and Turonian in other areas of the southern Alps of France (Sisteron area).

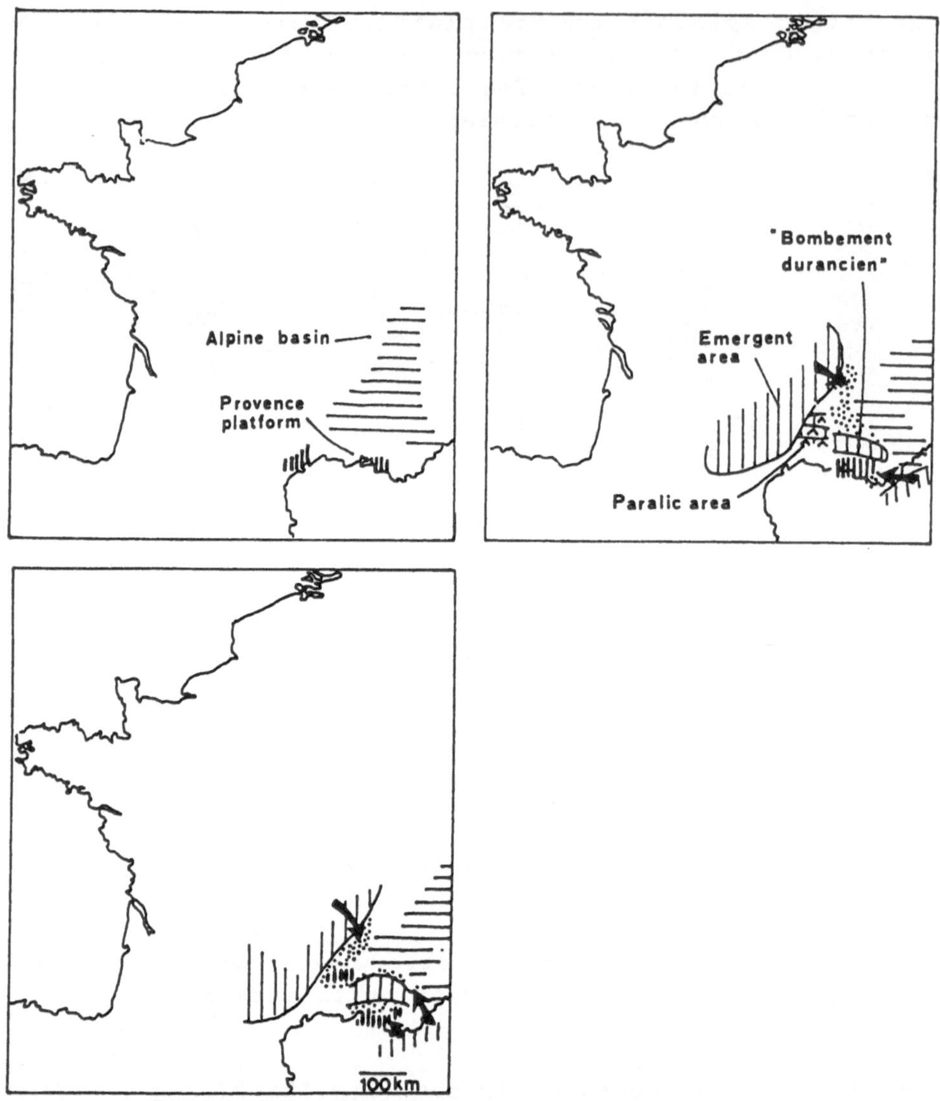

Fig.141 - Paleogeographic maps of the upper Albian (left),Cenomanian (center) and Upper Turonian (Masse and Philip,1981).

Ligurian Briançonnais (Fig.136: #6;Fig.142).

The Ligurian Birançonnais represents the prosecution of the Briançonnais domain, a paleogeographic unit located between the foreland and the alpine eugeosynclinal domain (Ligurian ocean). It extends from the Corsica throughout the western Alps. Probably it represents an intrageosynclinal ridge separating the miogeosyncline from the eugeosyncline. The differentiation of the Ligurian Briançonnais domain initiated since the Triassic. An uplift due to an oceanward tectonic tilting took place in the Sinemurian-Aalenian time. The Middle Jurassic to Lower Cretaceous interval was a time of shallow-water carbonate sedimentation. Successively, a deepening-upward trend, from the Upper Jurassic (Malm) led to a generalized development of a hardground in the Aptian-Cenomanian. Unfortunately, detailed analyses of this area are lacking (Vanossi and Gosso,1985). The persistence of a hardground horizon in an area subjected to a deepening-upward trend suggests that the Ligurian-Briançonnais underwent a relative uplift in the Turonian-Neocomian.

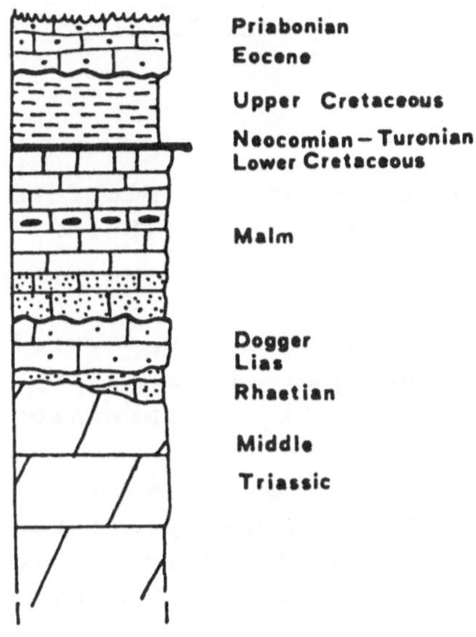

Fig.142 - Stratigraphic column of the Ligurian Briançonnais (Vanossi and Gosso,1985).

Abruzzi platform, central Italy (Fig.136: #3,4; Fig.143,144).

The Upper Cretaceous of central Italy, studied by Carbone and Sirna (1981) is represented by carbonate epioceanic platforms. Carbonate sedimentation was characterized by the development of rudist-coral communities, back-reef areas and foreslope sectors up to the Cenomanian. Progradation was interrupted in the Turonian: the shelf margin was subjected to disruption which led to the uplift of the shelf margin and to the formation of deep sea ways and intrashelf ramps (pers. obs.), and development of beaches (Fig.143). The subaerial exposure of the margin was attributed to tectonism. The transgressive trend continued in the Senonian, when the shelf margin was transformed into a seamount.

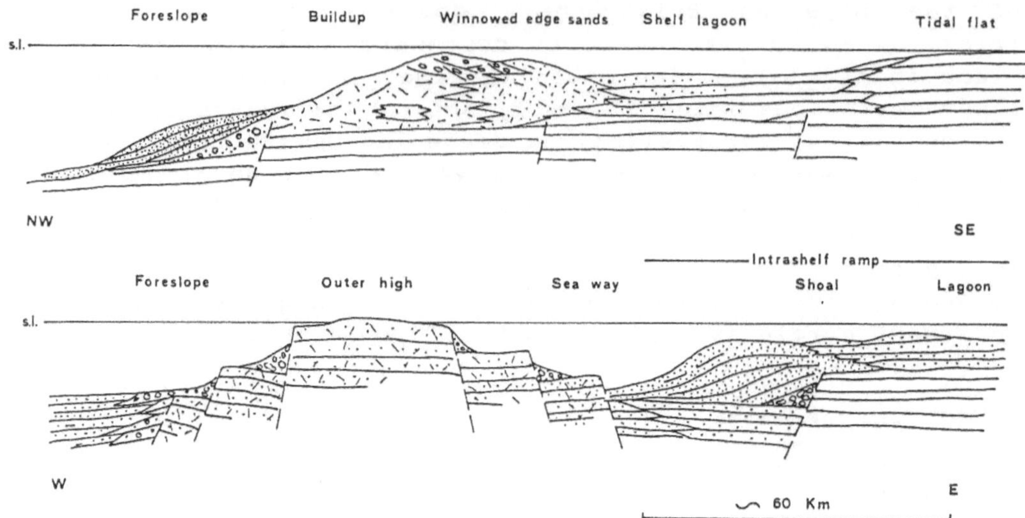

Fig.143 - Cross section showing the evolution of the platform margin in the Cenomanian (above) and Senonian time (below). From Carbone and Sirna (1981).

A detailed analysis of the Latium-Abruzzi platform was carried out by Colacicchi (1987) who distinguished three large sedimentary cycles, each characterized by a fast transgression followed by a more gradual progradational phase (Fig.144). He constructed a chart showing the evolution in time and space of various environmental zones of the platform (inner platform-margin-slope) from the Triassic to the Miocene. The first cycle (Lower Dogger to Valanginian) shows a strict correlation with the eustatic curve by Haq, et al.(1987). In the second (Valanginian - Cenomanian) and third cycles (Cenomanian - Maastrichtian) the three sectors of the platform follow different trends. A maximum landward shift of the inner platform took place in the Aalenian, Valanginian and Turonian, then in the Maastrichtian, Ypresian, Priabonian-Rupelian and Serravallian. These facts were taken as an evidence for tectonic interference: the drowning of peripheral areas was counterbalanced by uplift up to emersion of the inner platform areas.

Apulian platform, southern Italy (Fig.136:#12)

Iannone and Laviano (1980) and Sinni and Borgomano (1989) studied a shallow water Cretaceous succession. They found an angular unconformity and a discontinuity surface separating two shallow water Formations: the 'Calcare di Bari' from the

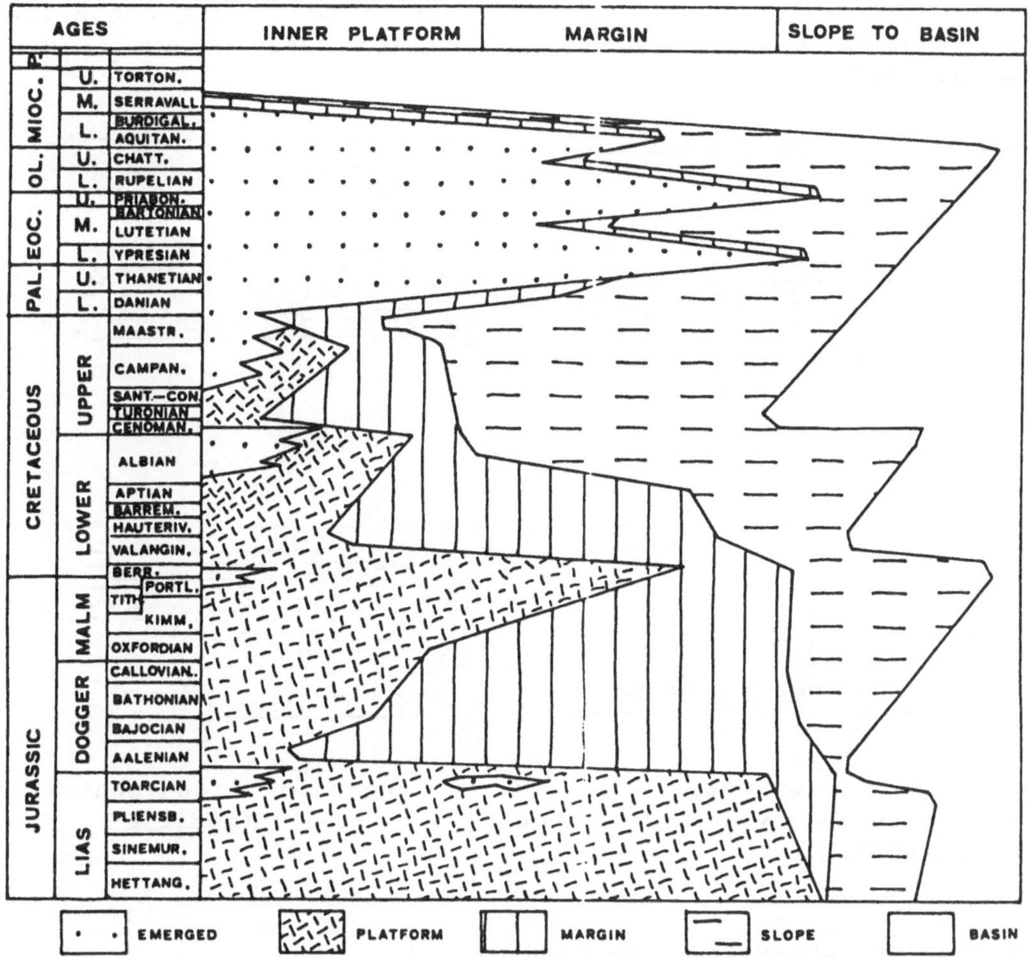

Fig.144 - Evolution of the Latium-Abruzzi carbonate platform margin (from Colacicchi,1987).

'Calcare di Altamura'. According to these authors this discontinuity corresponds to a generalized emersion of the carbonate platform due to a tectonic episode in the Middle and Upper Turonian.

Upper Cretaceous Rudist Reef complex,Central Tunisia (Fig.136: #10;Fig.145).

A shallow water carbonate platform, described by Negra, et al.(1987) was dissected in the Turonian by synsedimentary tectonics. A spectacular outcrop at Sidi Abdelkader (60 km SW of the city of Hajebel Aiöun) shows a recifal complex developed above uplifted parts of basinal deposits containing Ammonites, calcispheres and other undetermined biota. Faults are evident in the field.

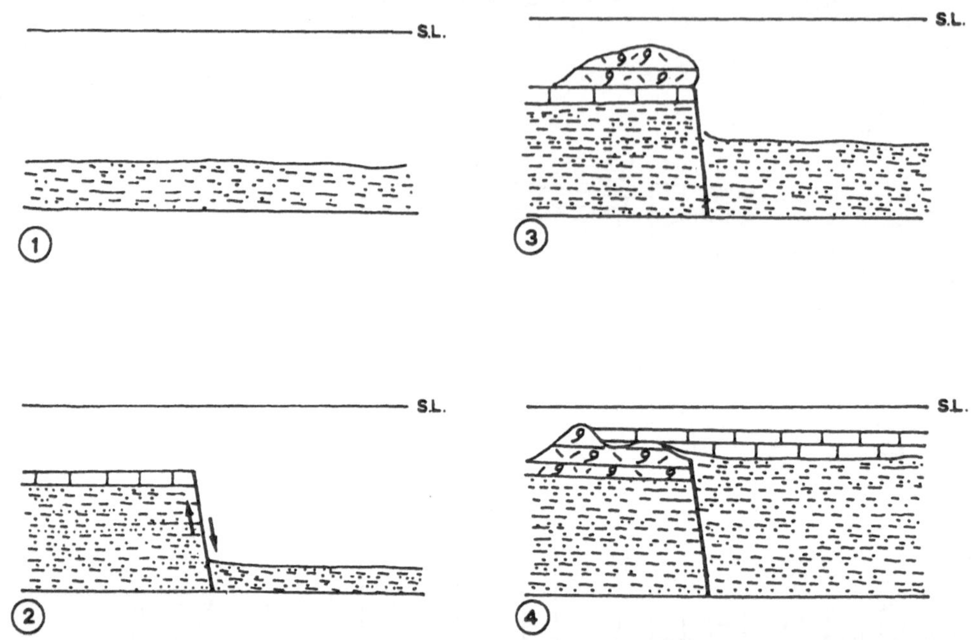

Fig.145 - Interpretative sketch of tectonic - sedimentation relationships at Sidi Abdelkader area (modified from Touir,1986).

Arabian carbonate platform margin,Oman

The Arabian carbonate platform,studied by Watts and Blome (1990) in the Oman area formed along the transitional zone between a passive continental margin and a deep oceanic basin in the east.The platform margin (Qumayrah and Mayhah Formations) was influenced by progressively increasing tectonism, as it approached a subduction zone. In the Cenomanian time, the Oman margin underwent uplifting and erosion; consequent to the oversteepening of the slope and deformation, a megabreccia was delivered into the basin. Watts and Blome (1990) argue that a 'notable sea-level lowstand at this time could have helped trigger mass movements downslope'. The above authors related the resedimentation of the Qumayrah and Mayhah Formations at this time to either initial deformation and thrust faulting of the slope, or to large gravity slides. However, they observed that 'the deformation of the Mayhah Formation and synorogenic sedimentation of the Qumayrah Formation preceded emplacement of the higher nappes'.

Seismoturbidites

Mutti and Ricci Lucchi (1984) suggested to term seismoturbidites the anomalous megaturbidites consisting of the delivery of exceptionally great volumes of sediment and of great lateral extent. They are the product of catastrophic gravity flows caused by shallow earthquakes.

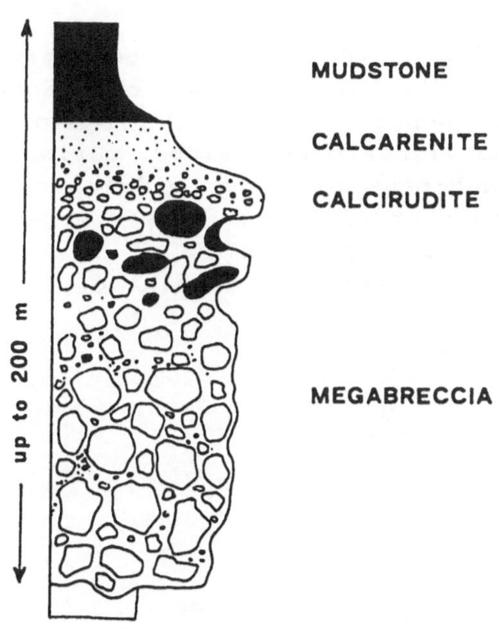

Fig.146 - Modal sequences of megaturbidites occurring in the northern Appennines and southern Pyrenees (Mutti and Ricci Lucchi,1984).

These seismoturbidites occur as scattered intercalations between fans or, more in general,in turbidite systems. Mutti and Ricci Lucchi (1984) proposed two idealized megaturbidite sequences,one reported in Fig.146.These megaturbidites,mostly calcareous in composition, are particularly frequent in the elongate turbidite basins of the northern Appennines.

The 'Contessa' megaturbidite layer,dated to the Serravallian (Middle Miocene) is a megaturbidite layer, 20 m thick, traceable for 140 Km along the axis of the Marnoso Arenacea foredeep, in the northern Appennines (Ricci Lucchi,1978). According to the authors working in the area, the 'Contessa' megabed was formed during a period characterized by a maximum rate of migration of the thrust front and high rate of uplift. It represents a transgressive phase in the basin,

counterbalanced by a tectonic tilting in the nearby carbonate platforms which led to a catastrophic collapse of the carbonate platform margin.
The Friuli platform and adjacent Slovenia basin crop out in the southern sector of the Julian Alps (NE of Italy). The evolution of the platform records a withdrawal of the margin (Tunis and Venturini,1987) which underwent a progressive dismantling. In the Maastrichtian the margin was affected by a strike-slip transcurrent tectonic control and listric faults.

During the Paleocene the narrowing of the basin was concomitant with the deposition of megabeds (Fig.147). In the Lower Eocene a huge amount of material was deposited into the basin. It consisted of redeposited shelf limestones, slope marls and fragments of megabeds. These megabeds crop out widely in the eastern Friuli region (Tunis and Venturini,1987).These megabeds are dated to the Thanetian and to the Ypresian.
In the Vernasso quarry, the Eocene megabeds consist of a tens

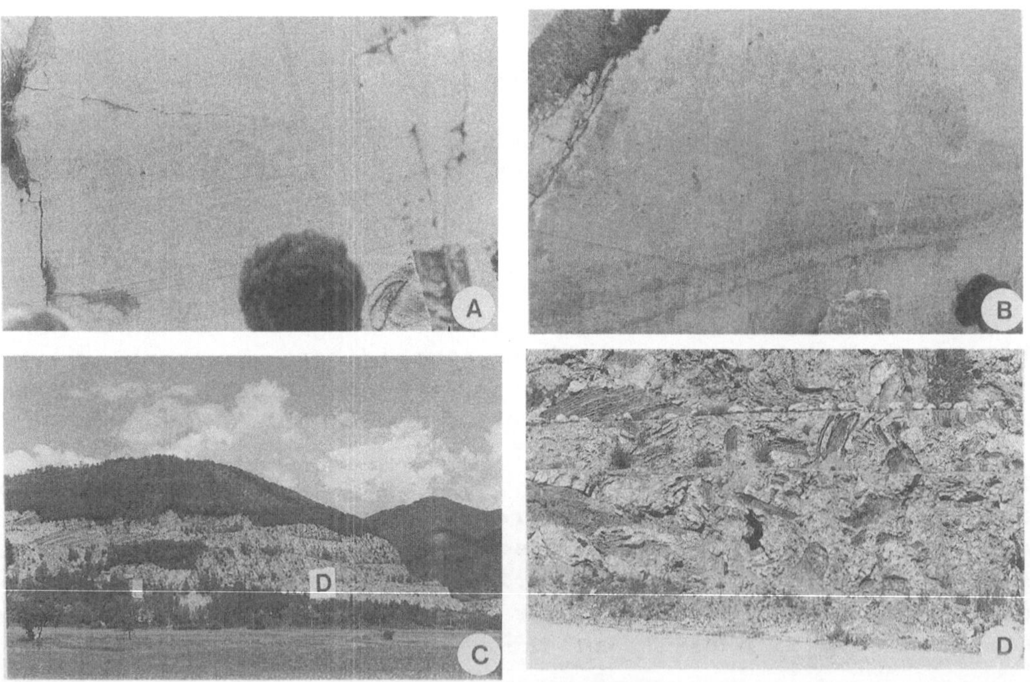

Fig.147 - A) In situ,autoclastic breccia of early lithified carbonate material, probably induced by seismic shocks (Vernasso quarry).B)Parabreccia formed by 'mud balls' floating in a fine-grained calcarenite matrix (Vernasso quarry).C) and D) Panoramic view of the Paleocene megabreccia occurring at Anhovo, Slovenia basin (Skaberne, 1987), transitional to the left to bedded turbidites;detail stressing the internal organization and size of megaclasts.

of m to hundred of m thick, fining-upward,stacked sequence. Individual beds contain features quite analogous to the intraclast parabreccias described by Spalletta and Vai (1984) from the Upper Devonian carbonate platform margin of the Carnic Alps, Italy, and interpreted as seismites (Fig.147). Some of these structures are also analogous to the 'mud balls' occurring in the shallow-water carbonate platform of the Venetian Alps (Trento platform; chapter #1) and interpreted as tsunami- generated features (Galli,1991).

Other interpreted seismoturbidite beds occur in the Eocene Echo Group, in the south Pyrenees, studied by Seguret, et al.(1984). This turbidite basin formed as a narrow, elongate basin dominated by strike-slip tectonics. Nine megabeds, 200 m thick, traceable for 150 Km throughout the basin (Fig.148),show a periodicity of deposition of about 5 x 10 to 1 x 10 years (Mutti and Ricci Lucchi,1984).
The triggering mechanism for the deposition of these calcareous megaturbidites was thought to result from superficial seismic activity due to thrust faults along the edge of the basin (Seguret, et al.,1984).

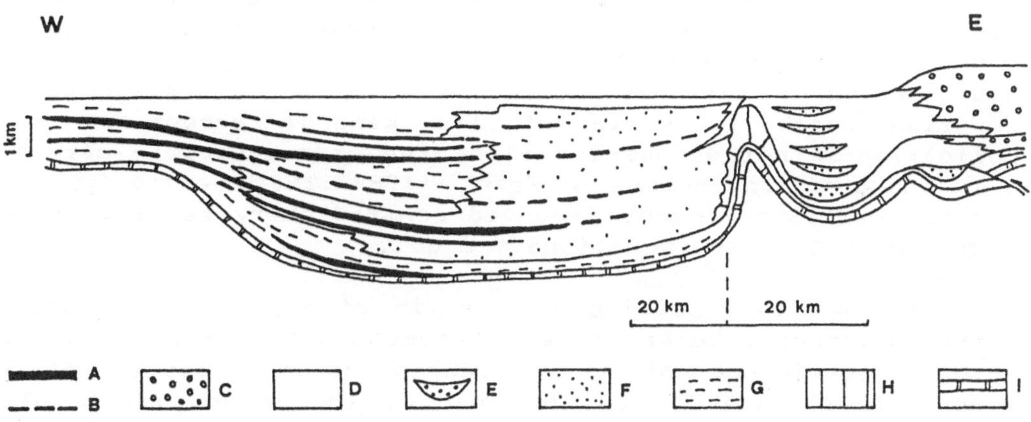

Fig.148 - Carbonate megaturbidites in the Eocene Echo Group, south-central Pyrenees (from Mutti and Ricci Lucchi,1984). A:complete,proximal megaturbidite sequences (including the basal magabreccia); B: incomplete,distal megaturbidites (calcarenite-mudstone couplets).

Are megabreccias the product of compressional or vertical tectonics?

A large number of authors tend to link the genesis of megabreccias to compressional orogenetic phases, linked to the emplacement of nappes which would be responsible for the formation of a peripheral bulge and oversteepening of the platform margin. For example, Abbate, Bortolotti and Sagri (1981) comment: " An analysis of available examples seems to indicate that the tectonic melanges are connected to accretionary wedges in convergent zones, where large sectors of oceanic crust and trench fillings have been or are being subducted. Olistostromes can be found in a wide range of structural settings: in more or less close connection with tectonic melanges, orogenic landslides and nappes in convergent and colliding margins and, also, in passive continental margins. Modern examples of the latter would be the afore mentioned mudflows off western Africa, earlier examples being olistostromes in the Scisti Policromi of the Tuscan sequence, and the megabreccias and olistoliths derived from the faulted rims of the carbonate platform of the Central Apennines. These ancient occurrences are related to the passive margins of the Adria plate. Owing to the complex structural pattern of the narrow Tethys seaway, there is no doubt that also these passive margins experienced since Cretaceous intense tectonism, induced by close convergent zones".
This intriguing view is based on a model of subduction-accretion which in spite of its general acceptance and application is not documented along the circum-Pacific margin, as has been demonstrated by Wezel (1988). The results of the Deep Sea Drilling Project produced contradictory results which failed to confirm an accretionary model.

An interpretation regarding the origin of megabreccias recurs to the 'peripheral bulge' model. Beaumont (1981) explained the tectonic subsidence of the foredeep as the result of a litospheric flexure due to the load of the thrust front. The peripheral bulge progressively uplifts towards the thrust front depending of the viscoelastic behavior of the crust. The model was based mainly on the study of seismic profiles and mathematical modelling on Wyoming and Appalachians, where the estimated thickness of nappes is of several kilometers.

Recently, Royden and Karner (1984) calculated the inflexion of the ramp beneath the Appennines and came down to the conclusion that the load of nappes is insufficient to produce such a bending: an additional load would be required to maintain the basement inflexion.

A consequence of the results of the work by Royden and Karner (1984) is that the model of peripheral bulge is not applicable

to the Appennines.
A peripheral bulge interpretation of megabreccias requires that 1) the nappe emplacement is concomitant with the formation of megabreccias; and 2) the load of the nappes is sufficient to produce a deformation of the foredeep.

Another problem concerns the direction of peripheral bulge migration which should in normal cases progress towards the subduction zone (basinward). In the Appennines the migration of the interpreted peripheral bulge goes in the opposite direction, as shown in the work by Vai and Castellarin (1988).

The Messinian 'Vena del Gesso' evaporite Formation described by Vai and Ricci Lucchi (1976;1977) and by Marabini and Vai (1985) consists of a gypsum lithosome, 150-170 m thick, cropping out as a homicline along a NW-SE strike,in the Appennines. Its original extent was of about 100 Km. The basin of sedimentation was a lagoon bounded by topographic highs, which became infilled with layers of microcrystalline gypsum interlayered with bituminous black shales. Tectonic sills separated different evaporite basins.
Vai and Ricci Lucchi (1976) described a regular repetition of six main facies types which compose a depositional modal cycle consisting of the following facies succession: black shales---> calcareous gypsum and algal laminite--->massive selenite---> banded selenite--->chaotic gypsum (pebbly mudstone) --->slump breccia (megabreccia). A great proportion of the banks is represented by mechanically reworked gypsum which underwent a basinward transportation by debris flow mechanisms in a subaerial environment. The authors proposed an autocyclic cannibalistic process whereby gypsum was eroded from the margins and redeposited basinward. The mechanical deposition was thought to have been the result of lowering sea-level causing a depositional regression.

The modal cycle can be split into two parts: a lower aggradational cycle (black shales to autochtonous and banded selenite), overlain by an upper coarsening- and thickening-upward cycle represented by allochtonous gypsum. The same organization is visible at a megasequence scale. This modal cycle is therefore analogous to the modal sequence described above.

An alternative interpretation to a purely autocyclical model of sedimentation would be a cannibalistic tendency resulting from the oversteepening of the relief due to a tectonic inversion. This cannibalistic trend was concomitant with a retrogradation of the cannibalized margin.

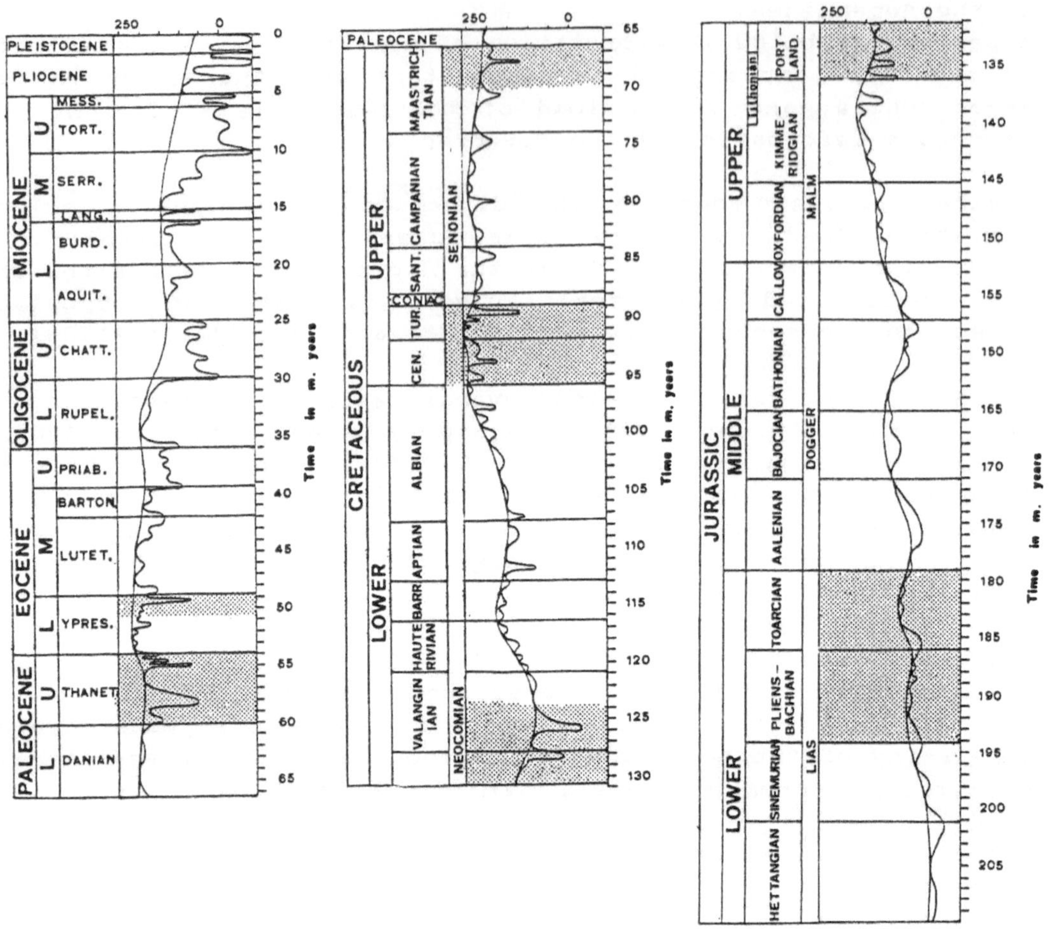

Fig.149 - Sequence of six significant time intervals (dotted areas) plotted on the global eustatic curve by Haq, et al.(1987) in order to show their correspondance with short-term sea-level falls. These time intervals may correspond to global periods of short-term krikogenetic rejuvenation or geoidal deformation. These time intervals represent 'event horizons', due to the clustering of several events within them. Geological processes are seen as the result of a non-linear punctuation of 'events' spaced by lag-times, the spacing being regulated by the omothetic proportion (see chapter 5). Other 'event horizons', not plotted here, occur in the 1) Upper Devonian - Lower Carboniferous, Anisian-Ladinian; and 2) Serravallian-Messinian - Pleistocene, respectively representing the end-terms and first terms of two other temporal sequences.

The structural setting of the northern Appennines consists of narrow, arcuate belts and overthrusts delimiting clastic wedges. The structure of the Appennines was interpreted by Van Bemmelen (1972) as the result of mantle diapirism and gravitational spreading of fluidized sub-crustal material. According to Wezel (1984) the Appennine frontal arcs are an indication of a tectonic deformation due to vertically rising diapiric domes. Krikogenesis (see page 1) results in the surface as transient crustal uplifts (see Wezel,1984,1985). The evolution of the northern Appennines implies a "paired migration and extension and compression towards east,linked to progression of transient orogenic arc systems" (Wezel,1984). The possible mechanism for the emplacement of thrust sheets, together with the foredeep migration consists of vertical (touche-de-piano) tectonics, as documented for example across the actual Java trench. This implies the migration of mantle diapirism through time and space ("megaundations" of Van Bemmelen).

In keeping with the concepts of Stille (1924), Vai (1987) showed that orogenetic deformation in the Appennines was discontinuous: he supported a punctuated migration of the deformation in the Appennines with a given recurrence time . In the Messinian and the Pleistocene the stages of maximum deformation and frontal advancement fall into the following ages: 18 m.y.;10 m.y.;5.5 m.y.;5 m.y.,3.5 m.y.; 2-1.5 m.y.; and 0.5 m.y. (Vai,198). The duration of the deformation episodes is brief (1-0.1 m.y.);the quiescence is of longer duration (5-0.5 m.y.).It seems that the deformation and advancement of the foredeep was synchronous with the extensional tectonics in the Tyrrenian Sea.

A crucial point is that most of the megabreccias, so common in the Mediterranean region,both in the Alps and Appennines, were emplaced during short-term sea-level falls.These occurrences (some of which described above) are confined within the following time intervals:

Anisian-Ladinian
Pliensbachian-Toarcian
Tithonian-Valanginian
Cenomanian-Turonian
Upper Maastrichtian
Thanetian
Ypresian
Serravallian
Messinian
Pleistocene

These time intervals correspond to short-term sea-level falls visible in the chart by Haq, et al.(1987),reported in Fig.149. The same ages coincide with the ages of unconformities which delimit the clastic wedges and Formations in the Appennines, as shown in Fig.150.

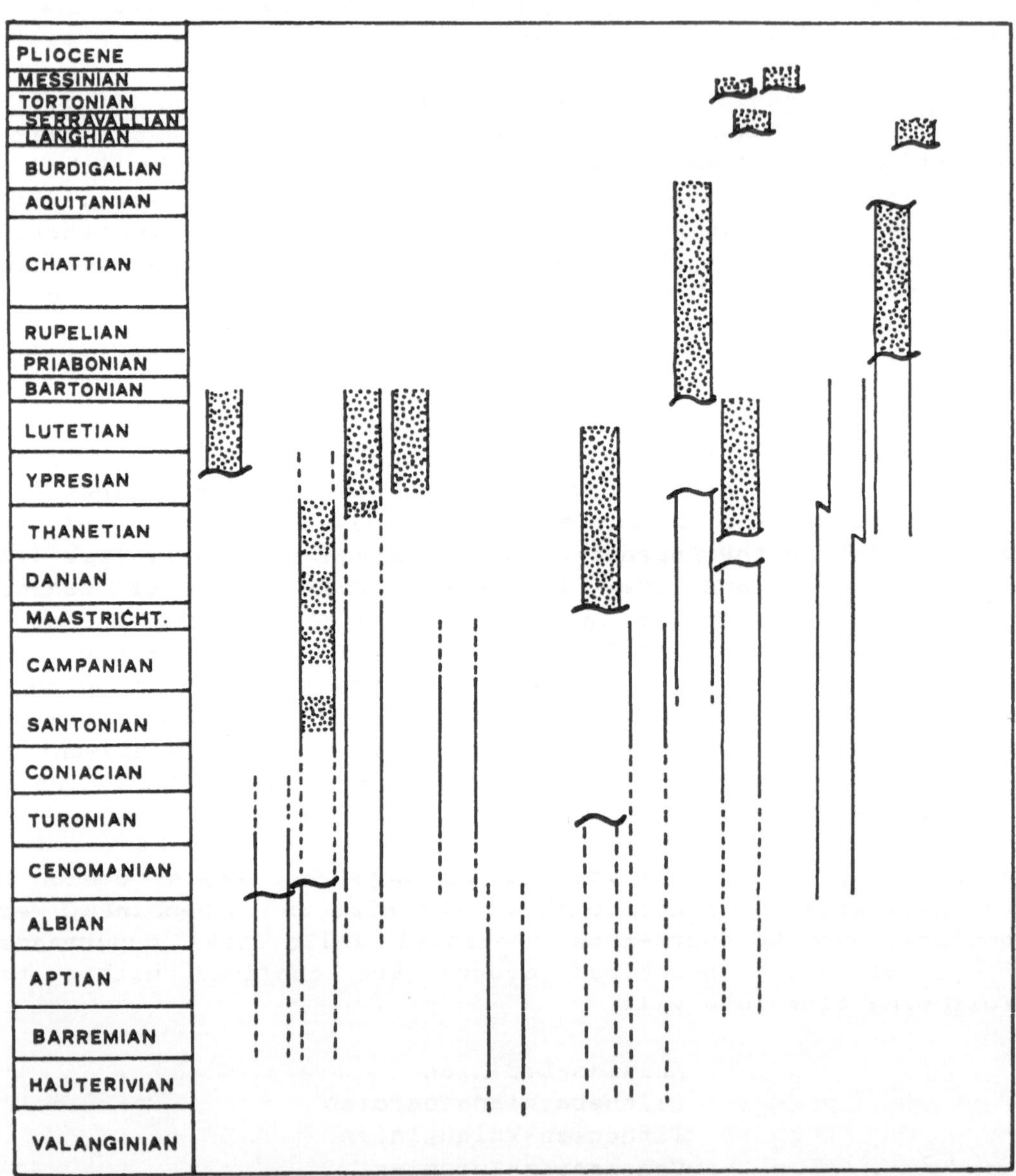

Fig.150 - Time-space distribution of sedimentary Formations in a sector of the Appennines,transversal to the Corsica-Tuscan-Umbro-Marchigiano Apennine belt). Stippled areas: synorogenic migrating basins; white areas: basinal turbidites, hemipelagites). The direction of migration is towards the right (East). From Principi and Treves (1984).

A possibility is that megabreccias and unconformities are the products of uplifts resulting from deep processes of vertical and lateral diapirism seated in the upper mantle.

This possibility allows for a reconsideration of piggy-back basins and denudation events described by Ori, et al.(1986), interpreted as the result of lateral tectonic compressional tectonics. The cross section described by Ori, et al.(1986) shows denudation complexes resulting from the erosion of the thrust sheets, overlain by prograding complexes (Fig.151).The geometrical relationships between fault scarp, denudation complex (megabreccias?) and progradation complex bear a striking resemblance to those of the second depositional sequence of the Gargano case history (Fig.106).

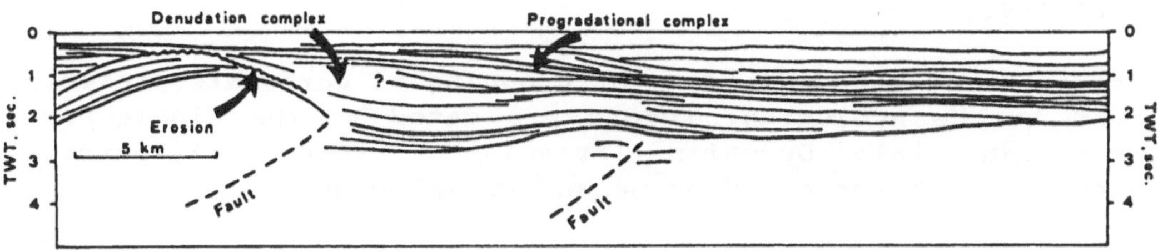

Fig.151 - Sketch from a seismic line showing relationships between fault scarps, denudation complex and progradational complex in the Plio-Pleistocene foredeep in the Central Adriatic Sea (from Ori, et al.,1986:Fig.6).

The above few examples,taken from the available literature, allow for the following working hypothesis concerning the origin of megabreccias:

1. Megabreccias occur in a wide range of settings and paleogeographic situations, ranging from the slope to a shallow lagoons.

2. They may be interpreted as lowstand system tract, being mostly associated with short-term sea-level falls.

3. Most of megabreccias encountered are the product of geoidal uplift, rather than of tectonic (orogenetic) or eustatic causes.

4. The abrupt sea-level falls punctuating the global eustatic

chart by Haq, et al.(1987) represent short-term phases of global geoidal deformation.

5. In some case megaturbidites appear to have been deposited during short-term sea-level falls, during short periods (less than 500.000 yrs.) of geoidal deformation, by catastrophic mechanisms induced by seismic shocks.
Therefore, it is not excluded that geoidal uplifts represent spasmodic phenomena.

Drowning of carbonate platforms

The drowning consists of the submergence of a carbonate platform below the photic zone and its termination.This phenomenon is dealth with in a thorough synthesis by Schlager (1981;1991,1989).

Most of world-wide phenomena of drowning of carbonate platforms found in the geological record are dated to the Middle-Lower Ordovician, Late Devonian (Frasnian - Famennian),Toarcian, Valanginian, Turonian, Miocene and Pleistocene.

As the age of worldwide drownings falls into periods of highstand of the sea-level, it is logical to accept the concept that carbonate platforms are drowned when the rate of sea level rise outpaces the carbonate accumulation rate (Kendall and Schlager,1981) On the other hand, estimated growth rates of reefs during the Late Holocene rise suggest that, in spite of a very rapid sea level rise, carbonate sediments in the late Holocene could recover and built to sea level.Schlager (1981) went to the conclusion that only an exceptional sea-level rise is capable to exceed the growth rate of a carbonate platform which is of about 1000 Bubnoff.

Among the causes of drowning, an increase in the rate of sea-floor spreading (Heezen,et al.,1973), a rapid eustatic rise produced by volcanism (Schlanger and Premoli Silva,1981), a dessication of a young ocean basin (Hsü and Winterer,1980) are discussed by Schlager (1981) as possibile causes of drownings.Also, an environmental change caused by environmental stress is a likely alternative, when sea-level rise is not sufficiently fast (Schlager,1991). Arthur and Schlanger (1979) observed the coincidence of anoxic events and drowning of carbonate platforms, and Hallock and Schlager (1986) proposed an excess of nutrient flood as a cause of demise of carbonate platforms.

Another possible cause of drowning is represented by a very rapid sea level fall followed by a transgression, as suggested by Mc Laren (1970), which may have led to drowning unconformities (Schlager and Camber,1986).

Most of anoxic events coincide with periods of drowning of carbonate platforms which in turn coincide with short-term sea-level falls in the chart of Haq, et al.(1987).Some of these rapid sea-level falls may not be lowstand (i.e. the Valanginian lowstand is probably a drowning unconformity mistakenly attributed to a lowstand: see Schlager,1989)

A further possibile cause of drowning would be a period of intense geoidal deformation,which may have determined in the platform some environmental change responsible for some reduction in its growth potential:for example, the uplift may have caused a discharge of great volumes of freshwater into the ocean, as suggested by Thierstein and Berger (1978).

In fact, the world-wide episodes of drowning of carbonate platforms coincide with periods of megabreccia shedding, formation of onlap ramps, divergent patterns, and seismoturbidites.

The Cenomanian-Turonian gives a good opportunity to attempt cross-correlations between seemingly independent phenomena. The drowning and retreat of carbonate platforms during the Upper Cretaceous is a worldwide event. Most of the drownings of carbonate platforms took place in the Cenomanian - Turonian period (Schlager,1981,1991,Schlager and Philip, 1990;Fig.152). Several seamounts in the western Pacific (Heezen, et al.,1973; Matthews ,et al.,1974), the Comanche shelf (Young,1977; Bebout and Loucks,1974), the south Florida platform (Worzel, et al.,1969) underwent abrupt drownings represented by the deposition of pelagic chalk deposits over shallow water sediments.The western Pacific seamounts dredged by the DSD Project probably underwent emergence prior to drowning: in fact, Heezen, et al.(1973) report that the top of seamounts and gujots in the Cenomanian - Turonian was covered by bioclast calcarenites: the well rounded grains evidence for a permanence in the surf zone; the flat surface of the gujots formed in the Turonian. The extermination of the reef fauna consequent to the shallowing was followed by a peneplanization of the reef surface, followed by rounding of the grains. Diagenetic features indicate a transition from marine phreatic to meteoric phreatic conditions.The Jordan Knoll in Florida was a shallow

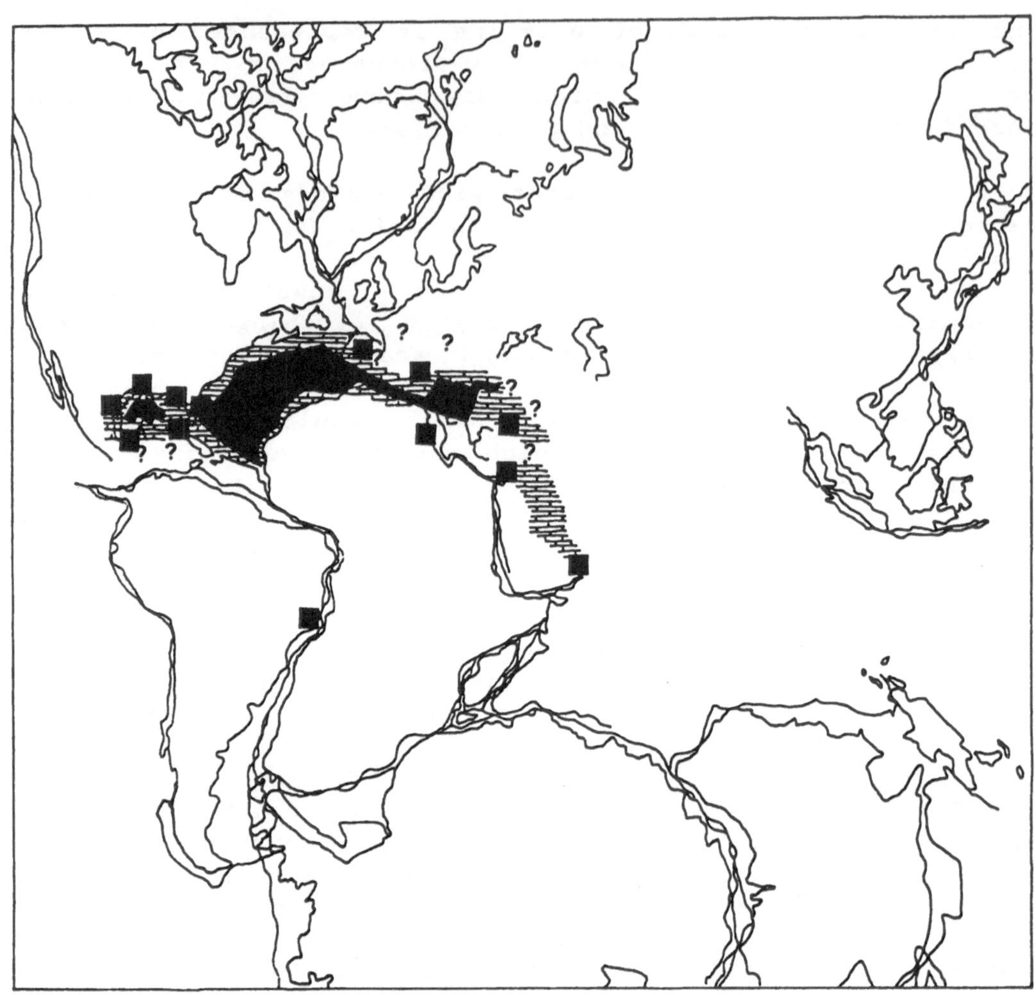

Fig.152 - Location of platforms drowned in the Aptian through the Cenomanian (modified after Schlager and Philip,1990;base map from Jenkins,1991).

water bank in the Albian - Cenomanian (Bryant, et al.,1969). It deepened from the Cenomanian to the Santonian. Megabreccias dated to this period may have formed by tilting of the knoll.According to Worzel, et al.(1973) the Middle Cretaceous was marked by a general period of emergence in the Gulf Coast Basin and East Texas. In the Albian - Cenomanian the Campeche shelf records pebbly mudstones interbedded or overlying calcarenite-calcisiltite beds (site #97: cores 7,8 and 9), delivered from the above shelf. In this case, the origin of the pebbly mudstones was thought to be linked to phases of the Laramide orogeny of Cuba.However, the Laramide orogeny took place in the Paleocene, not in the Cretaceous.

Relief inversions

The accentuation of the rotational tilting in some situations is responsible for relief inversions consisting of the emergence of some areas of the shallow-water carbonate platform or even the uppermost parts of the slope,close to the platformward terminations of clinoforms. The relief inversion is associated with the formation of divergent patterns,tectonic tilting, fault scarps, and shedding of megabreccias into the basin (Fig.153). It is seen here as an effect of rotational tilting associated with geoidal eustatic changes, as hypothesized by Mörner (1981). The effects of relief inversions in the platform interiors lead to the exposure with possible karstification, or a sudden transformation of a deep lagoon into a paralic swamp: the repeated mudstone interbeds within the deep lagoonal, deep ramp facies occurring in the 'Calcari Grigi' formation, and the sill sequences (chapter #1) are seen to have formed in response to relief inversion mechanisms.

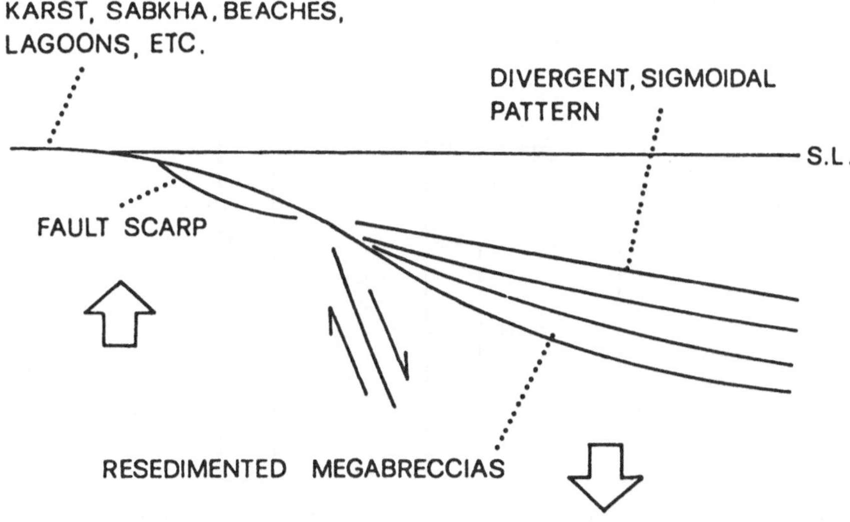

Fig.153 - Features associated with relief inversions of platform interiors and the margin of a carbonate platform.In a strike-slip situation the relief inversion is produced by transpression in which case the fault plane is either vertical or inclined towards the platform interiors.

In some cases the relief inversions take place by a touche-de-piano tectonics and a retrogradation of the margin. This process in the shallow-water platform is documented by the

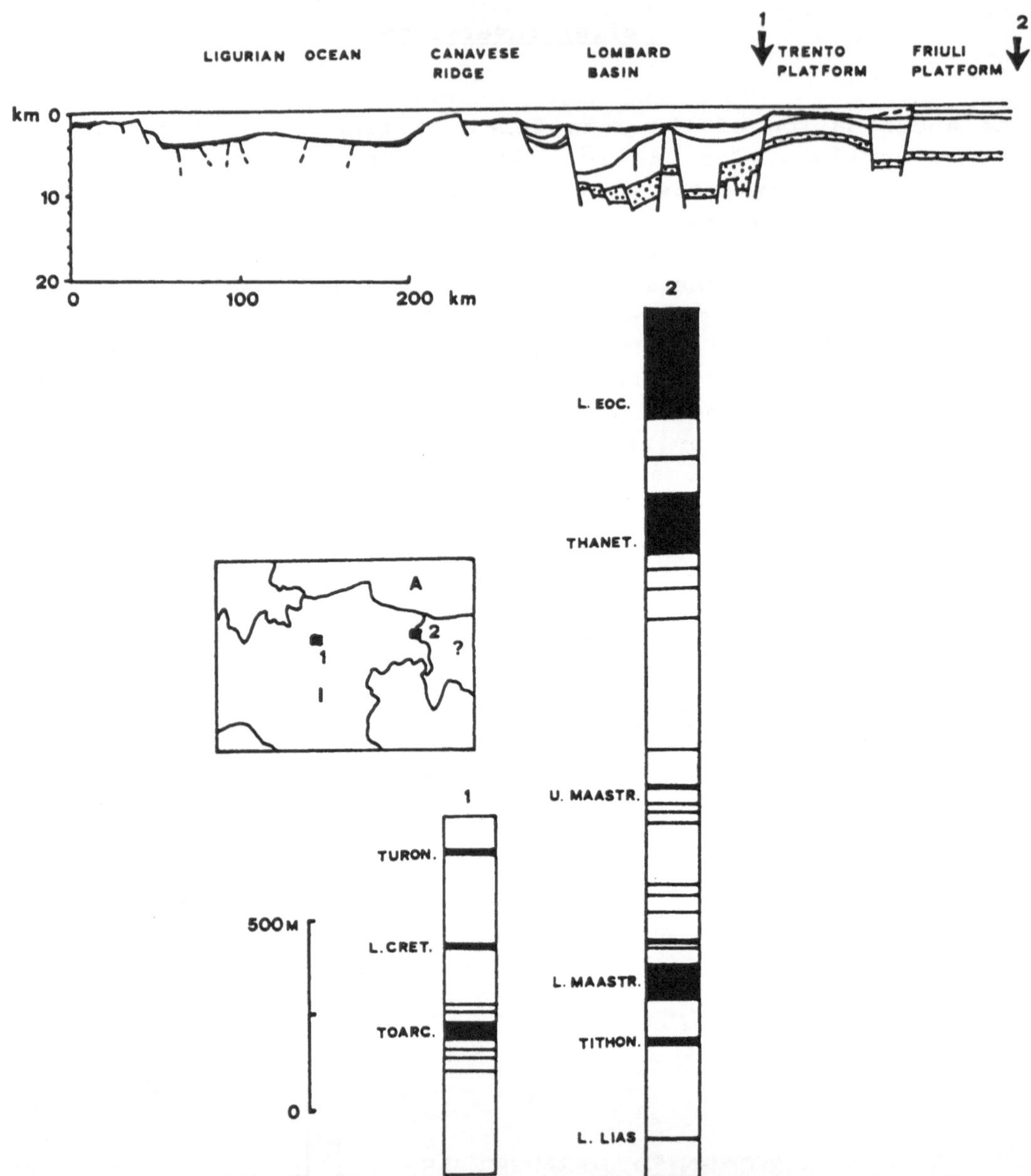

Fig.154 - Meridional Alps. Cross section from Winterer and Bosellini (1973). Below: 1): Castellarin (1972); 2): Tunis and Venturini (1987).Black zones in the stratigraphic columns denote megabreccias.

migration of the sill and flexure zones towards the hinge areas (see chapter #1).

In the Meridional Alps (Fig.154) retrogradation of the carbonate platforms was discontinuous, as testified by the deposition of megabreccias which are dated to the Pliensbachian

- Toarcian, Tithonian - Valanginian, Turonian, Maastrichtian, Thanetian and Lower Eocene.The megabreccia deposition in the western areas was accompanied by emersion of some areas in the east i.e. in the Toarcian, where deposition of megabreccias along the western border of the Trento platform was concomitant with emersion and karstification of some parts of the Friuli platform.The stratigraphic columns record an increase in the thicknesses of megabreccia bodies, probably related to a corresponding increase in the rate of tectonic tilting.

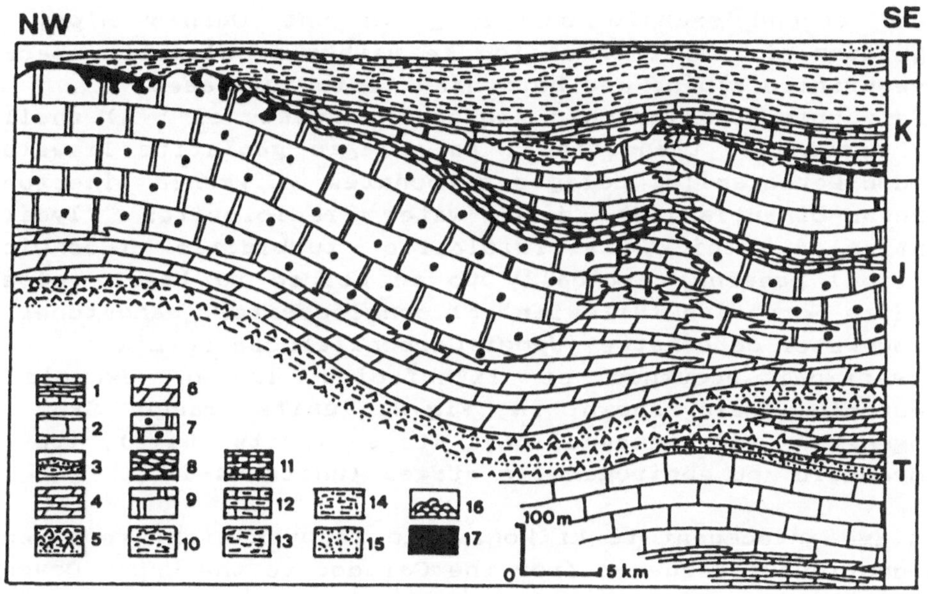

Fig.155 - Cross section of the Betic Cordillera, south Spain (after Martin Algarra,1988).1-5 : Hidalga Group (shallow-water Triassic carbonates and evaporites).6-9 : Libar Group (6-7: shallow-water carbonates; 8-9:shallow-water pelagic limestones. 10-16 : Espartina Group (pelagic, bituminous limestones, marls and flyschoid sandstones and clays).17: Jurassic and Cretaceous karst.

Relief inversions are common in convergent wrench zones, as a result of transpression mechanisms.Two examples come from the Betic Cordillera of south Spain and the Paleozoic sequence from the Carnic Alps (northern Italy).

In the first example (Fig.155) the NW sector of a Jurassic platform was raised by a tectonic tilting in the Lower Valanginian-Hauterivian, Lower Aptian and Lower Albian. Upon uplift, the NW sector was subject to karstification, whereas the SE sector was transformed into a 'shallow pelagic platform'.Later, the emerged area underwent drowning and became a submarine plateau.De Smet (1984) in a thorough study of the tectonic evolution of the Betic Cordillera, demonstrated that strike-slip movements followed by convergent motions occurred in the area comprised between the Spanish and the Maroccan Meseta. Strike-slip movements produced anticlinal and synclinal massifs; transpression led to the formation of 'flower' structures (Harding and Lowell,1979).

In the second example occurring in the Carnic Alps, the recognition of relief inversion is rather problematic, because of overprinted alpine and hercynian tectonic phases.
The Paleozoic stratigraphic sequence consists of : 1) shallow-water carbonates transitional to pelagic goniatite limestones (Caradoc-Devonian);2) pelagic carbonates (*Clymenia* limestones, Famennian-Dinantian);3) deep water radiolarites (lydites: Dinantian);4) thick, siliciclastic turbidites (Hochwipfel Formation) passing vertically to an acidic to basic volcanic turbidite Formation (Silesian);5) conglomerates, sandstones and shallow-water carbonates (Upper Carboniferous- Permian).
The vertical sequence of lithofacies is an example of geotectonic cycle shown in Fig.2: units range from the krikogenetic and geosynclinal stage (units #1÷3), to the tectogenetic and epeirogenetic stages (units #4÷5).

The stage antecedent to krikogenetic rejuvenation, represented by a continuous sequence from the Caradoc to the Upper Devonian (Vai,1980),is mainly constituted by shallow-water and pelagic carbonates which become the nearly exclusive lithofacies from the Emsian to the Frasnian.

The Frasnian records a worldwide establishment of restricted marine conditions; organic-rich black shales developed in the USA, for example in the Michigan Basin.In the Carnic Alps during this period anoxic conditions prevailed in the basin; corresponding changes in the platform are documented by a decrease in fossil diversity from the Givetian to the Frasnian and to the appearance of shallow-water black mudstone intercalated with other lagoonal facies. Semirestricted lagoonal facies (black mudstones and shales intercalated with other bioclast accumulations) are thought to represent the platformward extensions of anoxic pelagic facies occurring in the basin, which are widespread in Europe (see Preat,1988; Krebs,1974).The Frasnian records a change in platform facies organization and a transition from reef--back-reef-- lagoon--tidal flat to an inter-reef platform configuration with lagoons

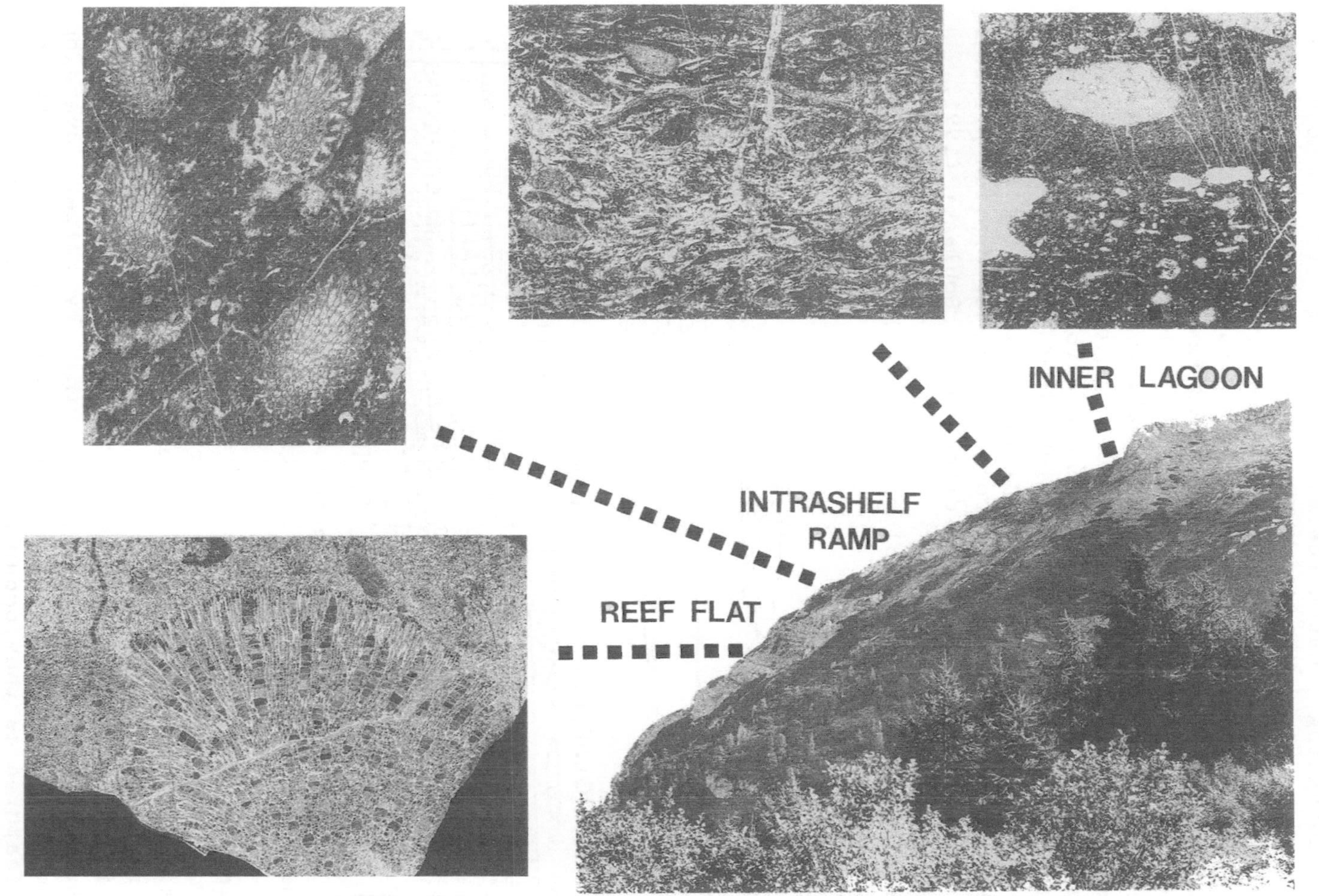

<u>Fig.156</u> - Cima Ombladet carbonate succession. The succession records a transition from a reef flat, high-diversity environment, to an intrashelf ramp constituted by oligotypic faunal assemblages, to a restricted lagoon.

populated by an oligotypic fauna (calcispheres and *Amphipora*) and intervening patch-reefs, composed of monospecific brachiopods (Vai,unpubl.data). The development of this lagoonal stress environment is more or less synchronous throughout the world and may be related to some drastic change in water paleocirculation (Copper,1986).A field example of this vertical decrease in faunal diversity (Fig.156) shows reef flat facies at the base containing crinoids, algae, stromatoporoids and corals, overlain by open lagoon facies with brachiopods, *Amphipora*,*Thamnopora* and *Trypanopora*,followed in turn by *Amphipora* meadows on top (Galli,1985,1986).
The evolution of the carbonate platform during the Frasnian may be taken as an example of environmental deterioration preceding the drowning of the platform, discussed by Schlager (1991).

The drowning of carbonate platforms at the Frasnian-Famennian

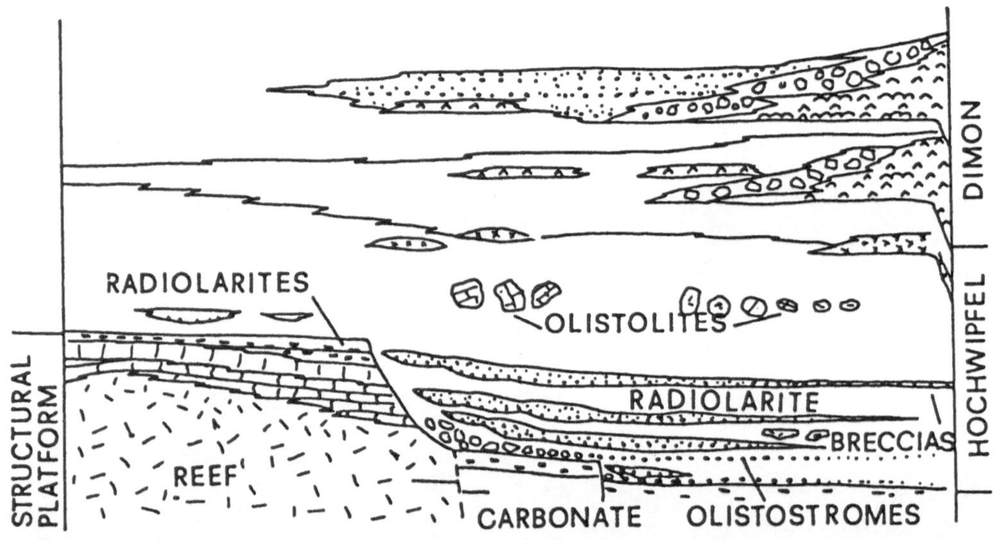

Fig.157 - Schematic cross section showing the evolution of the carbonate platform and the relationships with the hercynian flysch (after Spalletta, et al.,1979).This model assumes that radiolarite deposition draped the whole carbonate platform, with the exception of some emersion horizons dated to the Dinantian-Visean occurring at places above the shallow-water platform (Schönlaub,et al.,1991). Radiolarite megabreccias, amounting to about 100-200 m in thickness, were shedded into the basin either from the right (south), by relief inversion, or from the north, as results from this scheme. Olistoliths are local features as they occur only near Pizzo Collina area.

boundary is a global event concomitant with a general biotic

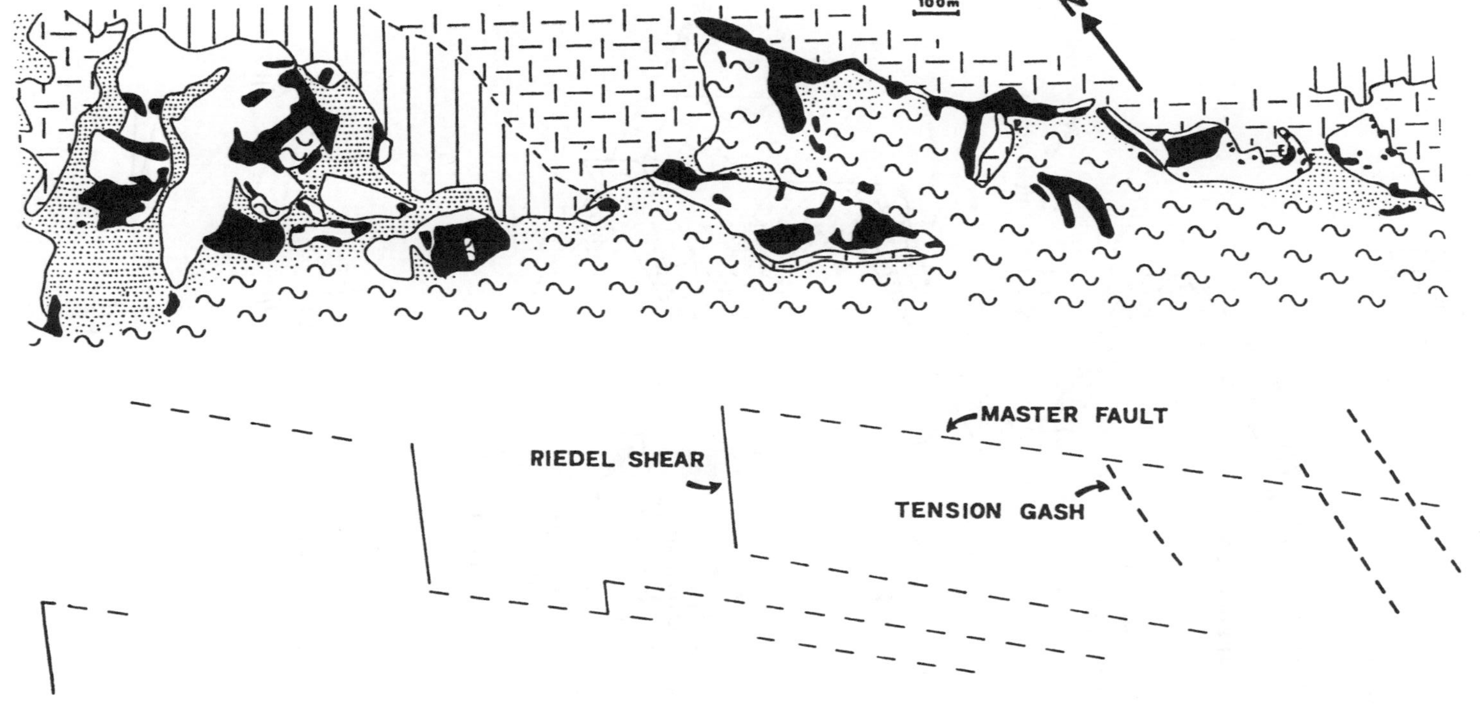

Fig.158 - Geological map of the Canale-Volaia area. Shown below is the synsedimentary hercynian structures.The fence diagram from the same area evidences fro the horst-graben geometry. The sketches below show the modalities of formation of the unconformities between the Formations (a:rotational tilt); b:relief inversions.

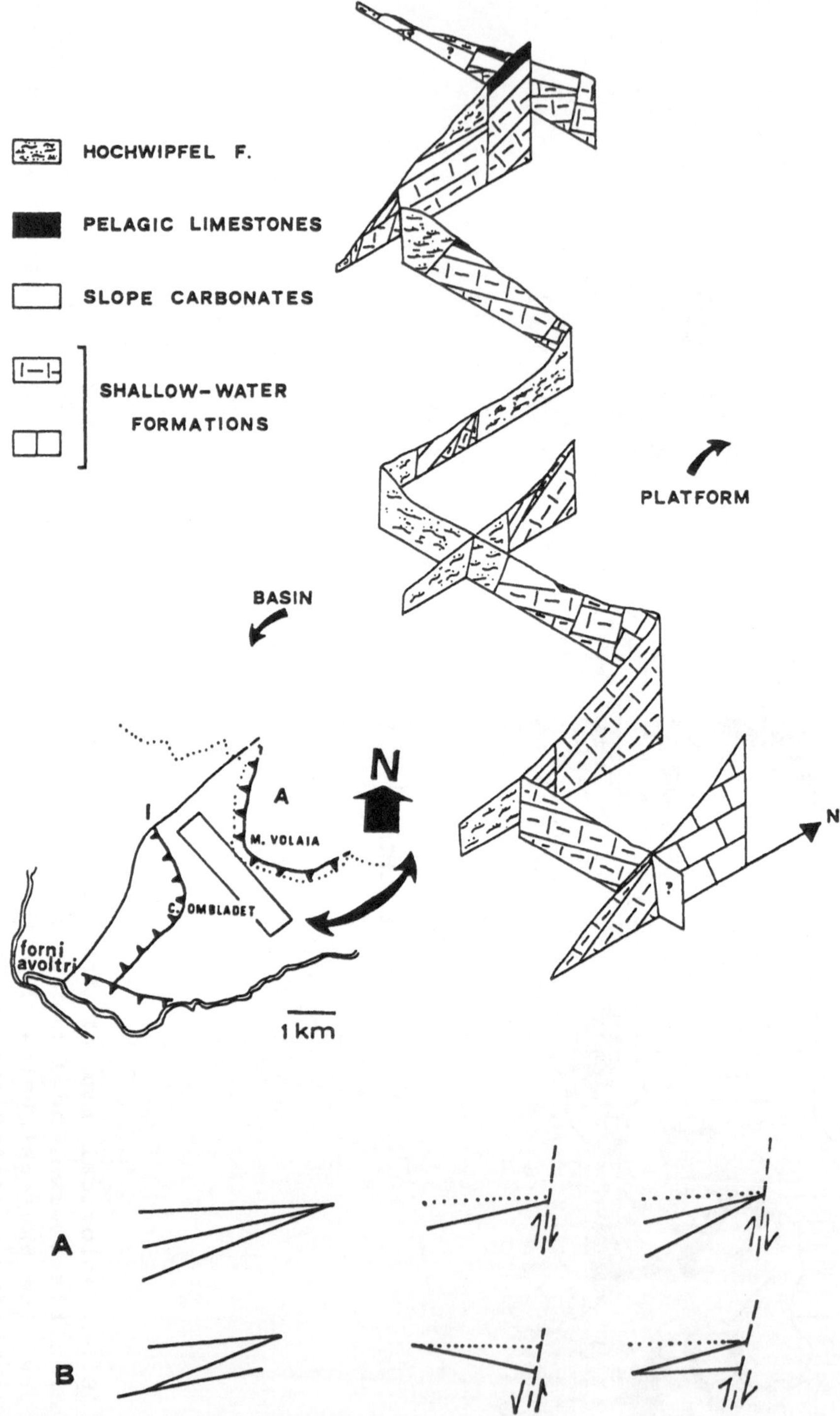
HOCHWIPFEL F.
PELAGIC LIMESTONES
SLOPE CARBONATES
SHALLOW-WATER FORMATIONS
PLATFORM
BASIN
N
A
M. VOLAIA
C. OMBLADET
forni avoltri
1 km
N
A
B

crisis and extinctions attributed by Mclaren (1970) and Playford, et al.(1984) to a bolide impact. In the Carnic Alps the drowning is recorded by the deposition of shallow-water pelagic limestones.

Tectonics became progressively important,as early as the Frasnian,when intrashelf onlap ramps were formed. In the Dinantian the deepening-upward trend was accentuated as a result of the fragmentation of the platform by extensional tectonics. Neptunian dykes, olistoliths, olistostromes, sedimentary hiathuses and unconformities evidence for the sedimentary tectonics. The cross section evidencing the geological evolution of the platform (Spalletta, et al.,1979; Fig.157) contains several features common to other platform margin transitions subjected to extensional tectonics and successive relief inversion such as: 1) coarsening-upward trends of megabreccias; 2) divergent wedges of pelagic-basinal sediments at the margin of the platform; and 3) synsedimentary faults.

The hercynian synsedimentary tectonics is evident from an examination of the fence diagram of Fig. 158 which reveals a horst-graben geometry. Angular unconformities amounting to 10°-20° separate some of the pelagic Formations. Two recurrent types of discordances can be observed in the study area. A first type consisting of fan-shaped discordances was the product of a rotational, basinward tilting.The second, more complex, was produced by basculatory movements or inversions in the direction of tilting which may be related to relief inversion mechanisms.

The development of the stratigraphic sequence was controlled by strike-slip tectonics from the Upper Devonian to the Lower Carboniferous,consequent to the activation of a transform system which determined a westward drift of continental blocks separated by narrow seaways.

The central areas of the hercynides display arched structural patterns, dome-like features, absence of ophiolites, widespread high-temperature metamorphism and a great number of granite intrusions which evidence for raised isotherms and thinned continental crust. No oceanic crust was present untill the Upper Devonian.The absence of ophiolites, nappe structures, island arcs and a lack of high-pressure metamorphism in the hercynides led Krebs and Wachendorf (1973) to infer a diapiric orogenesis as an alternative to a plate tectonic model.

Vai and Cocozza (1986) sustained that the deformation, progressing from west to east, was continuous and diachronous

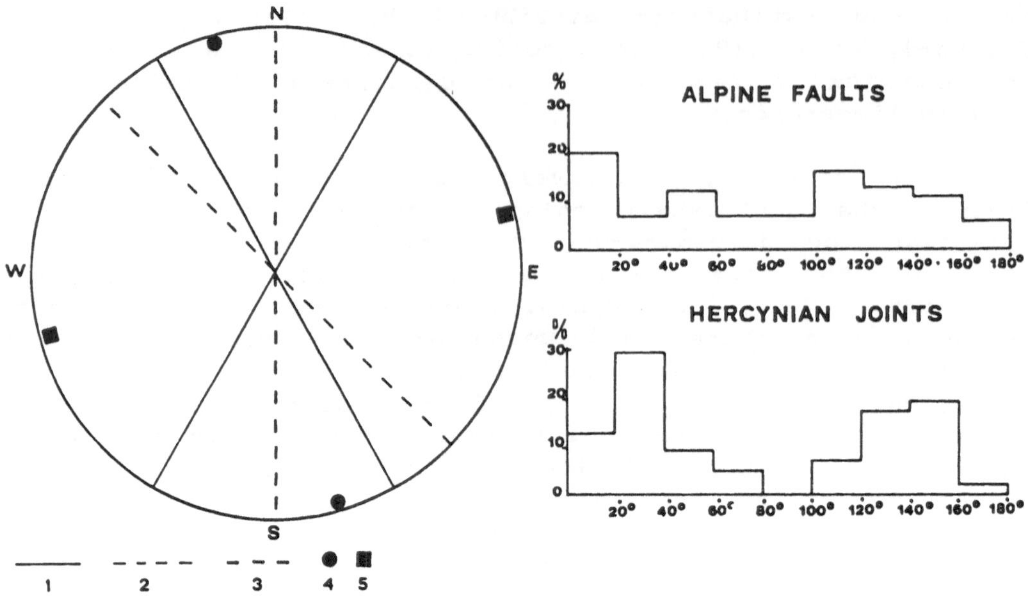

Fig.159 - Right. Equiareal projection showing the angular relationships between 'Riedel shears' (1), master faults (2), tension gashes (3), sigma 1 (4) and sigma 2 (5).Left. Canale-Volaia thrust sheet. Percent distribution versus direction of the bearings of 147 alpine faults and of 42 hercynian synsedimentary dykes and paleofractures.

in the Hercynian Belt; the deformation phases are dated to the following periods: 350 m.y. (Breton phase); 315 m.y. (Sudetic phase); and 300 m.y. (Asturic phase).According to this scheme the Sudetic phase did not affect the Carnic Alps.

In the study area the percent distribution of 42 strikes of tension joints, paleofaults,sedimentary dykes gives two modal directions (N 30° E; N 150° E) which represent Riedel planes (Wilcox,et al.,1973), as shown in the equiareal projection of Fig.159.The compressive hercynian and Alpine phases only produced a statistical remobilization of this joint system, as results from a comparison between the percent distribution of hercynian paleofractures and that of 147 alpine faults occurring in the same thrust sheet. At an early stage of wrenching a 'dog-legged' graben structure (Illies,1981) developed from an echelon system of structures as a consequence of motion of first-order shear zones and N 30°E second-order shear. Riedel shears initially formed under a low rate of axial shear motion; upon individuation, faulted blocks were successively remobilized and acted as high-angle shear planes.

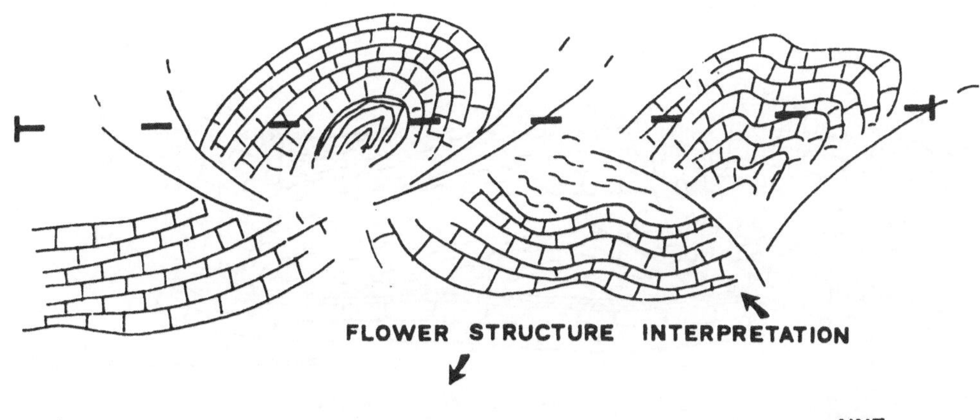
FLOWER STRUCTURE INTERPRETATION

SSW
NNE
UD
S
O
S
UD
H
H
UD
MD
0
1000
2000
2920 m
OLEODOTTO TAL (PSO MONTE CROCE CARNICO)

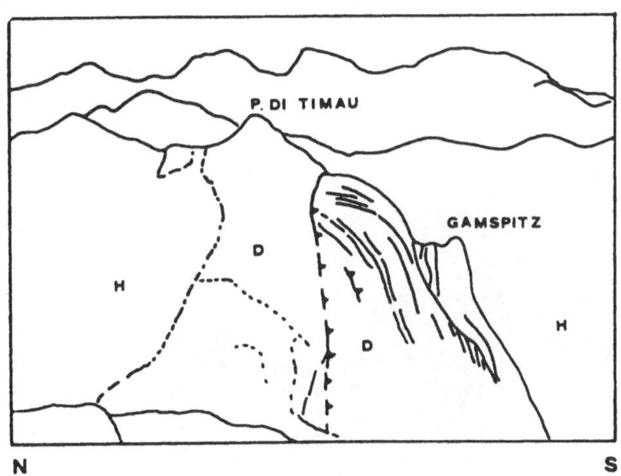
P. DI TIMAU
GAMSPITZ
D
H
D
H
N
S

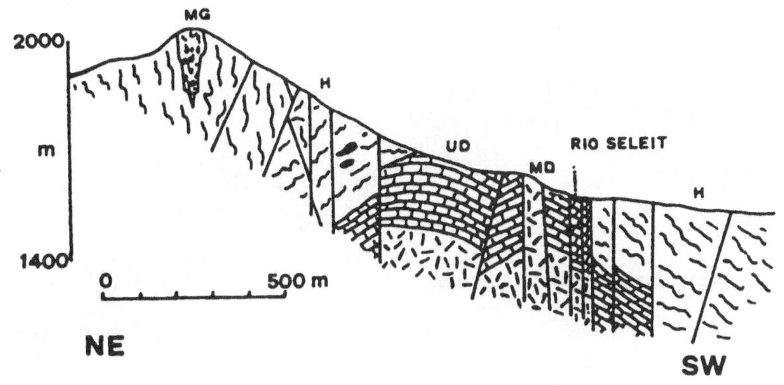
MG
2000
m
1400
H
UD
MD
RIO SELEIT
H
0
500 m
NE
SW

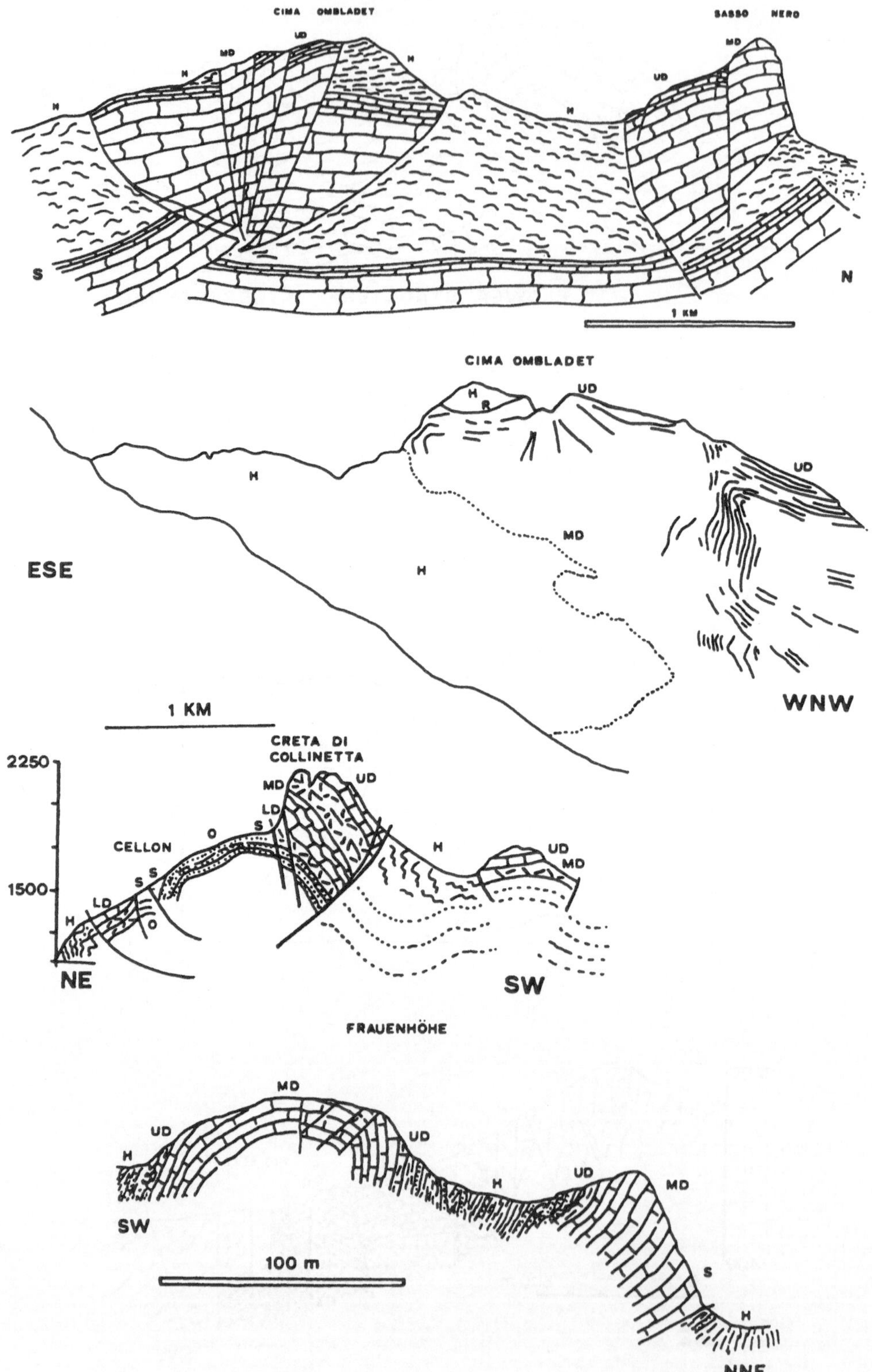
CIMA OMBLADET
UD
MD
H
H
H
H
SASSO NERO
MD
UD
H
S
N
1 KM
CIMA OMBLADET
H
R
UD
UD
H
MD
ESE
H
WNW
1 KM
2250
1500
CRETA DI COLLINETTA
MD
UD
LD
CELLON
O
S
S
S
H
UD
MD
LD
H
O
NE
SW
FRAUENHÖHE
MD
UD
UD
H
H
UD
MD
SW
100 m
S
H
NNE

Fig.160 - Examples of fold structures in the Paleozoic of the Carnic Alps. A: P.so di M.Croce Carnico (Cantelli et al.,1968). An anticlinal structure,however dismembered into slices, is clearly recognizable in the profile.This cross section was previously interpreted as a series of imbricated thrust sheets. Above :interpreted flower structure. B:Panoramic view towards east of the P.di Timau from Pizzo Collina (from Cantelli, et al.,1982) showing the southern limb of an anticlinal structure,close to the hinge area..C: Hinge of an anticlinal structure, overprinted by alpine tectonism (Spalletta, et al.,1979).D,E: Cima Ombladet and Sasso Nero (see Fig.73), interpreted as flower structures. The northern flank of the Cima Ombladet (sketch from Heritsch,1936) reveals complex box folds.The structure is unconformably overlain by the turbiditic Hochwipfel Formation, hence it formed in the Lower Carboniferous. F: Anticlinal structure at Creta di Collinetta (modified from Castellarin, 1965: in Desio,1973), interpreted here as a flower structure.G: Small-scale anticlinal and synclinal structures at Frauenhöhe,Lago Volaia (Vai,1963) (cf. with A).Surprisingly, these folded structures (and others not shown here) have been ignored or dismissed by italian authors working in the area, who made up a complicated style of imbricated underthrust sheets (cf.Fig.162) engulfed within the Carboniferous Hochwipfel Formation.
Symbols:H:Hochwipfel Formation; D: Devonian; S: Silurian; O: Ordovician; UD: Upper Devonian; MD: Middle Devonian; MG: megabreccias; LD: Lower Devonian; R: radiolarites.

The splaying of the divergent wrenching zone in the area was responsible for the northward block migration, retrogradation of fault planes, formation of the fan-shaped angular unconformities and the divergent pattern which can be perceived from an examination of Fig.158.

A successive stage in the strike-slip evolution in the area led to convergence, folding and rotation of faulted blocks,and the disintegration of the block mosaic leading to differential block movements and compressive structures squeezed out of adjacent areas,such as 'flower structures'. The second type of discordances visible in the fence diagram of Fig.158 were probably produced during this stage, by block uplifting and relief inversion.

The carbonate thrust sheets are intensively deformed and folded, as documented formerly by Heritsch (1936) who interpreted the tectonic structure as a folded style composed of brachianticlines and synclinal structures. Later, italian authors working in the area reconsidered such an interpretation and preferred a more sophisticated style of multiple, stacked thrust sheets engulfed within the flyschioid Hochwipfel Formation.This interpretation did not rule out the occurrence

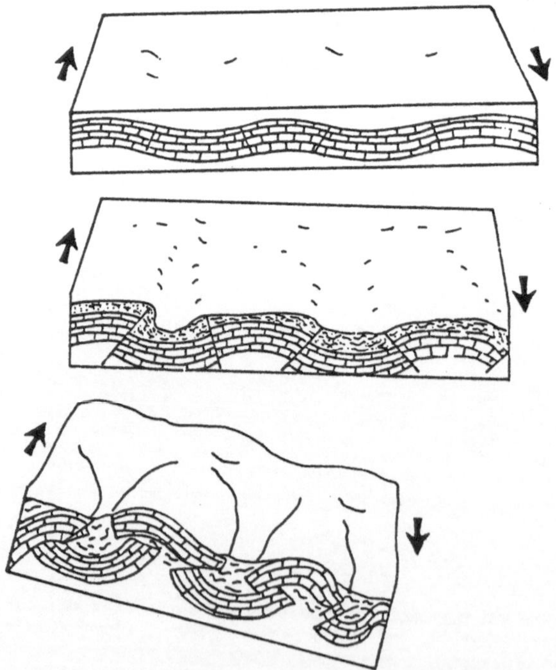

Fig.161 - Evolution of the flower structures produced by convergent wrenching (adapted from De Smet,1984).

of folds:there is no doubt in fact that most of carbonate sheets are strongly folded (Fig.160). Some structures,such as the Cima Ombladet (Fig.160), may be interpreted as 'flower structures'

The evolution of the area from the Frasnian to the Lower Carboniferous proposed here is shown in Fig.161 (adapted from De Smet,1984).Wrenching initially leads to folding and faulting of the carbonate platform. Successively the anticlinal areas are uplifted and squeezed out, with formation of 'flower structures'.The flower structure concept is a key in the interpretation of several tectonic structures in the area which display an opposite vergence, such as the structure shown in Fig.162, bracketed by arrows,which could be interpreted alternatively as a thrust.According to the 'flower structure interpretation', compressive phases (Asturic phase) may have reoriented the detached slabs of the carbonate platform, or overturned some blocks or limbs of flower structures.

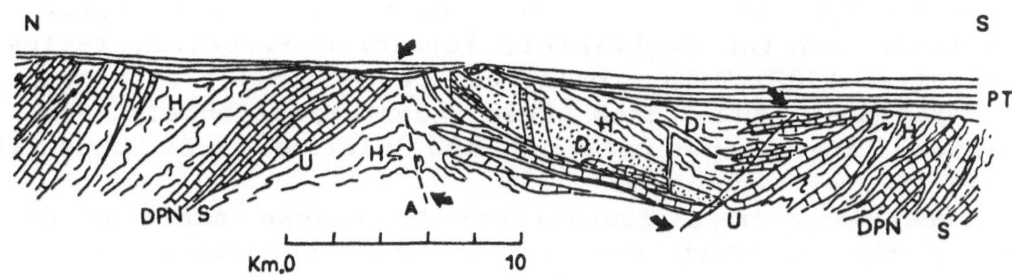

Fig.162 - Palinspastic reconstruction of the Paleocarnic chain (Upper Permian) (after Vai,1976).The structure of the chain is interpreted as a series of imbricated underthrusts. Observe the absence of anticlinal forms in the profile (cf. with Fig. 160),the opposed vergence of the area delimited by arrows interpreted here as a 'flower structure',the fault (a) floating within the flysch Hochwipfel Formation.According to the thrust sheet view, the volume of the Hochwipfel Formation is preponderant.Conversely, a flower structure interpretation leads to reducing such a volume,in keeping with stratigraphic data indicating an overall total thickness of the Hochwipfel Formation of about 800 m, in contrast with the thickness of the carbonate platform which amounts to about 1200 m. H: Hochwipfel Formation (Namurian- Lower Westfalian); DPN: neritic-pelagic carbonates (Gedinnian- Visean); S: Silurian carbonates and shales; U:Uqua Formation (Caradoc-Ashgill); D:Devonian platform carbonates; Di: Dimon Formation (Lower Westfalian).

A terrestrial - karst horizon is sandwiched between underlying pelagic and shallow-water carbonates and overlying turbidite deposits.It consists of silcrete deposits and various karst features.Recently, Schönlaub, et al.(1991) studied in great detail this horizon and conclusively demonstrated a karst origin.It was the object of much study by italian geologists working in the area, who considered and listed several interesting, alternative possibilities (Spalletta, et al.,1982 a,b,c,d),which served as a stimulus to workers in the area attempting at matching the occurrence of the karst horizon with the existing model of generalized deepening (cf. Fig.157).Some features of this horizon occurring in the study area are shown and described in Fig.163.

Interpretations concerning this terrestrial horizon have been contrasting, because of the difficulty in reconciling an uplift with the previous interpretation of the basin which assumed a generalized, deepening-upward trend (Spalletta, et al.,1979) and with the occurrence of some basinal sequence coeval with the terrestrial horizon and represented by pelagic carbonates (Famennian-Dinantian) overlain by deep water radiolarites (late Dinantian) and the Hochwipfel, turbidite Formation (Silesian) (Spalletta, 1982).This problem can be reconciled by a mechanism of rotational subsidence which contemporaneously sunk the basinward areas and uplifted the platformward zones which became 'flower structures'.

The occurrence of a Tournaisian to Visean emersion horizon draping the platform contradicts the theoretical scheme of diachronous deformation predicted by Vai and Cocozza (1986).The main implication of the occurrence of this terrestrial horizon is that compressional Bretonic (top Devonian) and Sudetic (top Dinantian) phases can not be excluded (according to italian authors only the Asturian phase -Westphalian- took place in the Carnic Alps).Mineralizations and karst horizons seem to be located preferentially upon flower structures (they are absent in the Devonian rocks cropping out in the Austrian area).This documents a localized tectonic control upon mineralizing agents.

The exact origin for these Bretonic and Sudetic phases is obscure.As a matter of fact,the Devonian- Carboniferous boundary (corresponding to the Bretonic phase in the Carnic Alps) records a worldwide emersion, as can be evaluated from an examination of the results of the 1986 symposium of Aachen on 'Late Devonian events around the Old Red Continent' (Ministry of Econ. Affairs,Adm.of Mines,1986).

The two phases correspond to short-term sea level falls in the

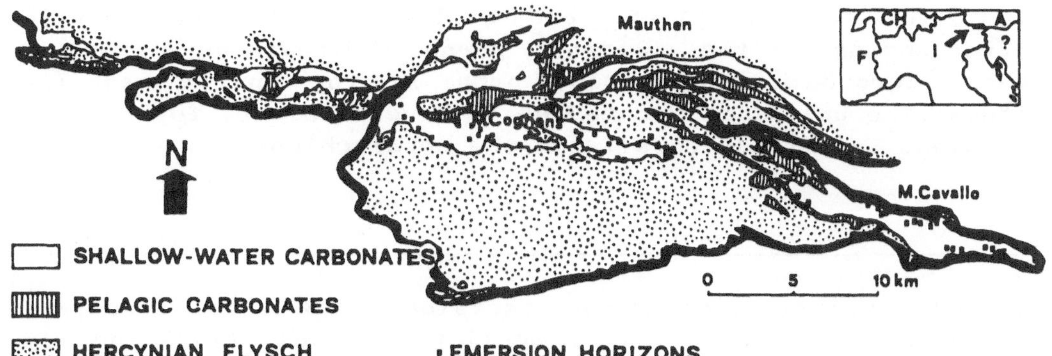
Mauthen
M.Coglians
M.Cavallo
N
CH
A
F
I
?
0
5
10 km
SHALLOW-WATER CARBONATES
PELAGIC CARBONATES
HERCYNIAN FLYSCH
• EMERSION HORIZONS

A
B
C
D

charts by Johnson (1985) and Stille (1924). The emersion corresponding to the Sudetic phase (Tournaisian-Visean) is less recorded in the world; probably the uplift was less intense.The uplift is documented in several areas, for example, in Poland, Armorican Massif and the Appalachians.

Fig.163 - Distribution of the emersion horizon and associated mineralizations (above: after Spalletta, et al.,1982) and some aspects of the shallow-water deposits sandwiched between pelagic carbonates and the turbiditic Hochwipfel Formation (A-E). According to the above scheme the mineralisations developed above pelagic and shallow-water carbonates are confined to a narrow belt trending east-west, which became individuated by relief inversion mechanisms produced by the convergent wrenching tectonics.
A): Silcrete deposits conformably overlying pelagic carbonates (Sasso Nero area).B) : Laminar and tubular bands reminiscent of algal tufas (photo courtesy of L.Brigo).C) : Dolomite layers alternating with scalenoedric calcite.D) : Low-angle lamination indicative of a shoreline environment.
These deposits, in addition to those studied by Schönlaub, et al.(1991),testify for an emersion horizon and uplift of the platform (relief inversion). Similar silica concentrations or silcrete deposits occur in other areas of the eastern Alps,for example in the Dolomites, above Ladinian buildups, and in the Hauptdolomite Formation in Switzerland.The general aspect of these silcrete deposits is also similar to silcretes reported by Smale (1973) from South Africa and Australia. In all such situations silicification took place close to synsedimentary faults, close to a Mg-Ca boundary, in a shallow environment resulting from spiculae concentrations.In the Hauptodolomite Formation (i.e. Munt de la Bascha, Engadine Dolomite) the site of silica deposition was subtidal (less than 10-20 m deep); probably subjected to a syndiagenetic instability (R. Trumpy,1987:pers.comm.).Barite, fluorite, scalenoedric calcite and to a lesser degree, sphalerite, are accessory components of the silcrete deposits.The occurrence of sedimentary structures such as rhythmites and the absence of contemporaneous volcanism rules out an epigenetic origin, as hypothesized by Spalletta, et al.(1982), or even a diagenetic transformation from overlying radiolarites (Spalletta, et al.,1982) which are never found associated with these deposits (Galli,1980).Sponge spiculae of the type found here flourish in shallow-water carbonate environments, such as the Bimini Lagoon (Hay, et al.,1987), and in other lagoonal areas in the Bahamas. Sponge

spiculae are abundant in marginal environments (Chouns and Elkerns,1974), as well as in sabkha facies. Concentrations of sponges may have taken place in a shoreline environment. The bitumen and the carbonaceous material produced in an anoxic lagoon by the decay of organic matter leads to an increase in the concentration of the carbon dioxide which in turn lowers the pH, therefore facilitating the silica replacement of carbonates (Walker,1960) and the precipitation along fissures. Silica is inorganically precipitating in the ephemeral lakes associated with the Coorong Lagoon in S. Australia (Peterson and Von Der Bork,1965).Nearby deposits to these silcretes consist of low-angle laminated,dolomite layers alternating with thin cm-thick horizons composed of scalenoedric calcite (B). Some low-angle laminated sets are composed of fragments of scalenoedric calcite.Oxygen-isotope data from crystals of this type (Schönlaub,et al.;1991) indicate a meteoric origin. Dolomite is commonly associated with beach deposits because the area landward of the beach is a discharge zone for continental groundwaters. An actual example occurs in the Coorong Lagoon, in south Australia. Algal tufas (C) consist of an alternation of tubular and laminar bands; they are quite similar to some components described from present-day terrestrial cyanobacterial stromatolites (Galli and Sarti,1989).The close association of bitumen, silcrete deposits, barite, fuorite, sphalerite, dolomite, beach deposits, algal tufas and organic matter points to a deposition in ephemeral, very shallow-water lagoons ranging from hypersaline to freshwater (cf. Galli,1983).Uplifts and block-faulting may have provided the diastrophic background for the formation of these marginal environments.

The Upper Devonian-Lower Carboniferous interval in the Carnic Alps is instructive because contains several features developed typically during krikogenetic rejuvenation periods; these are:

1)onlap intrashelf ramps in the platform;
2)anoxic basinal facies;
3)deposition of megabreccias and seismoturbidites;
4)strike-slip tectonics and formation of divergent patterns;
5)high-seismicity (Spalletta and Vai,1984);
6)paleokarsts and relief inversions;
7)accentuation of storm incidence on the platform (Galli,1986);
8)drowning of the carbonate platform at the Frasnian-Famennian boundary.

If one considers geoidal deformation as an effect of a global change in the shape of the geoid due to an increase in the rate of rotation (Whyte,1977), convergent wrenching results to be a phase of krikogenetic rejuvenation activity (Wezel,1988). The record of the punctuated, patchyly distributed geoidal deformation on the globe,or strike-slip tectonics, depends upon whether the deformed area is located within or outside a strike-slip megashear (cf. Carey,1988) which becomes activated during phases of geoidal deformation (change from prolate to oblate shape: Whyte,1977). The two case histories summarized above, documenting relief inversions in the Valanginian (Betic Cordillera) and Late Devonian-Carboniferous (Carnic Alps) were both localized within transcurrent megashears.

Part V

The evolution of geological processes, rather than the expression of a linear succession of facts, may be explained by a simple geometrical law regulated by the omothetic proportion, which is a fundamental geometrical property of living beings and neg-entropic processes.Geological 'events' such as storm accentuations, earthquakes, drownings of platforms, global uplifts, phases of increase in the evolution of organisms, and so on, are not distributed at random throughout the geological time. They are preferentially grouped within 'event horizons' which represent points of curvature increase in the relativistic space-time domain.
Based on this conceptual model, several geological facts may be reconsidered; cross-correlations between unrelated processes may reveal a 'logic' behind a tangle of processes obscured by models of linear evolution or gradualism.

Relativistic distribution of 'event horizons'

Introduction

The geological column is punctuated by rare 'events', such as drownings of platforms, faunal extinctions,global short-term episodes of relative se-level change, and so on.Several of these events are not distributed at random, but conversely are grouped within specific time intervals ('event horizons').

The growing interest in rare events, catastrophes and sudden changes interrupting a presumed gradually evolving system fosters the views of the forerunner of catastrophism, Cuvier (1769-1832) who first reported from several stratigraphic successions sudden events represented by appearances and disappearances of fossil species ("Life ... has often been disturbed on Earth by terrible events - calamities ... which at the beginning, have probably moved and overturned to a greater depth the entire outer crust of the globe" : cited from Raup ,1988: p.69).

Since the beginning of Earth Sciences, catastrophists together with their praecursor were regarded as eccentric people blinded by theological creed who resorted to irrational, supernatural causes to explain the evolution of natural processes.The reason of this view regarding catastrophism has been explained by Raup (1988): "... In time, the uniformitarian view of Earth history won decisively over the catastrophist view. It was a reflection of the abhorrence of rare, unpredictable events, that is common in many fields of science. In any event, uniformitarianism (or gradualism,as it is sometimes called) has dominated geology and the education of geologists for the century and a half since the original debate".

The paradigm of uniformitarian principles established by Lyell and Hutton conversely state that "processes that acted in the past are in no way different from processes acting today". The transformation of natural processes is thought to be a cumulative effect of small, trivial changes, since the remotest past. This concept is crystallized in the statement: "The present is the key to the past". Also the actual degree ,rate

and intensity of processes are considered to be the same as in the distant past.
Stratigraphic analysis has not escaped the application of the uniformitarian paradigm. The principles of Facies Models in sedimentology developed by Walther (1893) states that "only those facies and facies-areas can be superimposed primarily which can be observed beside each other at present time". Modal cycles are the product of the application of Walther's Law; in most cases however they are only an dealization of the reality because they show what the situation would have been in a scenario of gradual environmental shifting.

There are some basic problems with the Walther's Law. A first problem is the relation between the present-day accumulation rate of sediments and the geological preservation potential of a given sediment thickness. Sedimentologists studying the actual sediments are more concerned with the modality of deposition than with the preservation potential. A paradox in studying the stratigraphic sections is that most of the geological time is not recorded, due to non-deposition or/and erosion. This is particularly evident where fluctuations of the base level (level below which deposition takes place under any condition) are frequent. Beds usually represent a small time compared to that represented by bedding planes.Viewed under this perspective,beds are 'events'. When not produced by diagenesis, bedding planes represent changes in the environment.This means that the Walther's law is applicable correctly only to the sediments representing a small time interval separated by discontionuities. The Walther's Law works well only when sequences do not record major breaks in sedimentary processes, when facies boundaries are gradational. The application of stratigraphic gradualism has resulted in the definition of Formations and Members -diachronous rock units- which result from gradual environmental shifts. In several instances Formations do not have a genetic meaning: they are an artifact of stratigraphic gradualism; they have added confusion to the complexity of geological processes by leading to a complicated tangle of names and subdivisions which is familiar to every geologist starting the study of some new area.

Goodwin and Anderson (1985) chosed the 'PAC' as a fundamental unit of analysis of episodic punctuated sediment accumulation; their reults contrasted and led to different interpretations from those founded on a gradualistic approach: whereas the use of Formations and Members led to a disordered stratigraphy, the use of PAC led to a very ordered, layer-cake of stratigraphic units bounded by synchronous surfaces. The same approach is used in sequence stratigraphy, where sequence boundaries are synchronous surfaces of basinwide extent,produced by some global process.

The statement "The Present is the key to the past" needs some reconsideration. For example, just to cite few situations, the ancient epicontinental areas were characterized by a different water circulation and morphology; actual continental platforms underwent in the Quaternary atypical tectonic and glacial morphological modifications. Shelf regions are covered by relict sediments forming migrating sandwaves (i.e. along the eastern coasts of North America). Shelf margin models, as well as the Bahamian model must be used with caution,as stressed by the same carbonate sedimentologists working in the area. Paleozoic oceans had different geometries: they were undersaturated with respect to calcium carbonate. Genetic models for Mesozoic deep-water sediments advanced by Winterer and Bosellini (1975) for the Alpine - Mediterranean region can not be applied for recent oceans because of different calcium carbonate concentrations. The intensity of geological processes has also varied with time: in the distant geological past there was a prevalence of thalassocratic conditions whereas at present an epeirocratic regime is predominating (emergent areas are prevailing). Some classes of deposits and processes are better developed in the past that in the present: for example, the lack of vegetational cover in the distant past led to much higher rate of mechanical deposition. Orogeneses also increased in rate and volcanic activity since the Precambrian.Plate tectonics probably did not operate in the Paleozoic.

In short, the present represents only a very small fraction of evolution of geological processes compared to the span interval of time represented by the Phanerozoic. A blind application of uniformitarian principles may correspond in some cases to a fixist attitude.

The increasing interest in rare events in the last decades led to the need to accomplish a distinction between continuous and discontinuous processes. Recent analyses have shown that most of processes considered initially as continuous,such as those listed in Ager (1981):erosion of a meandering river, reef growth, subsidence, uplift,pelagic deposition, heat flow, seafloor spreading, magnetic field, cosmic rays, are punctuated by discontinuities. The distinction also depends upon the scale of observation.

The solar radiation for example undergoes discontinuities in intensity at various scales, related to changes in the magnetic

field below the photosphere. These discontinuities have a bearing on the terrestrial climate and weather such as an intensification of hurricane strength, changes in the Earth's spin rate, shift of climatic zones,and so on.
The deep sea drilling project results indicate synchronous variations in the sedimentation rate of pelagic deposition, formerly considered as continuous.Variations are related at various temporal scales to cooling episodes.

The rate of sea floor spreading is also punctuated by discontinuities; Schwan (1980) observed a coincidence between such discontinuities and unconformities of orogenic phases in orogenic belts. Orogenic deformation at a first approximation is a continuous process as an orogenic belt records a migration of the deformation towards the foreland (Vai,1987). Detailed studies conducted in several orogenic belts by Vai (1987) have shown that the deformation is concentrated within specific time intervals.According to Stille (1924) orogenic phases and unconformities are episodic and correspond to tectonic phases. Conversely, uniformitarian views assume that orogenic phases are continuous; a natural consequence of this view is the difficulty of interpreting the cause for major unconformities delimiting depositional sequences which are normally attributed to eustatic sea level changes or to an obscure,not clear reorganization phase of plate tectonic activity (Bally,1980).

Unconformities are considered by uniformitarians as local perturbations of an otherwise gradually evolving process.

There are two ways of considering events (Raup,1988):

1) events may occur as sudden happenings, totally unrelated to a process;
2) events may subtend a process operating before the onset of the event in which case they mark the overcoming of a threshold.

Based on this concept, Raupp (1988) made a distinction between <u>point events</u> and <u>threshold events</u>.

As natural processes become better understood, more and more point events shift into the threshold event category. For example,the theory proposed by Alvarez et al., (1980) according to which the dinosaur extinction was caused by a collision between an asteroid and the Earth, formerly considered as a random event, is now being reconsidered, in the light of

discoveries of other iridium anomalies associated with minor extinction periods (Lower Devonian;Middle Miocene;end of Cenomanian; Eocene-Oligocene boundary).
As our knowledge of phenomena increases, the idea of randomness tends to be replaced with cause-effect relationships, as pointed out by a mathematician of the beginning of the twentieth century: "randomness is a measure of our ignorance" (Poincaré,1910).

Hurricanes are normally considered as random events. If we consider some of the several maps of hurricane tracks hitting Florida peninsula, we can conclude without the slightest doubt that the possibility of some hurricane hitting the east Florida is regulated by chance. Conversely,the finding of a thickening-upward sequence generated by hurricanes and winter storm layers in the Holocene Florida Bay, and successively of various analogous or identical sequences of different ages and locations in Europe (Galli,1990; Fig.11,12) led me to reconsider on one side the randomness of hurricanes as random events,and,on the other side, the use of the term event as a synonymous of randomness. In other words, how is it possible that random events form a trend?

Likewise, the question I attempted at formulating in this work is the following: how is it possible that events that are spaced from each other several millions years form a temporal sequence of events regulated by a geometrical proportion? In fact, an important point observable from an examination of Fig.149 is the trend of successive lag times between the time intervals of short-term sea level falls (interpreted as phases of periods of geoidal deformation).

Such discontinuities are not distributed at random because each lag time between two temporal discontinuities is approximately the half of the preceding one.Is this pure chance or does it underline some hidden physical or geometrical law?

Relativistic concept of event

The study of 'events',however new in Earth sciences, is an old object of investigation in Physics.

Events were initially studied by Cusano, Leonardo da Vinci and Pacioli during the Renaissance period in Italy, in the XIV century.Leonardo da Vinci in particular studied the modalities of formation of impulse waves. He observed that sinusoidal waves under certain conditions change their configuration with their transformation into impulse waves.

The study of discontinuities and events was successively undertaken by Riemann, Gauss and Cantor in the XIX century, in Germany. It forms the basis for the development of the relativity theory of Einstein. Riemann demonstrated that under defined conditions of amplitude and wavelength the wave changes its configuration from a sinusoidal and forms a discontinuous front which represents a discontinuity. This transformation takes place through discrete amounts of energy,or infinitesimal quanta of action.

The concept of 'event' is one of the central concepts of the relativity theory. An 'event' is considered as the intersection point of the four space-temporal coordinates.The general theory of the relativity implies the hypothesis that nature, sometimes called the 'complex domain', can be analyzed through events. The space, rather than an absolute state, as assumed by the Newtonian physics, is determined by the event distribution which constitutes a non rigid system of reference, equivalent to a gaussian, curved quadridimensional system of coordinates.
The substitution of time and space with the time-space domain involves the replacement of the concept of matter with the event or event-particle concept. The reality of a phenomenon therefore represents a series of events. A group of events may be linked to each other by a determined law; their clustering may represent the arrival point of several geodesics (the path from one event point to another is determined by the least action principle: a body follows the path that correspond to the minimum action: material points move along geodesics).
A cluster of events, linked to a gravity center, close to a curvature of the space-time domain, may be related to other similar points in the chosed system of coordinates, which are the loci of similar events. All these events form a process. Matter is a mathematical construction based on the event distribution.Before Einstein Mach substituted the concept of space with that of summation of instantaneous distances between material points. The Universe results to have a discontinuous, corpuscular structure whose geometrical property is determined by the discontinuous distribution of event-matter. Several events clustered in the same space-time horizon are termed <u>event horizons</u>.

What emerged from the relativity theory was the possibility of developing a geometrical representation of the distribution of discontinuities.

Neg-entropy

Given a point 'x' in a natural system, or more generally in the Universe, characterized by a series of processes P1 ,P2 ,P3 ... Pn, any parameter is defined by the following function:

$$x = g\ (P1,P2...Pn) \quad (1)$$

According to this equation, organic processes control the development and geometrical properties of inorganic processes which represent discontinuities or 'events'.

Schrödinger associated the term neg-entropy to the process of formation and growth of organic processes, such as those characterizing living organisms. Neg-entropy is the opposite than entropy; it is a measure of the increase in the level of organization of a given process. Life is in fact associated to minimum values of entropy; it is in a constant disequilibrium state which contradicts the second principle of termodynamics equilibrium.

A measure of the neghentropy of a system may by accomplished on a geometrical basis; a geometrical approach stresses the relationship between the increase in the level of organization of the complex domain and the resulting increase in the associated neg-entropy.

The Gaia hypothesis by Lovelock (1979) which considers the Earth as an organic whole, is also based on the negentropy concept; this hypothesis assumes that all processes on the globe are regulated by Life through an omeostatic, autoregolatory principle capable of maintaining conditions suitable to Life.

This hypothesis has met with resistance, probably because of the actual orientation of science towards an opposite, reductive approach which may be schematized by the following equation:

$$Y = f(x1,x2...xn) \quad (2)$$

where Y is a process, f a function of a series of parameters x which correspond to a series of differential equations.

Inherent in the (2) is the tendency to explain organic processes as a complicated combination between inorganic processes. The reductive approach is dominant to overwelming in the science, which is concentrated on the study and description of structures which are considered as the reality of the complex domain.

The (1) conversely implies that structures or discontinuities are the product of a transformation of a neg-entropic process, mainly a process of organic growth.

Generation of singularities

According to Riemann (1854) the object of physical mathematics and geometry is the study of the process of transformation from n order series to a higher n+1 order series. The infinitesimal calculus developed by Leibniz is also based on the same principles of generation of numerical series.

One can consider a given series:

1,2,3,4,5,6 ... (3)

and two higher-order series, respectively the square and the cube of the first series:

1, 4, 9, 16, 25, 36...
1, 8,27, 64,125...

The differences between the terms of a series of a given order are the following:

1, 4, 9, 16, 25, 36, 49, 64...
3, 5, 7, 9, 11, 13, 15

The last series represents a process of transformation of the former, original series. The differences of the derived series can also be obtained:

3, 5, 7, 9, 13, 15...
2, 2, 2, 2, 2, 2...

The derived series of differences indicate that each term derives from the previous one by adding the same quantity (in this case the value '2').

This process is known as the invariant law of transformation.

The process of transformation can be reverted with the generation of a former series through integration or successive summations.
In keeping with Riemann ("The metric of a process implies the generation of singularities"),numbers originate from the counting of singularities. The interval between two successive terms of a series of differences is characterized by an infinitesimal increment dx to the system which becomes visible only through the formation of a new singularity.

A consequence of the above conceptual scheme is that a neg-entropic process of growth takes place only with the generation of singularities which correspond to quanta of action, that is,through discontinuous jumps.

Hierarchy of singularities

A neg-entropic process consists of a discontinuous growth of singularities which takes place at different levels. This can be schematized in the following way, by considering a system of series of first, second, third order.

```
x1 1, 2,  3,  4,  5,  6,...
x2 1, 4,  9, 16, 25, 36...
x3 1, 8, 27, 64,125,216...
x4 1,16, 81,256...
```

The horizontal rows belong to a class or order of singularity N characterized by a geometrical expansion. Vertical columns are exponential series N+1 corresponding to an exponential growth. Each of the series N+1 grows more rapidly than any other of the former series belonging to the class N. The series N+2 along the diagonal correspond to a function which grows even more rapidly than the series of transformation N and N+1.

The series of transformation N,N+1,N+2... may be defined by a higher-level procesds M which in turn generates M+1,M+2... processes.
According to Riemann the multiple domain is structured in this way; for example, the scale of length of physical processes follows an exponential geometrical increase in the order of magnitude.
Geological processes display such a hierarchical organization. The scheme of dynamic stratigraphy developed by Aigner (1985) is an example of hierarchical organization based on three levels of stratigraphic sequences (Fig.164).

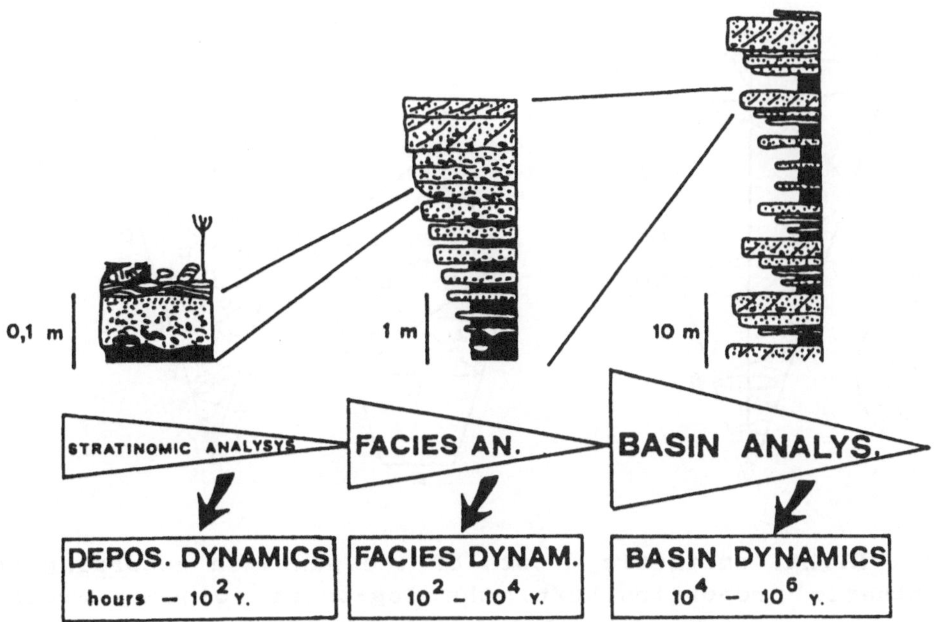

Fig.164 - Hierarchical analysis of three levels of stratigraphical sequences (after Aigner,1985,with permission).

Gemetrical distribution of singularities

The modality of growth of a living organism,or more generally,a neghentropic growth, is known as simple omothetic growth. The process may be visualized by means of a simple geometrical scheme.

In a pentagon - the typical symmetry of living organisms - an omothetic growth leads to an increase in size with a constant proportion , through the development of pentagons of progressively bigger size. The geometrical construction, quite simple, is obtained by prolonging the two sides of the triangles of Fig.165, until a side B and a diagonal A+B are constructed.

This growth must satisfy the following proportions:

$$C=A+B;\ D=B+C;\ E=C+D\ \ldots\ A:B=B:C=C:D\ldots$$

The omothetic proportion is defined as follows:

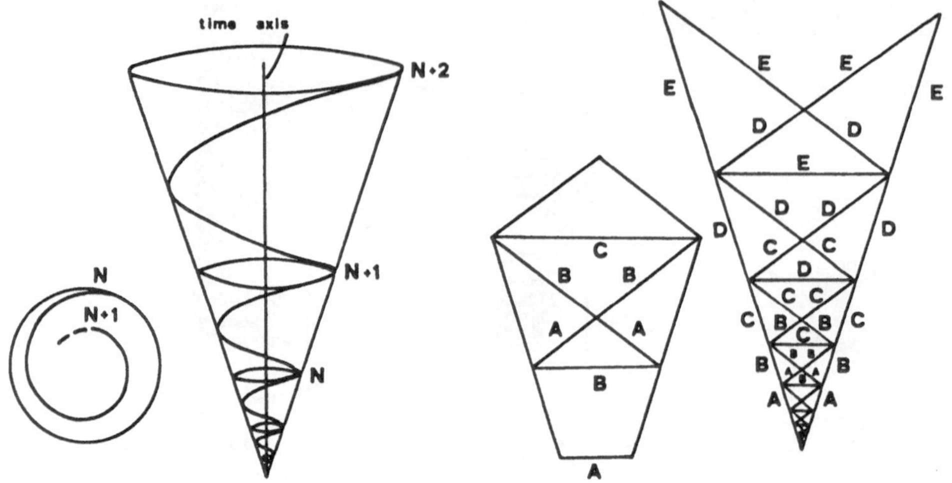

Fig.165 - Right: omothetic growth of the pentagon according to the omothetic proportion.Left: the log-spiral winding around the cones produces circular sections whose geometrical relationships vary according to the omothetic proportion.C) Projection of the log-spiral on the base of the cone.Upon a complete rotation the distance from the axis of the cone is halved.

A+B=B:(A+B)

The omothetic proportion, also known as golden section by ancient Greeks, is a fundamental character of living beings and organic processes, or even inorganic processes directly derived from organic processes. It occurs in DNA cells, microscopic organisms, trees, animals, shells, and so on. The simplest scheme of the omothetic growth of a population is given by the Fibonacci series (1,1,2,3,5,8,11...) whose successive rates (1/1;2/2;2/3;3/5;5/8...) rapidly converge towards the values of the omothetic proportion. Living beings differ from inorganic ones by their modality of growth which is regulated by the omothetic proportion and growth.

The concept of spiral conical action developed by Gauss at the beginning of the XIX century is useful in explaining the generation of singularities as the result of a projection of the relativistic space-time continuum onto the euclidean system of coordinates.

The geometrical construction of Fig.165 (right) can be visualized as a section of a spiral where the winding of the geodesic (or log-spiral) around a conic volume determines circular sections whose changes follow the omothetic proportion. A neg-entropic process can be thoght of as an outward and sideward expansion produced by a circulatory and rotational winding of the log-spiral around the conical volume. The rate of neg-entropic growth is expressed by the angle of the cone. Singularities are formed through a 180° rotation around the cone. The formation of a singularity corresponds to a phase change, or metrics, in the system (Fig.166).

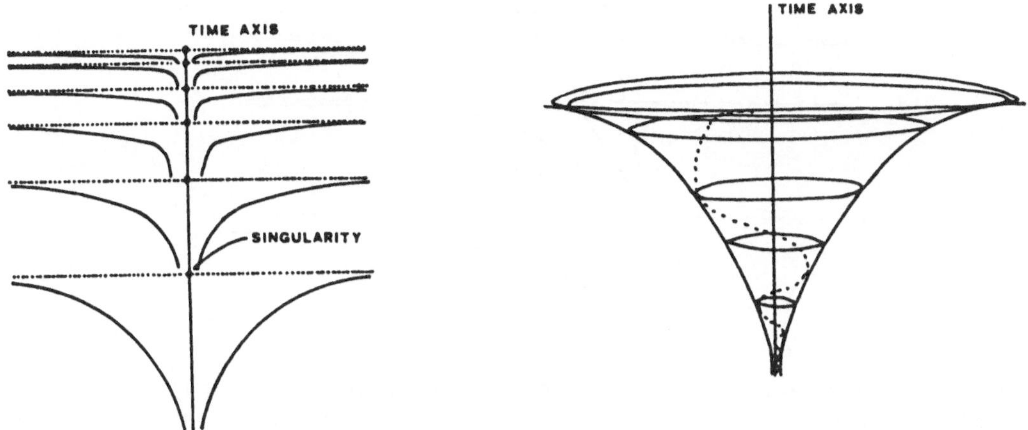

Fig.166 - Series of discontinuities or singularities ordered harmonically in the spiral conical action.

What is perceived as structures on a euclidean system of reference (i.e. Fig.166:left) is the projection of a process of continuous action from a non-euclidean system of reference (the space-time gaussian coordinate system by Einstein).This process can be visualized by the Fig.167 where the discontinuities, projected on the Riemann sphere,individuate a great circle; singularities are invariant points and are maintained in the stereographic projection onto a plane cut normal to the axis of the ellipse; other elements, such as the cartesian infinity disappear as they are not invariant points.A similar representation of neg-entropic process in conformity with the principles of the general relativity was produced by Weyl who conceptualized the expansion of the Universe as originating from a swarm of particles spreading out along geodesics from a point source.

It can be concluded that neg-entropic processess have a typical morphology of growth and functions derived from the process of expansion of an autosimilar spiral structure, congruent with the omothetic proportion. A neg-entropic action corresponds to

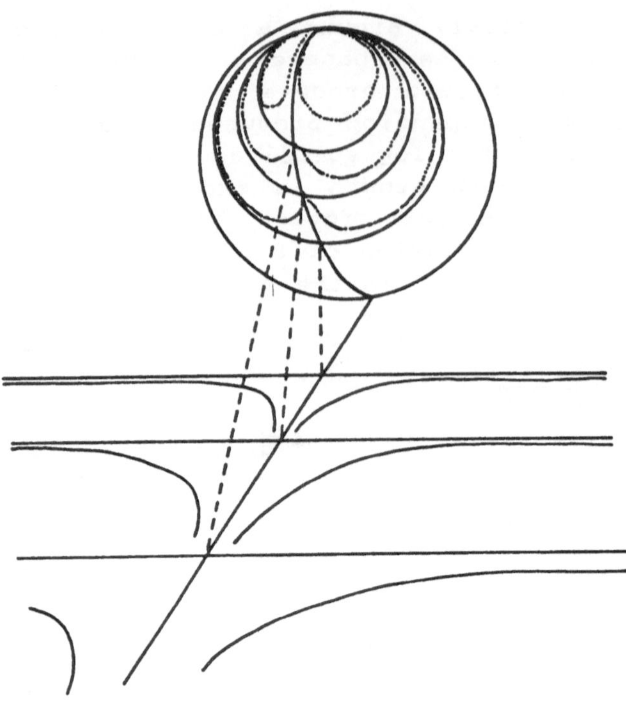

Fig.167 - Stereographic projection of hyperbolic singularities or discontinuities on the Riemann sphere. Observe that on the sphere the cartesian infinities disappear. Successive jumps to upper levels in the systems individuate in the sphere greater spheric volumes;the ratios between spheric volumes converge towards the omothetic proportion.

a spiral which produces a complete rotation between a series of circles progressively increasing in size. There is an exponential acceleration upwards and sidewards as a function of the angle of rotation around the conical action. The winding path is a log-spiral which defines on each rotation circular sections whose geometrical relations are defined by the omothetic proportion.
A plane cut normal to the axis of the cone contains the projection of the circular sections which appear as a series of concentrical circles whose distances from each other are defined by the omothetic proportion.

An example of such geometrical organization is shown by the Kepler's reconstruction of the Solar System.
In Kepler's system, the Solar System is divided into two main regions (excepting Pluto). There are inner planets, composed of heavier elements, devoids of rings and with a small number of satellites, and the outer planets characterized by a big size, a gaseous composition and a higher number of satellites. The division between the two zones is the asteroid belt, occupied

by 100.000 small bodies which correspond to a phase change, to a discontinuity between the two series of planets. The orbits of the planets are ordered harmonically in the way indicated by Gauss and Riemann: it is possible to construct two series of concentrical, slightly oblique sections for the two classes of planets which may be considered as the projections from a cone of the planes containing the orbits, normal to the axis of the cones, in each case the Sun and the Asteroid Belt. The planets appear to be located along a spiral (log-spiral) on the points that correspond to successive rotations of 180° of the log-spiral along the cone. The distance between the planets on the log-spiral is congruent with the omothetic proportion.

Biological evolution

At present the old concept of gradualistic evolution is being replaced by models of punctuated evolution. According to Dutuit (1986) the evolution of Life is a stepwise process progressing through discontinuous phase changes. The evolution of living organisms corresponds to a neg-entropic exponential growth process in response to an increase in the global energy budget of the Universe. Dutuit(1986) went back to the emergence of the first germs of Life, 3.5 to 4 billion years ago. Photosynthetic bacteria, composed of a primitive nucleous, appeared about 2.5 to 2 billion years ago, as is testified by the occurrence of microbial organo-sedimentary structures (stromatolites). Afterwards, eukaryotes appeared about 1 billion year ago. The Ediakara fauna in Australia records the first appearance of metazoans, approximately 750 million years ago.

Dutuit (1986) attempted at quantifying through phase changes the evolution of vertebrates: their evolution would have taken place through the acquisition of new planes of organization. The first stage is represented by advanced fished approximately 450 m.y. ago. Between the first and s the second stage,300 to 340 m.y. ago, there was the acquisition of carrying members, with the appearance of amphibians. The support and locomotion involved an increase in the energy expenses by a factor equal to 10. The third level involved a more advance in the organization marked by the evolution of advanced reptiles (200 m.y. ago) towards the mammalian development, and the transition to endothermy which involved a greater action upon the environment.The following level corresponds to the evolution of mammals (135 m.y. ago).
Here again, the temporal punctuation of the phase changes by

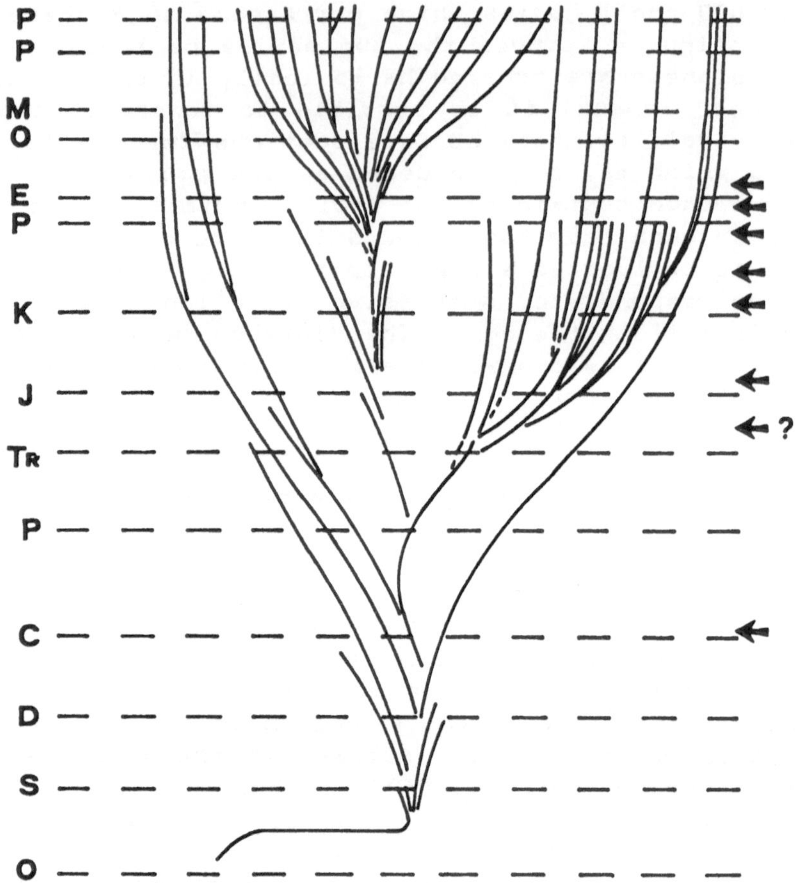

Fig.168 - Tree of Life of vertebrates. The arrows indicating the approximate location of the 'event horizons ' of Fig. 149, suggest a relation with phase changes in the evolution of vertebrates. New steps of the evolution take place close or just in correspondance with the event horizons.

Dutuit (1986) is congruent with the omothetic proportion.

The increase in size of individual phila (i.e. equids) may also be taken as a measure of the change in metrics of the universal evolution: in this case the increase in the energy is compensated by an increase in the mass of living organisms, as follows from the well known formula : $E=Mc^2$. In relativistic terms, the increase in size and/ or complexity of the organisms means that new mass is created because new energy is made available to the system.

Event horizons

The geological column is punctuated by short-term phases of global geoidal deformation (some of which considered in the above chapters), as shown in Fig.149.The intervals between these stratigraphic horizons are congruent with the omothetic proportion.

These time intervals may be thought of as singularities or 'events' of a spiral-type neg-entropic growth (Fig.169). There are two series (Fig.149),separated by a phase change dated approximately 38 m.y. ago which was a period of increase in tectonic deformation, global diastrophic events, acceleration of the Earth's flattening pulses and global cooling (Wezel, 1988).

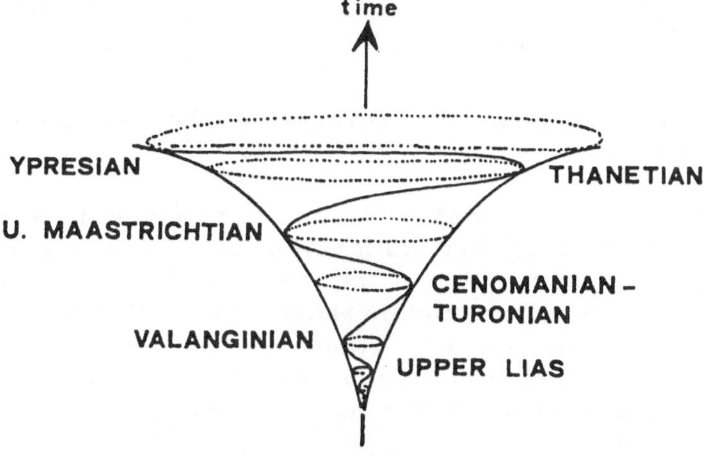

Fig.169 - Log-spiral arrangement of Jurassic-Tertiary event horizons of Fig.149.Event horizons are spaced 180° from each other (cf. Fig.1),and become more closely spaced with time,due to the upward widening of the spiral space-time domain.

It is not surprising to find that several, different types of events are clustered within these horizons, such as iridium anomalies, extinctions, storm accentuations, earthquakes, drownings of platforms, global uplifts, phases of increase in the evolution of organisms. These time intervals correspond to 'event horizons'.They may correspond to krikogenetic rejuvenation periods (see Introduction).

Event horizons may be thought of as the point sources of attraction of events which converge towards the event horizons.Each event may be considered as a point-component of a log-spiral process of winding around the cone and subjected to a deflection near the 'event horizon'.In fact, the event horizons correspond to points of curvature increase of the relativistic space-time domain.

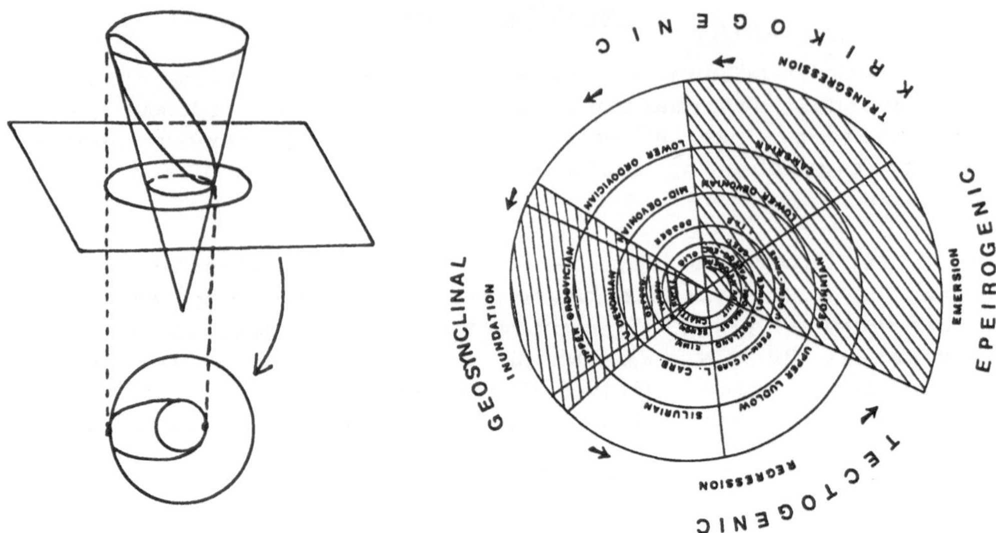

Fig.170 - Double singularity formed by the log spiral action around the cone.

The projection of the 'event horizons' onto the time spiral of Fig.1 shows that they are clustered into two zones with a 180° spacing from each other. The two zones may correspond to krikogenetic rejuvenation and krikogenetic quiescence periods. The 180° spacing may be explained by the concept of double discontinuity obtained by a projection of a conical spiral action on a plane normal to the axis of the cone (Fig.170).

In this respect, the modal sequence described in the above chapters (deepening-upward passing to shallowing-upward cycles; transitions from onlap to offlap geometry, divergent-convergent patterns; more in general all situations ranging from the geotectonic cycle: Fig.2 to 'PAC' cycles) may be thought of as double discontinuities produced by the projections onto our euclidean system of coordinates of a conical spiral action (Fig.171).

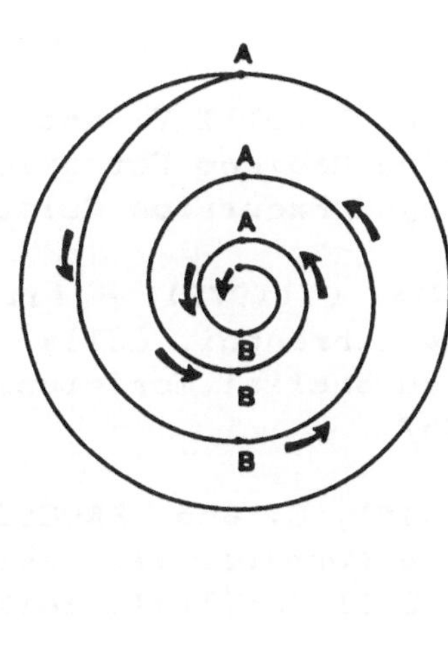

Fig.171 - Modal sequence seen as a double discontinuity or singularity of a spiral-type process.

References

ABBATE,E., BORTOLOTTI,V. and SAGRI,M.(1981) - Olistostromes in the Oligocene Macigno Formation (Florence area). 2nd Reg. Eur. Meeting,Bologna,Excursion Guidebook:163-203.

ACCORSI BENINI,C.(1979) - *Lithioperna*,un nuovo genere fra i grandi lamellibranchi della facies a *Lithiotis*.Morfologia, tassonomia ed analisi morfofunzionale.Boll.Soc. Paleont. ital., **18(2)**:221-257.

ACCORSI BENINI, C. and BROGLIO LORIGA, C. (1977) - *Lithiotis* Gümbel,1871 a *Cochlearites* Reis,1903. Revisione morfologica e tassonomica.Boll.Soc.Ital.,**16(1)**: 15-60.

AGER,D.V.(1981) - The nature of the stratigraphic record. Halsted Press,New York,122 p.

AHR,W.M.(1973) - The carbonate ramp: an alternative to the shelf model.Trans.Gulf Coast Assoc.Geol.,23rd Ann.Conv.: 221-225.

AIGNER,T. (1982) - Calcareous trempestites:storm-dominated stratification in Upper Muschelkalk limestones (Middle Trias,SW Germany). In:G.Einsele and A.Seilacher (Eds.),Cyclic and Event Stratification,Springer-Verlag:180-198.

AIGNER,T.(1985) - Storm depositional systems.Lecture notes on Earth Sciences.Springer-Verlag,174 p.

ALVAREZ,L.W., ALVAREZ,W., ASARO,F. and MICHEL,H.V. (1980) - Extraterrestrial cause for the Cretaceous-Tertiary extinction. Science,**208**: 1095-1108.

ANDERSON,E.J. (1971) - The interpretation of calcarenite paleoenvironments:the Coeymans Formation Lower Devonian of New York. S.E.P.M. Guidebook, Eastern Section, Temple University, Philadelphia, 67 p.

ARTHUR,M.A. and SCHLANGER,S.O. (1979) - Cretaceous 'Ocean Anoxic Events' as causal factors in development of reef-reservoired giant oil fields.A.A.P.G.Bull.,**63**:870-885.

AUBOIN,J., BOSELLINI,A. and COUSINS,M.(1965) - Sur la paléogéographie de la Vénétie au Jurassique.Mem.geopal.Univ.Ferrara,5 (1):147-158.

AUBOIN,J. and BROUSSE,R.(1977) - Compendio di geologia,2nd vol.,Casa editrice ambrosiana,654 p.

AZZAROLI,A. and CITA,M.B.(1967) - Geologia stratigrafica. 2nd vol.,Cisalpino Goliardica, Milan,353 p.

BAIN,R.J. and TEETER,J.W. (1975) - Previously undescribed carbonate deposits on Key Largo,Florida. Geology:137-139.

BALL,M.M. (1967) - Carbonate sand bodies of Florida and the Bahamas. J.Sedim. Petrol.,**37**:556-591.

BALLY,A.W. (1980) - Basins and subsidence-a summary. In: 'Dynamics of Plate Interiors', Geodynamics Series, vol.1, American Geophysical Union: 5-20.

BARBUJANI,C.,BOSELLINI,A. and SARTI,M. (1986) - L'oolite di San Vigilio nel Monte Baldo (Giurassico,Prealpi Venete). Annali Univ. Ferrara (NUova serie),**IX(2)**:19-47.

BEACH,D.K.(1982) - Depositional and diagenetic history of Pliocene and Pleistocene carbonates of NW Great Bahama Bank;evolution of a carbonate platform. PhD thesis, Univ. of Miami,660 p.

BEACH,D.K. and GINSBURG,R.N.(1981) - Facies succession of Pliocene-Pleistocene carbonates,Northwestern Great Bahama Bank. Amer. Assoc. Petrol. Geol. Bull.,**64/10**:1634-1642.

BEAUDRY,D. and MOORE,G.F. (1985) - Seismic stratigraphy and Cenozoic Evolution of West Sumatra Forearc Basin, A.A.P.G. Bull.,**69(5)**: 742-759.

BEAUMONT,C. (1981) - Foreland Basins. Geophys.Jl.R.Astr.Soc., **65**:291-329.

BEBOUT,D.G. and LOUCKS,R.G.(1974) - Stuart City trend,Lower Cretaceous,South Texas: Austin, Texas,Univ. of Texas of Econ. Geol. Report of Investigations,78,80 p.

BEMMELEN.R.W. van (1966) - On Mega-Undations: a new model for earth's evolution. Tectonophysics,**3(2)** :83-127.

BENINI,C. and BROGLIO LORIGA,C.(1974) - *Isognomon* (*Mytiloperna*) v. Ihering 1903 e *Gervilleioperna* Krumbek 1923 fra i grossi lamellibranchi della facies a "Lithiotis" del Veneto.Accad. Naz. Lincei,57(4): 233-245.

BERG,O.R. (1982) - Seismic Detection and Evaluation of Delta and Turbidite Sequences: their Application and Exploration for Subtle Trap. A.A.P.G.Bull.,66(9):1271-1288.

BERNOULLI,D. and JENKINS,H. (1974) - Alpine, Mediterranean,and Central Atlantic Mesozoic facies in relation to the early evolution of the Tethys.S.E.P.M.Sp.Publ.,19:129-160.

BERSEZIO.R. and FORNACIARI,M.(1987) - Cretaceous sequences in the Lombardy Basins:stratigraphic outline between the Lakes of Lecco and Iseo.Mem.Soc. Geol. It.,40:187-197.

BERTI CAVICCHI,A.,BOSELLINI,A. and BROGLIO LORIGA,C.(1971) - Calcari a *Lithiotis problematica* Gumbel o Calcari a "Lithiotis?". Mem.Geopaleontol. Univ.Ferrara,3(I/3):41-53.

BICHSEL,M. and HARING,M.O.(1981) - Facies evolution of Late Cretaceous Flysch in Lombardy (northern Italy).Ecl.Geol.Helv., 74(2):383-420.

BLENDINGER.W.(1984) - Late Ladinian strike-slip tectonics of the Marmolada-Costabella area (Dolomites).Jb.Geol. Bundesanst., 127:307-319.

BLENDINGER,W.(1985) - Middle Triassic strike-slip tectonics and igneous activity of the Dolomites (Southern Alps).Tectonophys., 113:105-121.

BLENDINGER,W.(1986) - Isolated stationary carbonate platforms: the Middle Triassic (Ladinian) of the Marmolada area, Dolomites, Italy. Sedimentol.,33:159-183.

BLENDINGER,W.,PARROW,A. and KEPPLER,F.(1982) - Paleogeography of the M.Cernera-Piz del Corvo area (Dolomites/Italy) during the Upper Anisian and Ladinian.Geol.Rom.,21:217-234.

BONAGA,G.,CANTELLI,C.,DE NUZZO,S.,GALLI,G., MONTANARI,R. and VANUCCI,F.(1989) - Cicli sedimentari nella Formazione del Dürrenstein nei dintorni di Cortina d'Ampezzo (Triassico, Dolomiti Orientali). Giorn. Geol.,ser3,51/1:33-43.

BOSELLINI,A. (1972) - Paleoecologia dei calcari a "Lithiotis" (Giurassico inferiore),prealpi Venete.Riv.ital.Paleontol.,78 (3): 441-464.

BOSELLINI,A.(1973) - Modello geodinamico e paleotettonico delle Alpi Meridionali durante il Giurassico-Cretacico. Sue possibili applicazioni agli Appennini. Accad. Naz. Lincei, Quad.,**183**:163-205.

BOSELLINI,A. (1984) - Progradation geometries of carbonate platforms:examples from the Triassic of the Dolomites,northern Italy. Sedimentology,**31**:1-24.

BOSELLINI,A.(1989) - Dynamics of Tethyan carbonate platforms. S.E.P.M. Spec. Publ.,**44**: 2-13.

BOSELLINI,A. and BROGLIO LORIGA,C. (1971) - I "Calcari Grigi" di Rotzo (Giurassico inferiore, Altopiano di ASiago) e loro inquadramento nella paleogeografia e nella evoluzione tettono-sedimentaria delle Prealpi Venete. Ann. Univ. Ferrara, n.s.,**IX,5(1)**: 1-61.

BOSELLINI,A. and FERIOLI,G.L.(1988) - Sequenze deposizionali e discordanze nel Gargano Meridionale. Atti del 74° congresso Soc. Geol. Ital.:A49-A54.

BOSELLINI,A. and HSU,K.J.(1973) - Mediterranean plate tectonic and Triassic paleogeography. Nature,**224**:144-146.

BOSELLINI,A.,MASETTI,D. and SARTI,M.(1981) - The Vajont Limestone: an oolitic deep sea fan, Middle Jurassic Venetian Alps. 2nd Regional Meeting IAS,Bologna (Excursion Guidebook): 308-342.

BOSELLINI,A. and WINTERER,E.L.(1975) - Pelagic limestone and radiolarite of the Tethyan Mesozoic. A genetic model. Geology,**3**:279-282.

BOSENCE,D. (1988) - Trends in shallow-water carbonate mounds, Florida. IAS 9th European Regional Meeting,Leuven (Abstr.): 26-28.

BOURROUILH-LE JAN and TALANDIER,J.(1985) - Sédimentation et fracturation de haute énérgié en milieu récifal:Tsunamis, ouragans et cyclones et leurs effects sur la sédimentologie et la géomorphologie d'un atoll: Motu et Hoa, à Rangiroa,Tuamotu, Pacifique SE. Mar. Geol.,**67**:263-333.

BRANDNER,R.(1984) - Meeresspiegelschwankungen und Tektonik in der Triass der NW-Tethys. Jb.Geol.Bundesanst.,**126**:435-475.

BRENCHLEY,P.J. and NEWALL,G.(1982) - Storm-influenced inner

shelf sand lobes in the Caradoc (Ordovician) of Shropshire,England. J.Sediment. Petrol.,**52(4)**:1257-1269.

BRETT,C.E.(1983) - Sedimentology, facies and depositional environment of the Rochester Shale (Silurian,Wenlokian) in Western New York and Ontario. J.Sediment. Petrol.,**53**:947-971.

BROADHEAD,R.F., KEPFERLE,R.C. and POTTER,P.E. (1982) - Stratigraphic and Sedimentologic Controls of Gas in Shale-Example from Upper Devonian in Northern Ohio. A.A.P.G.Bull., **66(1)**:10-27.

BROGLIO LORIGA,C. and NERI,C.(1976) - Aspetti paleobiologici e paleogeografici della facies a "Lithiotis" (Giurese inf.). Riv. ital. Paleont.,**82(4)**:651-706.

BROOKS,H.K.(1968) - The Plio-Pleistocene of Florida,with special reference to the strata outcropping in the Caloosahatche River. In: Perkins,R.D.(Ed.), Late Cenozoic Stratigraphy of southern Florida. Miami Geol.Soc.Annual Field Trip:3-42.

BROWN,L.F.,Jr. and FISCHER,W.L.(1977) - Seismic-stratigraphic interpretation of Depositional Systems: Examples from Brazilian Rift and Pull-Apart Basins.A.A.P.G.Memoir,**26**:213-248.

BRYANT,W.R.,MEYERHOFF,A.A.,BROWN,K.N.,FURRER,M.A.,PYLE,T.E. and ANTOINE,J.W.(1969) - Escarpments, reef trends, and diapiric structures. A.A.P.G.Bull.,**53**: 2506-2542.

BURTON,R.,KENDALL,C.G.St.C. and LERCHE,I.(1987) - Out of our Depth: on the Impossibility of Fathoming Eustasy from the Stratigraphic Record. Earth Sci. Reviews,**24**:237-277.

CANTELLI,C.,MANZONI,M. and VAI,G.B.(1965-1968) - Ricerche geologiche preliminari sui terreni paleozoici attraversati dalla galleria del Passo di M.Croce Carnico. Boll.Soc. Geol. It., **84**:27-36; **87**:183-193.

CANTELLI,C., SPALLETTA,C., VAI,G.B. and VENTURINI,C. (1982) - Sommersione della piattaforma e rifting devono-dinantiano e namuriano nella geologiadel Passo di M.Croce Carnico. In: Castellarin,A. and Vai,G.B.(Ed.),"Guida alla Geologia del Sudalpino centro-orientale". Guide Reg. Geol.S.G.I.:293-303.

CARBONE,F. and SIRNA,G.(1981) - Upper Cretaceous Reef Models from Rocca di Cave and Adjacent Areas in Latium,Central Italy.S.E.P.M.Sp.Publ.,**30**:427-445.

CASTELLARIN,A.(1972) - Evoluzione paleotettonica e sinsedimentaria del limite tra piattaforma veneta e bacino lombardo a nord di Riva del Garda.Gior.Geol.,38: 11-212.

CASTELLARIN,A.(1982) - Lineamenti ancestrali sudalpini. In:Castellarin,A. and Vai,G.B.(Eds.),'Guida alla geologia del Sudalpino centro-orientale',Guide Geol.Reg.S.G.I.:41-55.

CASTELLARIN,A., MORTEN,L. and BARGOSSI, G.M. (1976) - Conglomerati di conoide sottomarina nel flysch insubrico di Malè e Rumo (Trento). Boll.Soc.Geol.It.,95:513-525.

CASTELLARIN,A. and SARTORI,R.(1973a) - Dessication shrinkage and leaching vugs in the Calcari Grigi infraliassic tidal flat (S.Massenza and Loppio,Trento,Italy).Eclogae geol. Helv., 66 (2): 339-343.

CASTELLARIN,A. and SARTORI,R. (1973b) - I ciclotemi infraliassici di S.Massenza (Trento). Gior.Geologia, 39: 221-248.

CATALOV,G.A. (1983) - Triassic oncoids from central Balkanides (Bulgaria).In: T.M. Peryt (Ed.), Coated Grains, Springer-Verlag Berlin Heidelberg: 398-408.

CAUSARAS, C.R. (1987) - Geology of the surficial aquifer system, Dade County, Florida lithologiclogs.U.S.G.S.Water Resources Investigation Report, 86-4126.

CHOI,D.R. and GINSBURG,R.N. (1982) - Siliciclastic foundations of Quaternary reefs in the southernmost Belize Lagoon, British Honduras. Geol.Soc.Amer.Bull.,93:116-126.

CHOI,D.R. and HOLMES,C.W.(1982) - Foundations of Quaternary Reefs in south-central Belize Lagoon, Central America. A.A.P.G. Bull.,66(12):2663-2681.

CHOWNS,T.M. and ELKERNS, J.E. (1974) - The origin of quartz geodes and cauliflower cherts through the silicification of anhydrite nodules. J.Sediment.Petrol.,44(3) : 885-903.

CISNE,J.L. (1987) - Earthquakes recorded stratiphically on carbonate platforms. Nature,323 :320-322.

CLARI,P. (1975) - Caratteristiche sedimentologiche e paleontologiche di alcune sezioni dei Calcari Grigi del Veneto. Mem.Ist.Geol. Univ.Padova,31: 2-63.

COLACICCHI,R.(1987) - Sedimentation on a carbonate platform as controlled by sea level changes and tectonic movements. Mem.Soc.Geol.It.,40:199-208.

COOKE,C.W. (1945) - Geology of Florida. Florida Geol. Survey Geol. Bull., 29 :339 p.

COPPER P.(1986) - Frasnian/Famennian mass extinction and cold-water oceans. Geol.,14:835-839.

CRAIGHEAD,F.C.Sr.(1969)- Vegetation and Recent Sedimentation in Everglades National Park. Florida Naturalist (reprinted paper).

CREMONINI,G.,ELMI,C. and SELLI,R. (1971) - Note illustrative della carta geologica d'Italia. Fg.156 - San Marco in Lamis,Roma 1971.

CREVELLO,P.D. and SCHLAGER W.(1980) - Carbonate debris sheets and turbidites,Exuma Sound,Bahamas: J.Sedim.Petrol., 50:1121-1148.

CROS,P.(1974) - Evolution sédimentologique et paléostructurale de quelques plate-formes carbonatées biogénes (Trias des Dolomites italiennes).Sciences Terre, 19:299-379.

DAVIES,G.R.(1970) - Algal-laminated sediments, Gladstone Embayment,Shark Bay,Western Australia.Mem.Am.Ass.Pet. Geol., 13: 169-205.

DAVIES,T.D.(1980) - Peat formation in Florida Bay and its significance in interpreting the recent vegetational and geological history of the Bay area.PhD Thesis Pennsylvania State Univ.State College,Pennsylvania,316 p.

DAVIS,R.A.,Jr.,FOX,W.T.,HAYES,M.O. & BOOTHROYD,J.C.(1972) - Comparison of ridge-and-runnel systems in tidal and non-tidal environments.J.Sediment. Petrol.,32:413-421.

DE JONG,K.A. and SCHOLTEN,R.(Eds.) (1973) - Gravity and Tectonics. Wiley,New York,502 p.
DE JONG,K.A.(1973) - Mountain building in the Mediterranean region. In: K.A. de Jong and R.Scholten (Eds.),Gravity and tectonics,Wiley,N.York,502 p.

DESIO,A.(1973) - Geologia dell'Italia. UTET,1081 p.

DE SMET,M.E.M.(1984) - Investigations of the Crevillente Fault zone and its role in the tectogenesis of the Betic

zone and its role in the tectogenesis of the Betic Cordillera,southern Spain. Vrije Univ.,Amsterdam,174 p.

DIXON, O.A., NARBONNE, G.M. and JONES,B. (1981) - Event correlation in Upper Silurian rocks of Sommerset Island,Canadian Arctic. Bull.Am.Assoc.Petr. Geol.,65:303-311

DOGLIONI,C. and BOSELLINI,A.(1984) - Platform Break-downlap relationship in prograding carbonate buildups: a tool for the reconstruction of basin evolution. Boll.Soc.Geol. It.,108:175-182.

DROXLER,A.W. and SCHLAGER,W. (1985) - Glacial versus interglacial sedimentation rates and turbidite frequency in the Bahamas. Geology, 13 :799-802.

DOTT,R.H.Jr. & BOURGEOIS,J.(1982) - Hummocky stratification: significance of its variable bedding sequences. Geol.Soc. Amer. Bull.,93: 663-680.

EBERLI,G.P. (1987) - Carbonate turbidite sequences deposited in rift-basins of the Jurassic Tethys Ocean (eastern Alps, Switzerland). Sedimentol.,34: 363-388.

EBERLI,G.P. and GINSBURG,R.N. (1987) - Segmentation and coalescence of Cenozoic carbonate platforms,north-western Great Bahama Bank.Geology,15: 75-79.

ENOS,P.(1977) - Holocene sediment accumulations of the Florida shelf margin. In: P.Enos and R.D.Perkins, Quaternary sedimentation in south Florida. Geol.Soc.of Amer.Memoir., 147:1-130.

ENOS,P. and PERKINS,R.D. (1977) - Evolution of Florida Bay from island stratigraphy. Geol. Soc. Am. Bull.,90:59-83.

ERSKINS,R.D. and VAIL,P.R.(1988) - Seismic stratigraphy of the Exmouth Plateau. In: A.W.Bally (Ed.), Atlas of Seismic Stratigraphy, A.A.P.G. Studies in Geology,27(2): 163-173.

EVANS, C.C. and GINSBURG,R.N. (1987) - Fabric-selective diagenesis in the Late Pleistocene Miami Limestone. J.Sedim. Petrol.,57(2): 311-318.

EXON,N.F. and WILCOX,J.B. (1978) - Geology and Petroleum potential of Exmouth Plateau Area off Western Australia. A.A.P.G.Bull.,62 (1): 40-72.

EXON,N.F., VON RAD,U. and VON STACKELBERG,U. (1982) - The Geological Development of the Passive Margins of the Exmouth Plateau off Northwest Australia. Mar.Geol.,47:131-152.

FABIANI,R.and TREVISAN,L. (1939) - Foglio Schio.Note Illustr.Carta Geol.Tre Venezie, Uff.Idr.R.Mag. Acque:1-88, Padova.

FERIOLI,G.(1986-1987) - Geologia del Gargano Meridionale. Thesis, Univ. Ferrara,133 p.

FERRARI,A.and VAI,G.B.(1973) - Revision of the Fammennian *Rhynchonellid* genus *Plectorhynchella*. Giorn.Geol.,39: 163-200.

FISCHER,A.G.(1966) - The Lofer Cyclothems of the alpine Triassic. Kansas Geol.Surv. Bull.,109:107-149.

FISCHER,A.G. (1981) - Climatic oscillations in the biosphere. In:"M.H.Nitecki (Ed.), 'Biotic crises in ecological and evolutionary time'.Academic Press, New York:103-133.

FUGANTI,A.(1964) - Strutture geopete nei calcari mesozoici del Trentino e del Veronese (Alpi Orientali). Atti Accad. Roveretana Agiati, 4(6):5-9,Rovereto,Trento.

FUGANTI,A. and MOSNA,S.(1966) - Studio stratigrafico sedimentologico e micropaleontologico delle facies giurassiche del Trentino occidentale. Studi Trentini Sci. Nat.,S.A.,43(1): 25-105.

FROST,S.H.(1977) - Cenozoic reef systems of Caribbean - properties for paleoecological synthesis. Studies in Geology, A.A.P.G. (4):93-110.

GAETANI,M.,FOIS,E.,JADOUL,F. and NICORA,A.(1981) - Nature and evolution of Middle TRiassic carbonate buildups in the Dolomites(Italy). Mar.Geol.,44:25-57.

GALLI,G. (1980) - Rilevamento e stratigrafia della Cima Ombladet e dintorni (Alpi Carniche occidentali).Thesis, Univ. Bologna,197 p.

GALLI,G.(1983) - Marine evaporites with special reference to their relationship with base metal deposits.Msc. Thesis, Imperial College of Sci.& Thechnology, University of London, 187 p.

GALLI,G.(1984) - Studio sedimentologico preliminare dei

carbonati devoniani della Cima Ombladet (Alpi Carniche occidentali). Boll.Soc.Geol. ital.,**103**:341-347.

GALLI,G. (1984) - Hercynian synsedimentary tectonics: new analytical data.Mem.Scienze Geologiche,**XXXVI**:453-460,Padova.

GALLI,G. (1985a) - Depositional environments in the Devonian imestone succession of the Cima Ombladet (Carnic Alps,Italy). Facies,**12** :97-112

GALLI,G. (1985b) - Depositi litorali carbonatici devoniani (Alpi Carniche) (Applicazione dell'analisi della catena markoviana). Atti Soc.Ital.Sc.Nat. & Museo Civico Milano, **126**: 70-84.

GALLI,G. (1985c) - Ecology and dispersion of the fauna of the Cima Ombladet carbonate succession (Devonian,Carnic Alps, Italy). Palaeogeography, palaeoclimatology, Palaeoecology, **49**:265-275.

GALLI,G.(1986) - Facies analysis of a Devonian carbonate shoreline system (northern Italy).Sed. Geol.,**46**:91-110.

GALLI,G.(1986) - Analisi delle facies carbonatiche e terrigene della Formazione dcell'Auernig nel Permo-Carbonifero Pontebbano nella Alpi Carniche Orientali.Atti Soc.ital.Sci.nat. Museo Civ.Stor.nat.Milano,**127(1-2)**:3-12.

GALLI,G.(1988) - Struttura delle rampe di intrapiattaforma con esempi dalle Alpi Meridionali e dalla piattaforma della Florida (Applicazione di concetti di sismostratigrafia nella ricostruzione di geometrie carbonatiche di mare basso).PhD Thesis,Univ.Bologna,134 p.

GALLI,G. (1989a) - Depositional mechanisms of storm sedimentation in the Triassic Dürrenstein Formation,Dolomites, Italy.Sed. Geol.,**61**:81-93.

GALLI,G.(1989b) - Storm sedimentation in a Quaternary rocky shore sequence (southern Italy). Neues Jahrbuch Geol. Pal. Stuttgart,**H 10**: 590-602.

GALLI,G.(1989c) - Is Holocene storm-generated stratification in Florida Bay a reflection of solar storm cycles?Paleogeography, Palaeoecology, Palaeoclimatology,**76**: 169-185.

GALLI,G. (1990) - Origins of event beds in the Jurassic "Calcari Grigi" Formation, Venertian Alps, Italy.Geologie en Mijnbouw, **69** :375-390.

GALLI,G. (1991) - Mangrove-Generated Structures and Depositional Model of the Pleistocene Fort Thompson Formation (Florida Plateau). Facies, 25:297-314.

GALLI,G.,CANTELLI,C. and LAMBORGHINI,P. (1985) - Correlazione statistica di livelli ghiaiosi (torrente Idice,Bologna). Riv.Ital. Geotecnica, 4(XIX):199)207.

GALLI,G. and SARTI,C. (1989) - Freshwater stream stromatolites (Holocene, Villa Ghigi, Bologna, Italy).Revue de Paleobiologie, 8(1):39-49.

GALLOWAY,W.E. (1989) - Genetic stratigraphic sequences in basin analysis,I Architecture and genesis of flooding-surface bounded depositional units. A.A.P.G.Bull.,73:125-142.

GAUPP,R.,MOLLER,N.R. and BORSCHINSKY,R.(1981) - Epicontinental clastic sediments of the northern Calcareous Alps (Cenomanian-Turonian): examples of syntectonic sedimentation. IAS 2nd Reg. European Meeting,Bologna (Abstr.):69-72.

GEYER,O.F.(1977) - Die "Lithiotis-Kalke" im Bereich der unter jurassischen Tethys.N.Jb.Geol.Palaont.,Abh.,153(3): 304-340.

GINSBURG,R.N.(1956) - Environmental relationships of grain size and constituent particles in some south Florida carbonate sediments: A.A.P.G.Bull.,40 :2384- 2427.

GOHNER,D.(1980) -"Covel del'Angiolono"-ein mittelliassisches LIthiotis-Schlammbioherm auf der Hochebene von Lavarone (Provinz Trento,Norditalien). N.Jb.Paläontologie Abh.,10: 600-619.

GÖHNER,D.(1981) - Beitrage zur Kenntnis des Sudalpinen Juras mit besonderer Berücksichtigung der unterjurassischen Lithiotis - Fazies. I.Stratigraphie, Microfazies und Paläontologie, Universität Stuttgart, 163 p.

GILDHAMMER,R.K.,DUNN,P.A. and HARDIE,L.A.(1990) - Depositional cycles, composite sea-level changes,cycle stacking patterns, and the hierarchy of stratigraphic forcing: Examples from Alpine Triassic platform carbonates. Geol. Soc.Amer. Bull., 102: 535-562.

GOODWIN,P.W.and ANDERSON,E.J.(1985) - Punctuated aggradational cycles: a general hypothesis of episodic stratigraphic accumulation. J. Geol.,93:515-533.

GOODWIN, P.W.,ANDERSON, E.J.GOODMAN,W.M. and SARAKA, L.J.(1986) - Punctuated aggradational cycles: implications for stratigraphic analysis. Paleoceanography,1(4):417-429.

GROTZINGER, P. (1986) - Cyclicity and palaeoenvironmental dynamics, Rocknest platform, northwestern Canada. Geol.Soc. Amer. Bull.,97:1208-1231.

GUIDISH,T.M.,LERCHE,I.,KENDALL,C.G.St.C. and O'BRIEN,J.J.(1984) - Relationship between Eustatic Sea level changes and basement subsidence.A.A.P.G.Bull.,68:164-177.

GUTTERIDGE,P. (1987) - Dinantian sedimentation on the basement structure of the Derbyshire Dome.Geol.J.,22: 25-41.

HALLAM, A. (1978) - Eustatic cycles in the Jurassic. Paleogeography, Palaeoclimatology, Palaeoecology, 23:1-32.

HALLAM, A. (1981) - A revised sea-level curve for the early Jurassic. J. Geol. Soc. London,138:735-743.

HALLOCK, P. and SCHLAGER, W. (1986) - Nutrient excess and the demise of coral reefs and carbonate platforms.Palaios,1:389-398.

HAQ,B.U., HARDENBOL,J. and VAIL, P.R. (1987) - Chronology of fluctuating sea-level since the Triassic. Science, 235: 1156-1167.

HARDING, T.P. and LOWELL,J.D.(1979) - Structural styles,their plate tectonic habitats, and hydrocarbon traps in petroleum provinces. A.A.P.G.Bull.,63/7:1016-1058.

HARRIS, P. M. (1979) - Facies anatomy and diagenesis of a Bahamian ooid shoal. Sedimenta,7:1-163.

HAY,W.W., WIEDENMAYER,F. and MARSZALEK, D.S. (1970) - Modern Organism Communities of the Bimini Lagoon and their Relation to the Sediment.4th Annual FieldTrip of Miami Geol. Soc.: 19-30.

HEEZEN, B. C., MATTHEWS, J.L.,CATALANO,R.,NATLAND,J.,COOGAN,A., THARP, M. and RAWSON, M. (1973) - Western Pacific gujots.In: Heezen,B.C.,MacGregor,I.D. and others (Eds.),Initial reports of the DSDP, 20, Washington DC, US Government Printing Office:653-723.

HELLAND-HANSEN,W.,KENDALL,C.G.St.C., LERCHE,I. and NAKAYAMA,K. (1988) - A simulation of continental basin margin sedimentation in response to crustal movements,eustatic sealevel change and sedimentation accumulmation rates. Mathematical Geol.,**20**:777-802.

HINE,A. (1977) - Lilyu Bank,Bahamas:history of an active oolite sand shoal. J.Sediment. Petrol.,**47**:1554-1581.

HOFFMEISTER,J.E. and MULTER,H.G. (1965) - Geology and Origin of the Florida Keys. G.S.A Bull.,**709**:1487-1502.

ILLIES,J.H.(1981) - Mechanism of graben formation.Tectonophys., **73**:249-266.

JACKSON, P.C. (1984) - Paleogeography of the Lower Cretaceous group of Western Canada. In: Masters,J.A.(Ed.),Elmworth-Case study of a Deep Basin Gas Field.A.A.P.G.Memoir,**38**:49-77.

JAMES,N.P. (1972) - Holocene and Pleistocene calcareous crust (caliche) profiles: criteria for subaerial exposure.J.Sediment. Petrol.,**42/4**:817-836.

JANSA,L.F. (1981) - Mesozoic carbonate platforms and banks of the eastern North American margin. Mar.Geol.,**44**:97-118.

JENKINS, H.C. (1991) - Impact of Cretaceous Sea Level Rise and Anoxic Events on the Mesozoic Carbonate Platform of Yugoslavia. A.A.P.G.Bull.,**75(6)**:1007-1017.

JOHNSON, D.P., CUFF,C. and RHODES, E. (1984) - Holocene reef sequences and geochemistry,Britomart Reef,Central Great Barrier Reef,Australia. Sediment.,**31**:515-529.

JOHNSON, J.G., KLAPPER,G, and SANDBERG, C.A. (1985) -Devonian eustatic fluctuations in Euroamerica. Bull.geol.Soc.Am.,**96**:567-587.

KENDALL, C.G.St.C. and SCHLAGER, W. (1981) - Carbonates and relative changes in sea-level. Marine Geol.,**44**: 181-212.

KENDALL, C.G.St.C. and SKIPWITH, P.A.D.E. (1969) - Holocene shallow-water carbonate and evaporite sediments of Khor al Baxam, Abu Dhabi,Southwest Persian Gulf. A.A.P.G.Bull.,**53**: 841-869.

KIDWELL,S.M. (1991) - The stratigraphy of shell concentrations. In: Allison,A. and Briggs,D.E.G. (Eds.),'Topics in Geobiology', Plenum Press,N.York,9:211-290.

KINGSTON,D.R., DISHROON,C.P. and WILLIAMS,P.A. (1983) - Global Basin Classification.A.A.P.G.Bull.,**67**(**12**):2175-2193.

KREBS,W. (1974) - Depositional carbonate complexes of central Europe. SEPM Spec.Publ.,**18**:155-208.

KREBS,W. and WACHENDORF,H.(1973) - Proterozoic - Paleozoic and Orogenic Evolution of Central Europe. Geol.Soc.Amer.Bull.,**84**: 2611-2630.

KREISA,R.D. (1981) - Storm-generated sedimentary structures in subtidal marine facies with examples from Middle and Upper Ordovician of southwestern Virginia.J.Sediment.Petrol.,**51**:0823-0848.

KRUMBEIN, W.C. and Sloss, L.L. (1963) - Stratigraphy and Sedimentation.Freeman & Co. (II Ed.),660 p.

KUMAR,N. and SANDERS,J.E. (1976) - Characteristics of the shore face storm deposits: modern and ancient examples. J.Sediment. Petrol.,**46**:145-162.

LAPORTE,L.F. (1967) - Carbonate deposition near mean sealevel and resultant facies mosaic:Manlius Formation (Lower Devonian) of New York State. A.A.P.G.Bull.,**51**:73-101.

LECKIE,D.A. and WALKER,R.G. (1982) - Storm- and tide-dominated shorelines in Cretaceous Moosebar-Lower Gates interval: Outcrop equivalents of Deep Basin gas trap,western Canada:A.A.P.G.Bull. ,**66**:138-157.

LEMOINE,M.M.,BOURBON,M. and TRICART (1978) - Le Jurassique et le Cretacée pìrépiemontains à l'est de Briancon (Alpes occidentales) et l'évolution de la marge européenne de la Tethys: données nouvelles et consequences.Acad. Sci.Comptes Rendus,**286**(ser.D):1237-1240.

LEONARDI,P.(Ed.) (1967) - le Dolomiti.Geologia dei monti tra Isarco e Piave. Two voll..Cons.Naz.Ricerche,Roma,1019 p.

LOVELOCK,J.E. (1979) - Gaia. Oxford Univ.Press;,Oxford,176 p.

LOWENSTAM,H.A. (1950) - Niagaran reefs of the Great Lakes area. J.Geol.,**58**:430-487.

LOWENSTAM,H.A. (1963) - Biological problems relating to the

composition and diagenesis of sediments. In: T.W.Donnelly (Ed.),'The earth sciences', University of Chicago Press, Chicago:137-195.

MACINTYRE,I.G.,BURKE,R.B. and STUCKENRATH,R. (1977) - Thickest recorded Holocene reef section, Isla Perez core hole,Alacran Reef, Mexico. Geology,5:749-754.

MACINTYRE,I.G. and GLYNN,P.W. (1976) - Evolution of modern Caribbean fringing reef,Galeta Point,Panama.A.A.P.G.Bull.,**60**: 1054-1072.

MCLAREN,D.J.(1970) - Time, life and boundaries. J.Paleontol., **44**: 801-815.

MARABINI,S. and VAI,G.B.(1985) - Analisi delle facies e macrotettonica della Vena del Gesso in Romagna. Boll.Soc.Geol. It., **104**: 21-42.

MARSAGLIA,K.M. and KLEIN,G.De.V. (1983) - The paleogeography of Paleozoic and Mesozoic storm depositional systems. J.Geol.,**91**: 117-142.

MARTIN ALGARRA,A. (1988) - Paleogeographic evolution of the Pennibetic (Betic Cordillera,southeastern Spain). 9th IAS Reg.Meetg.of Sed.,Leuven (abstr.):138-139.

MARTINIS,B. and PAVAN,G. (1967) - Note illustrative della carta geologica d'Italia.Fg.156,San Marco in Lamis.

MASSE,J.P. and BORGOMANO,J. (1987) - Un modéle de transition plate-forme - bassin carbonatés controlé par des phénomènes tectoniques:le Crétacé du Gargano (Italia méridionale). C.R. Acad.Sc.Paris,**304(10)**:521-526.

MASSE,J.P. and PHILIP,J.(1981) - Cretaceous coral-rudist buildups of France.SEPM Spec.Publ.,**30**:399-426.

MATTAVELLI,L. and PAVAN,G. (1965) - Studio petrografico delle facies carbonate del Gargano.Rend.Soc.Mineral.Ital.,**XXI**,Pavia.

MATTHEWS,R.K. (1984) - Dynamic Stratigraphy.Prentice-Hall,New Jersey,2nd ed.,489 p.

MEISSNER,F.F. (1972) - Cyclic sedimentation in Middle Permian strata of the Permian basin,West Texas and New Mexico. In: J.C.ELam and S.Chuber (Eds.),"Cyclic sedimentation in the Permian Basin",2nd ed.,West Texas Geol.Soc.Midland Texas: 203-232.

MENARD,H.W. (1964) - Marine Geology of the Pacific. McGraw-Hill,New York,271 p.

MIALL,A.D. (1973) - Markov Chain Analysis applied to an ancient alluvial plain succession. J.Sediment. Petrol.,**20**:347-364.

MIALL,A.D. (1984) - Eustatic Sea Level Changes Interpreted from Seismic Stratigraphy: A Critique of the Methodology with Particular Reference to the North Sea Jurassic Record.A.A.P.G. Bull.,**70(2)**:131-137.

MILLIMAN,J.D. (1978) - Morphology and Structure of Upper Continental Margin off Southern Brazil.A.A.P.G.Bull.,**62(6)**: 1029-1048.

MINISTRY OF ECONOMIC AFFAIRS,ADM. OF MINES (1986) - Late Devonian Events around the Old Red Continent,Ann.Soc. Géol. Belgique, Spec. vol. "Aachen congr.1986,Avril1986,**109**.

MITCHUM,R.M. Jr.,VAIL,P.R. and SANGREE,J.B. (1977) - Seismic Stratigraphy and Global Changes of Sea Level,Part 6: Stratigraphic Interpretation of Seismic Reflection Patterns in Depsoitional Sequences.A.A.P.G.Memoi,**26**:117-133.

MITTERER,R.M. (1975) - Age and diagenetic temperatures of Pleistocene deposits of Florida based on isoleucine epimerization in *Mercenaria*.Earth and Planetary Sci.Letters,**28**: 275-282.

MOLENAAR,C.M. (1983) - Depositional relations of Cretaceous and Lower Tertiary Rocks,Northeastern ALaska.A.A.P.G.Bull.,**67(7)**: 1066-1080.

MOUGENOT,D.,BOILLOT,G. and REHAULT,J.P. (1983) - Prograding Shelfbreak types on passive continental margins:some european examples.SEPM Sp.Publ.,**33**:61-77.

MOUNT,J.F. (1982) - Storm-surge ebb origin of hummocky cross-stratified units of the Andrews Mountain Member,Campito Formation (Lower Cambrian), White-Inyo Mountains, Eastern California. J.Sediment.Petrol.,**52(3)**:0941-0958.

MORNER,N.A. (1976) - Eustasy and Geoid changes. J. of Geol.,**84**: 123-151.

MORNER,N.A. (1981) - Revolution in Cretaceous sea-level analysis.Geology,**9**:344-346.

MULLINS,H.T.,GARDULSKI,A.F. and HINE,A.C.(1986) - Catastrophic collapse of the West Florida carbonate platform margin. Geology,**14**: 167-170.

MULLINS,H.T. and HINE,A.C. (1989) - Scalloped bank margin: Beginning of the end for carbonate platforms. Geology, **17**: 30-33.

MULTER,H.G.(Ed.) (1977) - Field Guide to some carbonate rock environments - Florida Keys and Western Bahamas -Kendall/Hunt Publ. Co.,Iowa,415 p.

MUTTI,E. RICCI LUCCHI,F.R.,SEGURET,M and ZANZUCCHI,G. (1984) - Seismoturbidites: a new group of resedimented deposits. Mar.Geol.,**55**:103-116.

NEGRA,M.H.,RABET,A.M.,BISMUTH,H. and ABDALLAH,H. (1987) - Upper Crecateous rudist reef complexes and shallow marine carbonates in central Tunisia.IAS 8th Reg.Meetg.of Sed.,Tunis, Excursion Guidebook:45-88.

ORI,G.G.,ROVERI,M. and VANNONI,F. (1986) - Plio-Pleistocene sedimentation in the Apenninic-Adriatic foredeep (central Adriatic Sea,Italy).Spec. Publs.IAS,**8**:183-198.

PARKER,G.G.,FERGUSON,G.E.,LOVE,S.K. and OTHERS (1955) - Water resources of southeastern Florida with special reference to geology and ground water of the Miami area. U.S.G.S.Water Supply Paper **1255**,965 p.

PARKINSON,R.W.(1989) - Decelerating Holocene sea-level rise and its influence on southwest Florida coastal evolution: a transgressive-regressive stratigraphy. J.Sedim.Petrol., **59/6**: 960-972.

PARONA,C.F.(1924) - Trattato di geologia. Vallardi ed.

PATTARA,E. (1966-1967) - Rilevamento geologico della dorsale a Est di Monte San.Angelo (Foggia).Thesis, Univ. of Bologna.

PAVAN,C. and PIRINI,C. (1966) - Stratigrafia del Foglio 157,M.S.Angelo. Boll. Serv. Geol. Ital.,**LXXXVI**,Roma.

PERKINS,R.D.(1977) - Depositional framework of Pleistocene Rocks in South Florida. In: P.Enos and R.D.Perkins (Eds.), Quaternary Sedimentation in South Florida,Geol.Assoc.Am. Memoir,**147**:131-198.

PERYT,T.M. (Ed.) (1983) - Coated Grains. Springer-Verlag, Berlin.

PETERSON,M.N.A. and VON DER BORK,C.C.(1965) - Modern chert in a carbonate-precipitating locality. Science,**149**:1501-1503.

PISA,G.,MARINELLI,M. and VIEL,G.(1980) - Infraraibl Group: a proposal (southern Calcareous Alps,Italy). Riv.ital.Paleont. Stratigr., **85/3/4**: 983-1002.

PLAYFORD,P.E.(1980) - Devonian "Great Barrier Reef" of Canning Basin,Western Australia:A.A.P.G.Bull.,**64**:814-840.

PLAYFORD, P.E., MACLAREN, D.J., ORTH, C.J., GILMORE, J.S. and GOODFELLOW, W.D.(1984) - Iridium anomaly in the Upper Devonian of the Canning Basin,Western Australia.Science,**226**:437-439.

POINCARE',H. (1913) -Derniéres pensées,ch.2.

POSAMENTIER,H.W., JERVEY,M.T. and VAIL,P.R. (1988) - Eustatic controls on clastic deposition.I.Conceptual framework.S.E.P.M. Spec. Publ.,**42**: 109-124.

PRATT,B.P. and JAMES,N.P.(1986) - The St.George Group (Lower Ordovician) of western Newfoundland: tidal flat island model for carbonate sedimentation in shallow epeiric seas. Sedimentol.,**33**: 313-343.

PRAY,L.C.(1966) - Hurricane Betsy (1965) and nearshore carbonate sediments of the Florida Keys.Geol.Soc. of America Spec. paper **101**:168-169.

PREAT,A.(1984) - Etude litostratigraphique et sédimentologique du Givétien belge (Bassin du Dinant).PhD Thesis,ULB ,Bruxelles, 466 p.

PRINCIPI,G. and TREVES,B. (1984) - Il sistema corso-appennino come prisma di accrezione. Riflessi sul problema generale del limite Alpi-Appennino. Mem.Soc.Geol.It.,**28**:549-576.

PURDY, E.G. (1974) - Karst-determined facies pattern in British Honduras:Holocene carbonate sedimentation model.A.A.P.G. Bull., **58**:825-855.

RAUP,D.M. (1988) - Changing views of natural catastrophe. The Great Ideas Today,Encyclopaedia Britannica,Inc.,Chicago:54-77.

READ,J.F.(1985) - Carbonate platform facies models. A.A.P.G.Bull.,**69**:1-21.

REY,J.,ANDREO,B.,GARCIA-HERNANDEZ,M.,MARTIN-ALGARRA,A.and VERA, J.A. (1990) - The Liassic "Lithiotis" facies north of Vélez Rubio (Subbetic zone).Rev.Soc.Geol.Espana,3(1-2): 199-212.

RIDING,R. and WRIGHT,V.P.(1981) - Paleosols and tidal-flat, lagoon sequences on a Carboniferous carbonate shelf: sedimentary associations of triple disconformities.J.Sedim. Petrol., **51**(4): 1323-1339.

RICCI LUCCHI,F.R. (1978) - Miocene paleogeography and basin analysis in the Periadriatic Apennines. In: C.Squires (Ed.), 'Geology of Italy',Earth Sci Soc. of Libya:129-162.

RIEMANN,B. (1854) - Uber die Hypothesen, welche der Geometrie zu Grunde liegen. In: 'Gesammelte matematische Werke' (1876).

ROBERTS,D.G. and CASTON,V.N.D.(1975) - Petroleum potential of the deep Atlantic Ocean.9th World petroleum cong. Proc.,**V**,II. Applied Sci. Publisher,Ltd.:281-289.

ROYDEN,L. and KARNER,G.D.(1984) - Flexure litosphere beneath Apennine and Carpatian foredeep basins: evidence for an insufficient topographic load. A.A.P.G.Bull.,**68**:704-712.

RUPPEL, S.C. and WALKER,K.R. (1982) - Sedimentology and distinction of carbonate buildups: Middle Ordovician, East Tennessee. J.Sedim. Petrol.,**52(4)**:1055-1071.

RYAN,W.B.F.and CITA,M.B.(1978) - The nature and distribution of Messinian erosional surfaces - indicators of a several kilometer deep Mediterranean in the Miocene.Marine Geol.,**27**: 193-230.

SANDBERG,P.A. (1983) - An oscillating trend in Phanerozoic non skeletal carbonate mineralogy. Nature,**305**:19-22.

SARG,J.F.(1988) - Carbonate sequence stratigraphy. S.E.P.M. Special Publ.,**42**:155-181.

SCATURO,D.M., STROBEL,J.S., KENDALL,C.G.St.C, WENDTE,J.C., BISWAS,G., BEZDEK,J. and CANNON,R. (1989) - Judy Creek: a case study for a two-dimensional sediment deposition simulation. S.E.P.M.Special Publ.,**44**:63-76.

SCHLAGER,W. (1981) - The paradox of drowned reefs and carbonate platforms. Geol.Soc.Am.Bull.,**92**:197-211.

SCHLAGER,W.(1989) - Drowning unconformities in carbonate platforms.In: 'Controls on carbonate platform and basin development'.SEPM Spec. Publ.,**44**:15-25.

SCHLAGER,W.(1991) - Depositional bias and environmental change - important factors in sequence stratigraphy. Sed. Geol.,**70**: 109-130.

SCHLAGER,W. and CAMBER,O. (1986) - Submarine slope angles, drowning unconformities, and self-erosion of limestone escarpments. Geology,**14**:762-765.

SCHLAGER,W. and PHILIP,J. (1990) - Cretaceous carbonate platforms. In: R.N.Ginsburg and Beaudoin,B.(Eds.), "Event and Rhythms. NATO ASC Ser.,**304**:173-195.

SCHLANGER,S.O. JENKINS,H.C. and PREMOLI SILVA,I. (1981) - Volcanism and vertical tectonics in the Pacific Basin related to global Cretaceous Transgressions. Earth Planet. Sci. Lett.,**52**:435-449.

SCHOLL,D.W.(1963) - Sedimentation in modern coastal swamps, southwestern Florida.A.A.P.G.Bull.,**47**:1581-1603.

SCHOLL,D.W., CRAIGHEAD,F.C.Sr. and STUIVER,M.(1969) - Florida submergence curve revised: its relation to coastal sedimentation rates. Science,**163**: 562-564.

SCHONLAUB,H.,KLEIN,P.,MAGARITZ,M.,PANTITSCH,G. and SCHARBERT,G and S. (1991) - Lower Carboniferous Paleokarst in the Carnic Alps (Austria,Italy). Facies,**25**:91-118.

SCHWAN,W. (1980) - Geodynamic peaks in Alpinotype orogeneses and changes in ocean-floor spreading during Late Jurassic-Late Tertiary time.A.A.P.G.Bull.,**64**:359-373.

SCHWARZACHER,W.(1975) -Sedimentation models and quantitative stratigraphy. Elsevier,382 p.

SCOFFIN,T.P. (1977) - Sea-level features on reefs in the northern province of the Great Barrier Reef.Proc.3rd.Int.Coral Reef Symp.,Miami,Fla.:319-324.

SEGURET,M., LABAUME,P. and MADARIAGA,R. (1984) - Eocene seismicity in the Pyrenees from megaturbidites of the south Pyrenean Basin (Spain). Mar.Geol.,**55**:117-131.

SELLI,R. (1962) - Le Quaternaire marin du versant Adriatique-Jonien de la Péninsula Italienne. Quaternaria, 6: 391-413.

SELLI,R.(1967) - The Pliocene - Pleistocenen Boundary in italian marine sectors and its relationship to continental stratigraphy. Progr. in Oceanogr.,4:67-86.

SINNI,E.L. and BORGOMANO,J. (1989) - Le Crétacé supérieur des Murges sud-orientales (Italie Méridionale): stratigraphie et évolution des paléoenvironments. Riv.It. Paleont.Str., **95** **(2)**: 95-136.

SINNI-LUPERTO,E. and MASSE, J.P. (1986) - Données nouvelles sur la stratigraphie des calcaires de plateforme du Crétacé inferieur du Gargano (Italie méridionale). Riv. ital.Paleont. Strat.,9:33-66.

SKABERNE,D.(1987) - Megaturbidites in the Paleogene Flysch in the region of Anhovo (W Slovenia,Yugoslavia). Mem.Soc. Geol. It., **40**:231-239.

SLOSS,L.L. (1984) - Comparative anatomy of cratonic unconformities.In:J.S.Schlee(Ed.),'Interregional Unconformities and Hydrocarbon Accumulation'.A.A.P.G.Mem.,**36**:7-36.

SMALE,D.(1973) - Silcretes and associated silica diagenesis in southern Africa and Australia.J.Sedim. petrol.**43**(4):1077-1089.

SMITH,D.B.,HARWOOD,G.M.,PATTISON,J. and PETTIGREW,T.H.(1986) - A revised nomenclature for the UpperPermian strata in eastern england. In: G.M.Harwood & D.M.Smith (Eds.), The English Zechstein and related topics. Spec. Publ. Geol. Soc. London, **22**: 9-17.

SOMMERVILLE,J.D.(1979) - A cyclicity in the early Brigantian (D2) limestones east of Clwydian Range,North Wales and its use in correlation. Geol. J.,**14**:68-86.

SPALLETTA,C.(1982) - Concordanza stratigrafica tra carbonati radiolariti e flysch ercinico nelle Alpi Carniche (Devoniano-Silesiano). Mem. Soc. Geol.Ital.,**24**:11-21.

SPALLETTA,C. and VAI,G.B. (1984) - Upper DEvonian intraclast parabreccias interpreted as seismites. Mar.Geol.,**55**:133-144.

SPALLETTA,C.,VAI,G.B. and VENTURINI,C.(1980) - Il Flysch ercinico nella geologia dei Monti Paularo e Dimon (Alpi Carniche). Mem. Soc. Geol. Ital.,**20**: 243-265.

SPALLETTA,C.,VAI,G.B. and VENTURINI,C.(1982) - Controllo ambientale e stratigrafico delle mineralizzazioni in calcari devono - dinantiani delle Alpi Carniche. Mem.Soc.Geol. Ital.,**22**:101-110.

SPALLETTA,Vai,G.B. and VENTURINI,C. (1982) - La Catena Paleocarnica. In: A.Castellarin and G.B.VAI (Eds.),"Guida alla geologia del sudalpino centro-orientale",Guide regionali SGI,Bologna Technoprint:281-292.

STACKELBERG,VON U.,EXON,N.F.,VON RAD,U.,QUILTY,P.,SHAFIK,S., EIESDORF,H.,SEIBERTZ,E. and VEEVERS,J.J. (1980) - Geology of the Exmouth and Wallaby Plateaus off northwest Australia: sampling of sieismic sequences. BMR J of Australian Geology and Geophysics,**5**: 113-140.

STAMP,L.D.(1922) - An outline of the Tertiary geology of Burma. Geol. Magazine,**59**:481-501.

STANLEY,S.M. (1966) - Paleoecology and diagenesis of Key largo Limestone,Florida. A.A.P.G.Bull.,**50(9)**:1927-1947.

STILLE,H. (1924) - Grundfragen der Vergleichenden Tectonik. Gebr. Bornt.,Berlin,443 p.

SWIFT,D.J.P.,HUDELSON,P.M.,BRENNER,R.L. and THOMPSON,P.(1987)-helf construction in a foreland basin:storm beds,shelf sandbodies, and shelf-slope depositional sequences in the Upper Cretaceous Mesaverde Group,Book Cliffs,Utah. Sediment.,**34**: 423-457.

THIERSTEIN,H.R. and BERGER,W.H. (1978) - Injection events in ocean history. Nature,**276**:461-466.

TOUIR,J. (1986) - Etude stratigraphique et tectono-sédimentaire des séquences du Crétacé supérieur du Jebel Mrhila (Tunisie Central). Thesis, Fac. Sci. Tunis, Earth Sci Dpt.:1-163.

TUNIS,G. and VENTURINI,S. (1986) - Nuove osservazioni sul Mesozoico delle Valli del Natisone (Friuli Orientale). Gortania. Atti Mus. Friuli St. Nat.:17-68.

TUNIS,G. and VENTURINI,S. (1987) - New data and interpretation on the geology of the southern Julian Prealps (Eastern Friuli). Mem.Soc. Geol. It.,**40**:219-229.

VAI,G.B.(1963) - Ricerche geologiche nel Gruppo del Monte

Coglians e nella zona di Volaia (Alpi Carniche).Giorn. Geol.,**30**:137-198.

VAI,G.B.(1976) - Stratigrafia e paleogeografia ercinica delle alpi. Mem.Soc. Geol. Ital.,**13**(1): 7-37.

VAI,G.B.(1980) - Sedimentary environments of Devonian pelagic limestones in the southern Alps. Lethaia,**13**:79-81.

VAI,G.B. (1987) - Migrazione complessa del sistema fronte-deformativo - avanfossa - cercine periferico: il caso dell' Appennino settentrionale. Mem.Soc.Geol.It.,**38**:95-105.

VAI,G.B. and CASTELLARIN,A. (1988) - Southalpine versus Po Plain Apennine Arcs. IN: Wezel,F.C. (Ed.), 'The origin of Arcs',Elsevier Pub.l.:253-278.

VAI,G.B. and COCOZZA,T. (1986) - Tentative schematic zonation of the Hercynian chain in Italy. Bull.Soc.géol. France, **2**(**1**):95-114.

VAI,G.B. and RICCI LUCCHI,F. (1976) - The Vena del Gesso in northern Apennines: growth and mechanical breakdown of gypsified algal crusts. Mem.Soc.Geol. It.,**16**:217-249.

VAI,G.B. and RICCI LUCCHI,F. (1977) - Algae Bearing and clastic gypsum in a 'cannibalistic' evaporite basin: a case history from the Messinian of northern Apennines. Sedimentology, **248**:211-244.

VAIL, P.R. (1987) - Seismic stratigraphy interpretation procedure. In:A.W.Bally (Ed.), Atlas of seismic stratigraphy (vol.1),A.A.P.G.Studies in Geol., **27**:1-10.

VAIL,P.R. and HAQ,B.U. (1988) - Sea level history - reply. Science, **241**:599.

VAIL,.R.,HARDENBOL,J.and TODD,R.G. (1984) - Jurassic unconformities, chronostratigraphy,and sea-level changes from seismic stratigraphy and biostratigraphy.In: J.S.Schlee (Ed.),"Interregional unconformities and hydrocarbon accumulation", A.A.P.G.Memoir **36**:129-144.

VAIL,P.R.,MITCHUM,R.M.Jr.,TODD,R.G.,WIDMIER,J.M.,THOMPSON,S.III, SANGREE,J.R.,DUBB,J.N. and HALSLID,W.G. (1977) - Seismic stratigraphy and global changes in sea-level.In: C.E.Payton (Ed.),Seismic Stratigraphy- Applications to Hydrocarbon Exploration,A.A.P.G.Memoir,**26**:49-205.

VAIL,P.R. and TODD,R.G.(1981) - Northern North Sea Jurassic unconformities, chronostratigraphy and sea-level changes from seismic stratigraphy. In: L.V. Illing & G.D. Hobson (Eds.),Proceedings of the Petroleum Geol. of the Continental Shelf of NW Europs Conf., march 4-6,1980,London,Heydon and Sons Ltd.,London:216-235.

VANOSSI,M. and GOSSO,G.(1985) - Introduzione alla geologia del Brianzonese Ligure. Mem.Soc.Geol.It.,**26**:441-461.

VAN STEENWINKEL,M.(1988) - The sedimentary history of the Dinant platform during the Devonian-Carboniferous transition. PhD thesis,Katholieke Universiteit Leuven, Belgium,173 p.

VAN STEENWINKEL,M. (1990) - Sequence stratigraphy from 'spot' outcrops - example from a carbonate-dominated setting:Devonian - Carboniferous transition,Dinant synclinorium (Belgium). Sed. Geol.,**69** (3/4):259-280.

VENERANDI PIRRI,I.(1978) - Le paragenesi a Zn,Cu,Pb,Sb, Hg,Ni,As,fluorite, barite nel Devoniano della Catena Carnica. Rend. Soc. It. Min. Petr.,**33**(2):821-844.

VENZO,G.A.(1963) - La Formazione dei "Calcari Grigi" in Valterragnolo - Trentino.Giorn. Geol.,**31**: 1-25.

VIEL,G. (1979) - Litostratigrafia ladinica: una revisione. Ricostruzione paleogeografica e paleostrutturale della area Dolomitico - Cadorina (Alpi Meridionali). Parte I. Riv.ital. Paleont. Strat.,**85(1)**:85-126;**(2)**:297-352.

VON DER BORCH,C.C. and JONES,J.B. (1976) - Spherular modern dolomite from Coorong area, South Australia.Sedim.,**23**:587-591.

YOUNG,K.(1977) - Middle Cretaceous rocks of Mexico and Texas. In: Bebout,D.G. and Loucks,R.G. (Ed.),"Cretaceous carbonates of Texas and Mexico;applications to subsurface exploration",Univ. of Texas Bureau of Economic Geology Report of Investigations, **89**:325-332.

WALKDEN,G.M.(1982) - Field Guide to the Lower Carboniferous rocks of the south east margin of the Derbyshire Block, Wirksworth to Grangemill.Publ. Dep.Geol. and Mineral., Univ. of Aberdeen, **3**.

WALKDEN,G.M. and GUTTERIDGE,P.(1987) - Field Excursion to the Derbyshire Carbonate platform. Publ.Dep.Geol. and Mineral. Univ. of Aberdeen, **5**.

WALKER,K.R. (1974) - Community patterns: Middle Ordovician of Tennessee.In: Ziegler,A.M.,Walker,A.M.,Anderson,E.J., Kauffman, E.G., Ginsburg,R.N. and James,N.P. (Eds.)"Principles of Benthic Community analysis",Sedimenta IV,Comparative Sed. Lab.,Univ. of Miami:9.1-9.9.

WALKER,R.G., DUKE,W.L. and LECKIE,D.A. (1983) - Hummocky stratification, significance of its variable bedding sequences: discussion. Geol.Soc.Amer.Bull.,**94**:1245-1251.

WALKER, T.R. (1960) - Carbonate replacement of detrital crystlline silicate minerals as a source of authigenic silica in sedimentary rocks.Geol.Soc. Amer. Bull.,**71**:145-152.

WANLESS,H.R.,TYRRELL,K.M.,TEDESCO,L.P. and DRAVIS,J.J.(1988) - Tidal-flat sedimentation from hurricane Kate, Caicos platform,British West Indies.J.Sed.Petrol.,**58**(4): 724-738.

WEISS,M.P. (1969) - Oncolites paleoecology and Laramide tectonics, central Utah.A.A.P.G.Bull.,**53**:1105-1120.

WESLEY,A.(1956) - Contributions to the knowledge of the Grey Limestones of the Veneto. Part I. A revision of the flora fossilis formationis oolitichaeof the Zigno.Mem.Ist.Geol.Min. Univ. Padova,**19**:1-69.

WENDT,J. and FURSICH,F.T. (1980) - Facies analysis and paleogeography of the Cassian Formation, Triassic, southern Alps. Riv.It.Paleont. Stratigr.,**85**:1003-1028.

WEZEL,F.C.(1984) - The Tyrrenian Sea: a rifted krikogenic-swell basin.Mem.Soc.Geol. It.,**24** (1982):531-568.

WEZEL,F.C.(1985) - Facies anossiche ed episodi geotettonici globali.Giorn.Geol.,**47(1-2),ser.3°**:281-286.

WEZEL,F.C.(1988) - Earth structural patterns and rhythmic tectonism. Tectonophysics,**146**:1-45.

WHYTE,M.A.(1977) - Turning points in Phanerozoic history. Nature, **267**: 679-682.

WILCOX,R.E.,HARDING,T.P. and SEELEY,D.R.(1973) - Basic wrench tectonics.A.A.P.G.Bull.,**578**:74-96.

WILLIAMS,B.G.and HUBBARD,R.J.(1984) - Seismic Stratigraphic framework and depositional sequences in the Santos Basin, Brazil. Marine and Petrol. Geology,1:90-104.

WINTERER, E.L. and BOSELLINI,A. (1981) - Subsidence and sedimentation on the Jurassic passive continental margin, southern Alps,Italy.A.A.P.G.Bull.,**65**: 394-421.

WORZEL,J.L.,BRYANT,W. and OTHERS (1969) - Initial reports of the Deep Sea Drilling Project.,**10**,Washington D.C.,US Government Printing Office,748 p.

WRIGHT,V.P., PLATT,N.H. and WIMBLEDON,W.A. (1988) - Biogenic laminar calcretes: evidence of calcified root-mat horizons in paleosols.J.Sedim.Petrol.,**35**:603-620.

Springer-Verlag and the Environment

Lecture Notes in Earth Sciences

Vol. 1: Sedimentary and Evolutionary Cycles. Edited by U. Bayer and A. Seilacher. VI, 465 pages. 1985. (out of print).

Vol. 2: U. Bayer, Pattern Recognition Problems in Geology and Paleontology. VII, 229 pages. 1985.

Vol. 3: Th. Aigner, Storm Depositional Systems. VIII, 174 pages. 1985.

Vol. 4: Aspects of Fluvial Sedimentation in the Lower Triassic Buntsandstein of Europe. Edited by D. Mader. VIII, 626 pages. 1985.

Vol. 5: Paleogeothermics. Edited by G. Buntebarth and L. Stegena. II, 234 pages. 1986.

Vol. 6: W. Ricken, Diagenetic Bedding. X, 210 pages. 1986.

Vol. 7: Mathematical and Numerical Techniques in Physical Geodesy. Edited by H. Sünkel. IX, 548 pages. 1986.

Vol. 8: Global Bio-Events. Edited by O. H. Walliser. IX, 442 pages. 1986.

Vol. 9: G. Gerdes, W. E. Krumbein, Biolaminated Deposits. IX, 183 pages. 1987.

Vol. 10: T.M. Peryt (Ed.), The Zechstein Facies in Europe. V, 272 pages. 1987.

Vol. 11: L. Landner (Ed.), Contamination of the Environment. Proceedings, 1986. VII, 190 pages.1987.

Vol. 12: S. Turner (Ed.), Applied Geodesy. VIII, 393 pages. 1987.

Vol. 13: T. M. Peryt (Ed.), Evaporite Basins. V, 188 pages. 1987.

Vol. 14: N. Cristescu, H. I. Ene (Eds.), Rock and Soil Rheology. VIII, 289 pages. 1988.

Vol. 15: V. H. Jacobshagen (Ed.), The Atlas System of Morocco. VI, 499 pages. 1988.

Vol. 16: H. Wanner, U. Siegenthaler (Eds.), Long and Short Term Variability of Climate. VII, 175 pages. 1988.

Vol. 17: H. Bahlburg, Ch. Breitkreuz, P. Giese (Eds.), The Southern Central Andes. VIII, 261 pages. 1988.

Vol. 18: N.M.S. Rock, Numerical Geology. XI, 427 pages. 1988.

Vol. 19: E. Groten, R. Strauß (Eds.), GPS-Techniques Applied to Geodesy and Surveying. XVII, 532 pages. 1988.

Vol. 20: P. Baccini (Ed.), The Landfill. IX, 439 pages. 1989.

Vol. 21: U. Förstner, Contaminated Sediments. V, 157 pages. 1989.

Vol. 22: I. I. Mueller, S. Zerbini (Eds.), The Interdisciplinary Role of Space Geodesy. XV, 300 pages. 1989.

Vol. 23: K. B. Föllmi, Evolution of the Mid-Cretaceous Triad. VII, 153 pages. 1989.

Vol. 24: B. Knipping, Basalt Intrusions in Evaporites. VI, 132 pages. 1989.

Vol. 25: F. Sansò, R. Rummel (Eds.), Theory of Satellite Geodesy and Gravity Field Theory. XII, 491 pages. 1989.

Vol. 26: R. D. Stoll, Sediment Acoustics. V, 155 pages. 1989.

Vol. 27: G.-P. Merkler, H. Militzer, H. Hötzl, H. Armbruster, J. Brauns (Eds.), Detection of Subsurface Flow Phenomena. IX, 514 pages. 1989.

Vol. 28: V. Mosbrugger, The Tree Habit in Land Plants. V, 161 pages. 1990.

Vol. 29: F. K. Brunner, C. Rizos (Eds.), Developments in Four-Dimensional Geodesy. X, 264 pages. 1990.

Vol. 30: E. G. Kauffman, O.H. Walliser (Eds.), Extinction Events in Earth History. VI, 432 pages. 1990.

Vol. 31: K.-R. Koch, Bayesian Inference with Geodetic Applications. IX, 198 pages. 1990.

Vol. 32: B. Lehmann, Metallogeny of Tin. VIII, 211 pages. 1990.

Vol. 33: B. Allard, H. Borén, A. Grimvall (Eds.), Humic Substances in the Aquatic and Terrestrial Environment. VIII, 514 pages. 1991.

Vol. 34: R. Stein, Accumulation of Organic Carbon in Marine Sediments. XIII, 217 pages. 1991.

Vol. 35: L. Håkanson, Ecometric and Dynamic Modelling. VI, 158 pages. 1991.

Vol. 36: D. Shangguan, Cellular Growth of Crystals. XV, 209 pages. 1991.

Vol. 37: A. Armanini, G. Di Silvio (Eds.), Fluvial Hydraulics of Mountain Regions. X, 468 pages. 1991.

Vol. 38: W. Smykatz-Kloss, S. St. J. Warne, Thermal Analysis in the Geosciences. XII, 379 pages. 1991.

Vol. 39: S.-E. Hjelt, Pragmatic Inversion of Geophysical Data. IX, 262 pages. 1992.

Vol. 40: S. W. Petters, Regional Geology of Africa. XXIII, 722 pages. 1991.

Vol. 41: R. Pflug, J. W. Harbaugh (Eds.), Computer Graphics in Geology. XVII, 298 pages. 1992.

Vol. 42: A. Cendrero, G. Lüttig, F. Chr. Wolff (Eds.), Planning the Use of the Earth's Surface. IX, 556 pages. 1992.

Vol. 43: N. Clauer, S. Chaudhuri (Eds.), Isotopic Signatures and Sedimentary Records. VIII, 529 pages. 1992.

Vol. 44: D. A. Edwards, Turbidity Currents: Dynamics, Deposits and Reversals. XIII, 175 pages. 1993.

Vol. 45: A. G. Herrmann, B. Knipping, Waste Disposal and Evaporites. XII, 193 pages. 1993.

Vol. 46: G. Galli, Temporal and Spatial Patterns in Carbonate Platforms. IX, 325 pages. 1993.

Vol. 47: R. L. Littke, Deposition, Diagenesis and Weathering of Organic Matter-Rich Sediments. IX, 216 pages. 1993.

Vol. 48: B. R. Roberts, Water Management in Desert Environments. XVII, 337 pages. 1993.

Vol. 49: J. F. W. Negendank, B. Zolitschka (Eds.), Paleolimnology of European Maar Lakes. IX, 513 pages. 1993.

Vol. 50: R. Rummel, F. Sansò (Eds.), Satellite Altimetry in Geodesy and Oceanography. XII, 479 pages. 1993.